CHEMISTRY COUNTS

GRAHAM HILL

HODDER AND STOUGHTON
LONDON SYDNEY AUCKLAND TORONT

Contents

Preface

This book is about chemistry and the part that chemistry plays in our lives, in industry and in society. It is specially written for the new GCSE courses and examinations in chemistry. I hope that you will find it lively, colourful and interesting.

The book is divided into twelve major sections. These sections are divided into two-page units, each of which ends with several short questions to help you understand the topic you have just studied. At the end of each section, there are longer study questions. Many of these are taken from specimen GCSE papers and from recent joint O level/CSE examinations. Try to answer as many as possible of these questions because asking questions and answering them is a good way of learning.

If you are looking for information on a particular topic look it up in the contents list at the beginning of the book and in the index at the back of the book.

Many people have influenced the planning and writing of 'Chemistry Counts'. I am particularly grateful to Maggie Hannon, Campbell Grant, my daughter Clare and my wife Elizabeth. It has been a privilege and a pleasure to work on a project with such able, enthusiastic and supportive colleagues.

Graham Hill
(August 1986)

To Aunt Clara

SECTION A

Introducing Chemistry

1. What is Chemistry?
2. Why Study Chemistry?
3. Elements—Building Blocks
4. Classifying Elements
5. The Uses of Elements
6. Elements and Compounds
7. Separating Mixtures
8. Separating Solutions
9. Separating Liquids
10. Pure Substances

Black gold—crude oil is an important source of chemicals and fuels

1 What is Chemistry?

Mining iron ore

Iron ore can be made into steel and used to build structures like the Humber Bridge

Iron ore, like that in the photograph above, is useless. We can't eat it, wear it or grow things in it. But, if we heat it with limestone, coke and air we can turn it into iron, and from iron we can make steel.

> *This is what chemisty is about:*
> - *finding out what the Earth is made of, and*
> - *changing useless things like iron ore into useful things like iron and steel.*

These different things like iron, iron ore, limestone, coke and air are called **substances**.

Raw materials from the Earth

Cooks are restricted in the meals they can make by the food they have in their kitchen. In the same way, chemists are restricted in what they can make by the raw materials, like iron ore, which they can get from the Earth. The chemical and mining industries produce valuable substances like metals, fuels, fertilizers and plastics, from raw materials in the sea, in the air or in the earth.

The table on the opposite page shows the six most important sources of raw materials and the substances we get from them.

	Raw material	Substances obtained from the raw material
1	Plants	Timber, foods (e.g. flour, oats, sugar, cooking oil), clothing (e.g. cotton, linen), rubber
2	Coal	Fuels (e.g. coke, coal gas), dyes and plastics from coal tar
3	Oil and natural gas	Fuels, chemicals (e.g. plastics, pesticides, perfumes)
4	The sea	Salt (sodium chloride), magnesium, bromine
5	The air	Oxygen, nitrogen, argon
6	Rocks	Metals (e.g. iron, copper, gold), limestone, sand, aggregate for building

Important raw materials and their products

Over 6 billion glass bottles and jars are used each year in Britain. Throwing these away produces 2 million tonnes of wasted glass

Finite Earth

In our desire for more possessions and easier living, we have spoilt large areas of the Earth with mines, quarries, motorways and pylons. Large buildings have turned our cities into 'concrete jungles' and forests have been destroyed forever. As a result of this, important raw materials, such as copper and oil, are being used up rapidly.

Today there are 5000 million people on the Earth and the world's population will be 7000 million by the year 2000. Unfortunately, *more* people need *more* food, *more* water, *more* raw materials and *more* land.

During the last 200 years, society has benefited greatly from the discoveries of science and technology. But *the Earth's resources will not last forever.* Some raw materials, like coal, will last for centuries but others, like oil and natural gas, will only last a few decades unless we use them more sparingly. As raw materials run out, chemists will try to develop new substances so that reserves will last longer than predicted.

A bottle bank—recycling materials like glass is becoming more important

Questions

1 How would you describe to someone of your age what chemistry is?

2 Make a list of 4 substances in your kitchen and give the chemical name of their main constituent (e.g. salt—sodium chloride). Say which substances occur naturally and which are manmade.

3 Coal is an important raw material. List 4 substances made or extracted from it.

4 (a) Which metals are obtained from the following ores: *bauxite; copper pyrites; haematite; galena; tinstone?*
(b) What does the word 'ore' mean?

5 Look at today's newspaper. What news or adverts does it contain about chemists or the chemical industry?

2 Why Study Chemistry?

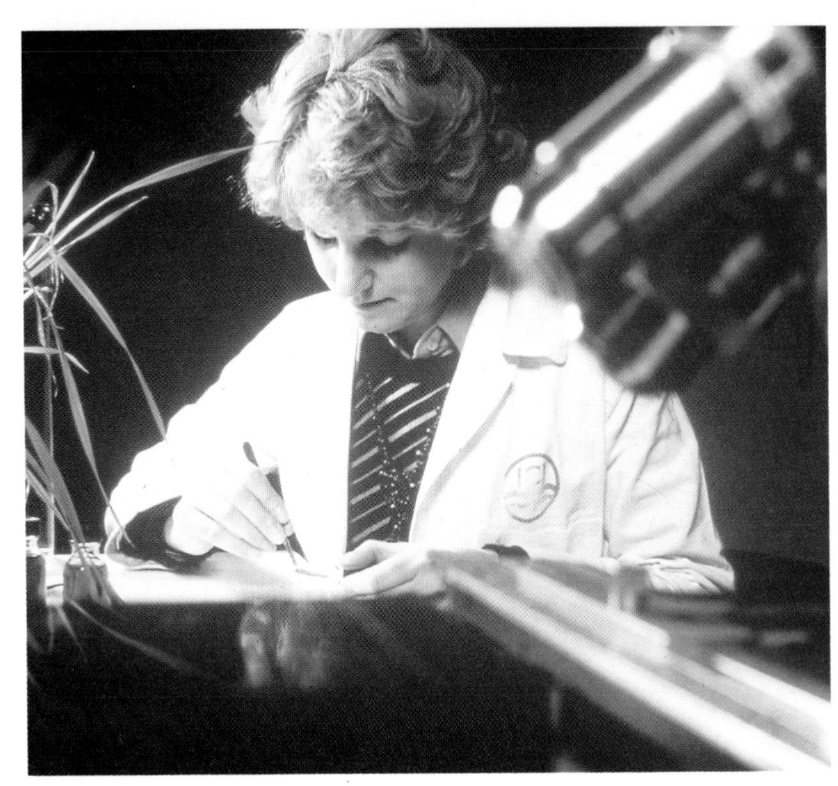

Most chemists use their knowledge and skills to help others and benefit mankind

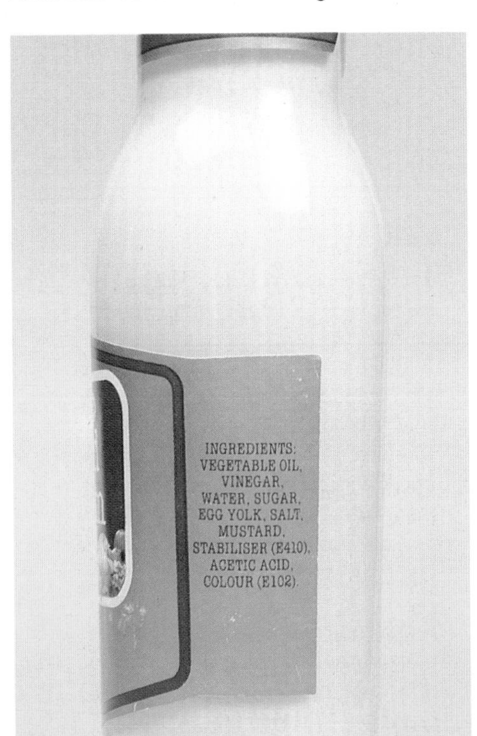

Chemistry is concerned with the food we eat. This label shows the ingredients (chemicals) in salad cream

In unit 1 we met some important raw materials and the substances that chemists can obtain from them. These substances, like metals, fuels and oxygen, have important uses in everyday life. This shows that chemistry is not just something that goes on in laboratories in schools, universities and industry. *Chemistry is an essential part of our lives.* It is concerned with the food we eat, the clothes we wear, the fuels we burn and the way in which we live.

Look around your home. Everywhere you will see the results of chemistry.

1 *Metals* used to make knives and forks, pans, ornaments and other items. These materials have been manufactured by chemists and metallurgists from rocks and minerals in the Earth's crust.

2 *Curtains, carpets, clothes and other textiles* which have been made by chemists. Some textiles like cotton and wool occur naturally as part of animals and plants, but others like polyester (Terylene) and nylon have been made from the chemicals in coal, crude oil and plants. Have a look at the labels on your clothes. They will tell you what the clothes are made of. There are probably large amounts of polyester and nylon in the clothes you are wearing.

3 *Plastics* such as polythene, perspex and PVC used for such different articles as clingfilm (polythene) and records (PVC).

4 *Building materials*—even the materials used in building our homes have been studied by chemists and manufactured in the chemical industry. These materials include bricks (from clay), cement (from limestone) and steel in the girders and supports.

These examples show two things:
- *chemistry affects the way we live,*
- *chemists and the chemical industry play an important part in turning raw materials into new and more useful substances.*

Some of these substances like foods, drinks and medicines are essential to our health and well being. Other substances make our lives easier, more comfortable and more enjoyable. Next time you go out for the evening, think of the pharmaceutical chemists who tested your make-up, the industrial chemists who made the fibres in your clothes, the metallurgists involved in manufacturing the steel in the bus or train which you use and the food scientists and dieticians who checked the quality of your food.

None of this could have happened without the study of chemistry. This is why it is important to study chemistry. In order to produce new materials and to improve our lives we need to study and understand the chemicals which make up the universe, and the chemical processes which can be used to change one substance into another. This is what chemistry is all about.

Marie Curie dedicated her life to the study of science. She won two Nobel prizes—one for chemistry and one for physics. Her study of radium has led to the treatment of various cancers by radiotherapy

Next time you enjoy a family meal, think of the chemists, food scientists and dietitians involved in preparing our food and checking its quality

Questions

1 Give *three* reasons why it is useful to study chemistry.

2 What branch of chemistry are the following particularly interested in? *biochemists; metallurgists; pharmacists; chemical engineers; geologists.*

3 Some scientific discoveries can be used either for helpful purposes or for harmful purposes. Write two paragraphs on three of the following to show first their helpful uses, and second their harmful uses: *detergents; atomic energy; insecticides; explosives; Aspirin.*

3 Elements—Building Blocks

All this can be made from a few basic pieces of Lego. In the same way, all substances are made from a few different elements

Everything in the universe is made up of elements. So far, we know of 105 elements. These include aluminium, iron, copper, oxygen, nitrogen and carbon.

> *In 1661, Robert Boyle suggested the name* ***element*** *for a substance that cannot be broken down into a simpler substance. This is still a useful description of an element.*

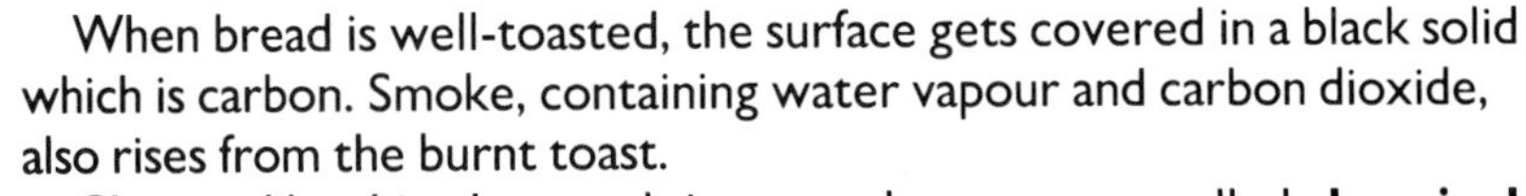

When bread is well-toasted, the surface gets covered in a black solid which is carbon. Smoke, containing water vapour and carbon dioxide, also rises from the burnt toast.

Changes like this, that result in new substances, are called **chemical reactions** and the new substance is called the **product** of the reaction. The chemical reaction which produced carbon from bread is caused by heat from the cooker. Many of the foods which we eat are prepared by chemical reactions caused by heat.

We can summarize the reaction which takes place when bread is heated by writing a **word equation**:

$$\text{bread} \xrightarrow{heat} \text{carbon} + \text{carbon dioxide} + \text{water} + \text{toast}$$

No matter how the black carbon is treated, it cannot be broken down into a simpler substance. Therefore, carbon is an element. Substances like water and carbon dioxide are not elements because they can be broken down into simpler substances. Substances like water and carbon dioxide which contain two or more elements are called **compounds**.

Robert Boyle—the first scientist to use the word 'element'

When electricity is passed through water containing a little sulphuric acid, it breaks up forming hydrogen and oxygen (figure 1).

$$\text{water} \xrightarrow{\textit{electricity}} \text{hydrogen} + \text{oxygen}$$

The hydrogen and oxygen cannot be made any simpler: they are elements.

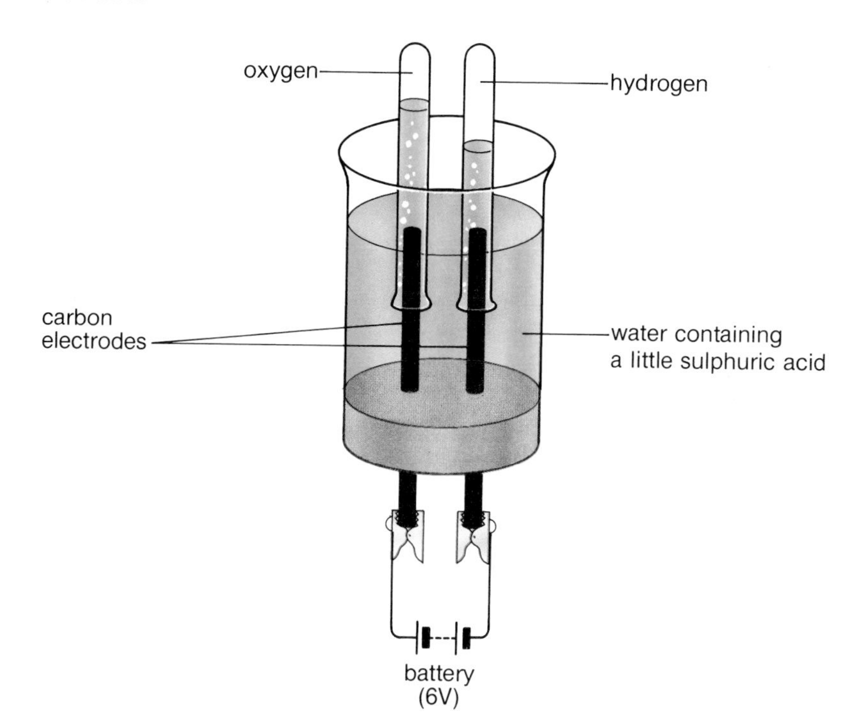

Figure 1
When electricity is passed through water containing a little sulphuric acid, it splits up into hydrogen and oxygen

Although there are millions and millions of different substances in the universe, they can all be split up into one or more of the 105 known elements. For example, grass is made of carbon, hydrogen and oxygen. Sand is made of silicon and oxygen. So elements are *the building blocks for all substances.* In the same way, bricks are the building blocks for houses. Using a few different types of brick it is possible to build millions of different houses.

A reaction in which a compound splits up into two or more simpler substances is called **decomposition**. An example of this is the splitting of water into hydrogen and oxygen.

Water can be decomposed into hydrogen and oxygen, but if hydrogen is exploded with oxygen, these two simpler substances react to form water again. A reaction like this in which two or more substances join together to form a single product is called **synthesis**.

Notice that synthesis is the reverse of decomposition. Synthesis is a 'building up' whereas decomposition is a 'breaking down'.

$$\text{water} \underset{\textit{synthesis}}{\overset{\textit{decomposition}}{\rightleftarrows}} \text{hydrogen} + \text{oxygen}$$

Gold is an element which occurs in the Earth's crust. This photograph shows the gold in Tutankhamun's funeral mask

Questions

1 Explain the following:
element; chemical reaction; product; decomposition; synthesis.

2 Write down the names of all the elements mentioned in this unit.

3 A gas was produced when a solid was heated. How would you test to see if the gas is (i) carbon dioxide, (ii) water vapour, (iii) oxygen?

4 When red mercury oxide is heated, shiny beads of mercury are produced and oxygen gas is given off.
(a) Write a word equation for the reaction.
(b) Is this decomposition or synthesis? Explain your answer.

5 (a) Is it possible to decompose an element? Explain your answer.
(b) Is it possible to decompose a compound? Explain your answer.
(c) Which of the following can be decomposed?
sugar; zinc; hydrogen; salt.
(d) Explain why a synthesis reaction *always* produces a compound.

4 Classifying Elements

Why are metals ideal for making bells?

Element	Melting point /°C	Boiling point /°C	Electrical conductivity
Aluminium	660	2450	Good
Carbon (graphite)	3730	4830	Moderate
Copper	1083	2600	Good
Gold	1063	2970	Good
Iron	1540	3000	Good
Oxygen	−219	−183	Poor
Sulphur	119	445	Poor

Table 1: the properties of some elements

Look at the properties of the elements listed in table 1.

1 Which elements have a melting point above 500° C?
2 Which elements have a boiling point above 1000° C?
3 Which elements are good conductors of electricity?
4 Which of these elements can be polished to a shine?
5 Which of these elements can be bent or hammered into different shapes?
6 Which of these elements are metals?
7 Do the properties in table 1 separate metals from non-metals?

Dividing elements into metals and non-metals is very useful. One of the best ways of checking whether an element is a metal or a non-metal is to see if it conducts electricity. This can be done by using the apparatus in figure 1. The bulb lights when the element tested is a metal. When non-metals (except graphite) are tested, the bulb does not light because these are poor conductors.

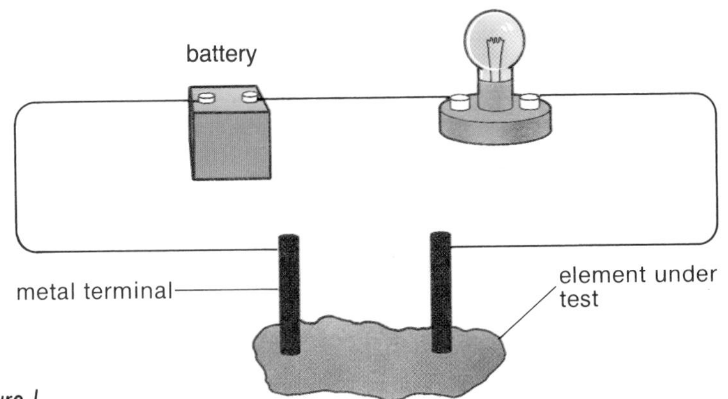

Figure 1

Although one or two elements, such as graphite, are difficult to classify, the classification of elements as metals and non-metals is very helpful. There are about 80 metals, about 20 non-metals and five or six elements that are difficult to place. Table 2 compares the properties of metals and non-metals. There is more about classification in section E.

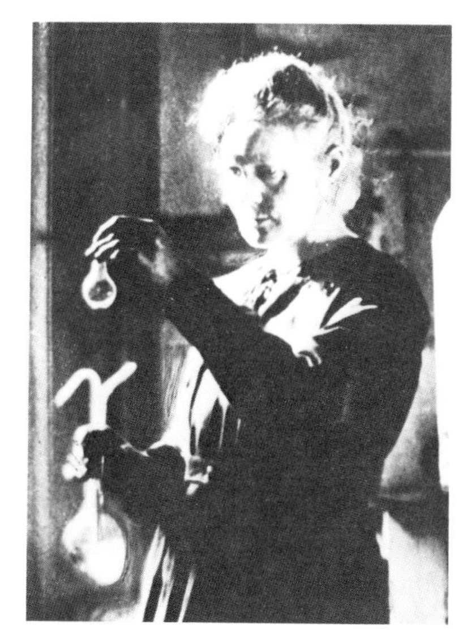

Marie Curie

Enrico Fermi

Dmitri Mendeléev

Albert Einstein

Which elements have been named after these four scientists?

Property	**Metals**	**Non-metals**
State	Usually solids at room temperature	Solids, liquids or gases at room temperature
Melting point and boiling point	Usually high	Usually low
Density	Usually high	Usually low
Appearance	Shiny	Dull
Effect of hammering (malleability)	Can be hammered into shape (malleable)	Brittle (non-malleable) and soft when solid
Thermal and electrical conductivity	Good	Poor

Table 2: comparing the properties of metals and non-metals

Questions

1 Make a list of the properties of metals and non-metals.

2 (a) Make a table showing the names of 20 metals and 10 non-metals.
(b) Write down the names of 2 elements that are difficult to classify as either metal or non-metal. Explain why they are difficult to classify.

3 How would you find out whether a chunk of a solid element is a metal or a non-metal?

4 A colourless gas *A* relights a glowing splint:
(i) *A* reacts with copper to give a black solid *B*;
(ii) *A* reacts with a hot dark grey solid *C* to give a colourless gas *D*;
(iii) *D* turns limewater milky;
(iv) *B* and *D* are both produced when the green powder *E* decomposes on heating.
(a) Identify the substances *A* to *E*.
(b) Write word equations for the reactions (i), (ii) and (iv).

5 The Uses of Elements

Making a stained glass window using strips of lead

Concorde—20 tonnes of aluminium

Metals, like gold and silver, and the non-metal diamond are attractive and unreactive. They can be used in ornaments and jewelry. This photograph shows the Imperial State Crown

Look at the photographs above.

1 Why is aluminium used for aircraft construction?
2 Why is lead used to hold the pieces together in stained glass windows?

The uses of an element depend on two things:
- *how easily it can be obtained and*
- *its properties.*

Aluminium is hard, it has a low density and it does not rust. These properties explain why it is used in aircraft construction and for window frames. Aluminium is also a good conductor of heat and electricity. Because of these properties it is used for pans and other cooking vessels. Lead is a soft, malleable metal which does not rust. It is useful for roofing and for stained-glass windows. It also has a high density and because of this it is used to protect hospital workers from X-rays. Copper is a good electrical conductor. It is cheap compared to most metals and it can be drawn into wires. Because of these properties, it is used for electrical wires and cables.

Metals can be mixed with other elements to form **alloys**. Alloys have properties that are different from those of the elements they contain. So it is possible to make alloys with particular properties. The most common alloy is steel which is mainly iron with about 0.15% carbon. The carbon makes steel harder and tougher than pure iron. Stainless steel also contains chromium and nickel to prevent it from rusting. Iron, aluminium and copper are the most widely used metals. All three are used in large quantities in alloys. There is more about alloys in section F.

What about the uses of non-metals? Silicon and germanium conduct electricity but not as well as metals. Thus, these two elements are known as **semiconductors** (see section D). They are used to make 'silicon chips' for electronic circuits. Chlorine, a reactive gas which is poisonous in large quantities, is used at low concentrations to kill bacteria in drinking water and swimming pools.

The bands of rocks in the Grand Canyon are mainly sandstones and limestones. The main elements in sandstone are silicon and oxygen. The main elements in limestone are calcium, carbon and oxygen

Elements in the Earth's crust

Although there are 105 known elements, they occur in very different proportions on the Earth. Figure 1 shows the abundance of the commonest five elements in the Earth's crust. Notice that these five elements make up 91% by mass. The remaining 100 elements account for only 9%. The most abundant element in the Earth's crust is oxygen. This forms 23% by mass of the air and 89% of water. Most oxygen, however, is combined with other elements in rocks and soils. Sand and sandstone, for example, contain roughly equal proportions of oxygen and silicon, which is the second most abundant element. Clays also contain silicon and oxygen, together with aluminium. Aluminium is the third most abundant element and the most abundant metal in the Earth's crust. Iron is the fourth most abundant element. Most of it is combined with oxygen in iron ore (rust-coloured iron oxide). Chalk and limestone are made of calcium carbonate which is 40% calcium, 12% carbon and 48% oxygen. This accounts for most of the calcium in the Earth's crust.

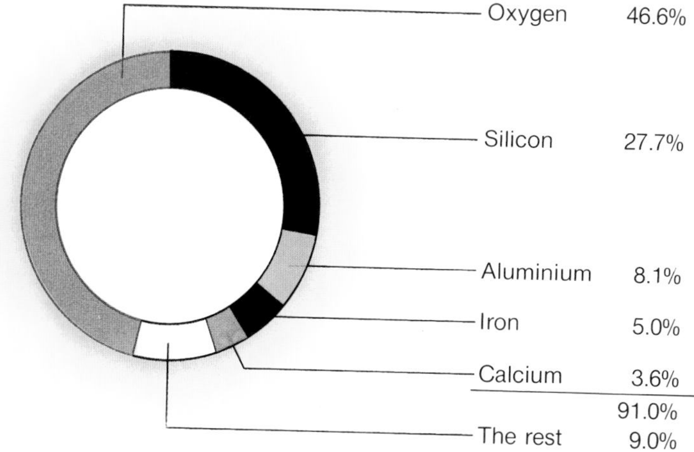

Figure 1

Questions

1 What determines the uses of an element?

2 (a) What is an alloy?
(b) Why are alloys important?

3 Platinum is a rare, hard and unreactive metal with a high melting point. What uses does platinum have because of these properties?

4 Mercury is the only liquid metal at room temperature. It has a high boiling point, a high density and it is a good conductor of heat and electricity. What uses does mercury have on account of these properties?

6 Elements and Compounds

When aluminium pans and window frames are first made, they look shiny. Gradually, they become dull as a layer of white aluminium oxide forms on the surface. The aluminium and oxygen have not just *mixed* together. They have joined together to form a new substance called aluminium oxide. We say that the aluminium and oxygen have **reacted** or **combined** to form white aluminium oxide which is a compound. Salt and sugar are also compounds. Salt contains the elements sodium and chlorine chemically combined (figure 1). Sugar is a compound of carbon, hydrogen and oxygen.

- *A compound is a substance that contains two or more elements combined together.*

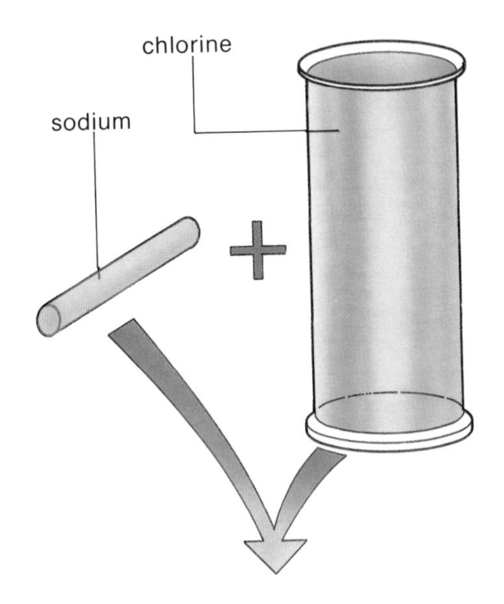

Figure 1
The properties of a compound (salt) are very different from those of the elements in it

Sugar is a compound containing the elements carbon, hydrogen and oxygen

When *two* elements react together to form a compound, the name of the compound ends in **-ide**. For example,

aluminium + oxygen → aluminium ox**ide**
sodium + chlorine → sodium chlor**ide**
hydrogen + sulphur → hydrogen sulph**ide**

When a metal reacts with a non-metal, the non-metal forms the -ide part in the name of the compound. When two non-metals react, the more reactive non-metal forms the -ide part to the name of the compound.

Mixtures and compounds

There are important differences between mixtures and compounds.

1 When two or more elements combine to form a compound, a new substance is formed and a chemical reaction occurs. This does *not* happen when two substances are just mixed together to form a mixture.

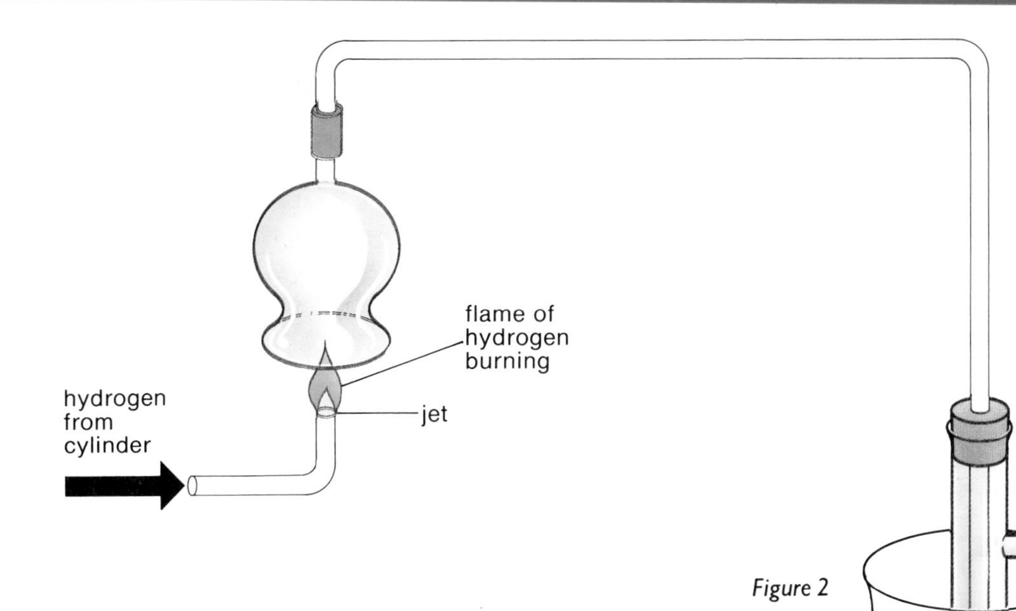

Figure 2

2 A compound has very different properties from those of its elements. For example, white aluminium oxide has a different set of properties from aluminium metal and colourless oxygen gas. When a mixture is formed, the separate constituents still have their own properties.

3 The elements in a compound can only be separated by a chemical reaction but the components in a mixture can often be separated easily. A mixture of aluminium powder and oxygen separates by itself, but aluminium oxide needs a great deal of heat and electricity to separate the aluminium and oxygen.

4 If we analysed aluminium oxide, we would find that it *always* contains 52.9% aluminium and 47.1% oxygen. Water always contains 11.1% hydrogen and 88.9% oxygen. Salt always contains 39.3% sodium and 60.7% chlorine. The composition of a compound is always the same but the composition of a mixture can vary. Aluminium oxide always has 52.9% aluminium and 47.1% oxygen, but a mixture of aluminium and oxygen could have 0.1% aluminium, 99.9% aluminium or any percentage in between.

The table below summarises the main differences between pure compounds and mixtures.

	Compound	**Mixture**
1	A new substance is produced when the compound forms.	No new substance is produced when the mixture forms.
2	The properties are different from those of the elements in it.	The properties are similar to those of the elements or compounds in it.
3	The elements in it can only be separated by chemical reaction.	The elements or compounds in it can often be separated easily.
4	The composition is always the same.	The composition can vary.

Differences between pure compounds and mixtures

Questions

1 What is meant by the words: *compound; mixture; reacted?*

2 What are the important differences between compounds and mixtures?

3 Figure 2 shows how the compound water can be made from hydrogen and oxygen. Mixtures of hydrogen and oxygen explode if a lighted splint is put into them. The 'pop' test for hydrogen is the result of hydrogen exploding with oxygen in the air. Because of this, the gas at the jet must not be lit until all the air has been pushed out of the tube and the gas coming from the jet is pure hydrogen.

(a) Write a word equation for the reaction taking place in figure 2.

(b) Why is the test tube surrounded by cold water?

(c) Why is the pump necessary?

(d) How would you show that the liquid which collects in the test tube is water?

(e) Why is water a compound and not a mixture of hydrogen and oxygen?

7 Separating Mixtures

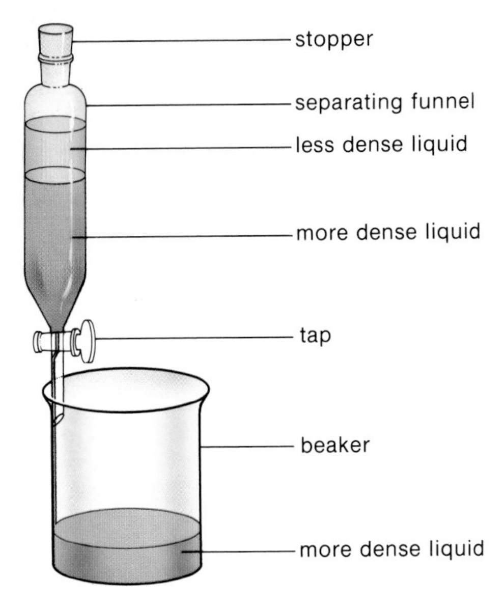

Figure 1

Most substances which occur naturally are mixtures. Very often these mixtures have to be separated before we can use them. Just imagine what might happen if we used untreated, muddy water for cooking and cleaning, or if we tried to run our cars on crude oil rather than petrol.

Muddy water contains particles of solid 'floating' in it. If the water is left for some time, heavier particles of solid sink to the bottom of the vessel as a **sediment**. The clearer water can then be poured off easily and separated from the sediment. This process is called **decantation**. Decantation is important in cooking. It is used to separate water from vegetables after cooking and also to separate two liquids which do not mix, like cream and milk. These liquids which do not mix are called **immiscible liquids**. Have you noticed how difficult it is to separate the cream from milk really well? Chemists use a separating funnel to separate two immiscible liquids (figure 1).

If water is decanted from its sediment of mud, it still appears cloudy due to very small particles of solid 'floating' in it. We say that the particles are suspended in the water and that the mixture of fine solid particles and water is a **suspension**. In a suspension, some particles are so small that they do not sink. Smoke is a suspension of fine solid particles in air; milk is a suspension of tiny droplets of oil (cream) in a watery liquid.

In order to separate the suspension of fine particles from cloudy water, we must use **filtration** (figure 2). There are millions of tiny holes in filter paper. Liquids can pass through these holes, but solid particles are too large to do so. The solid that remains in the filter paper is called the **residue** and the liquid that trickles through is the **filtrate.**

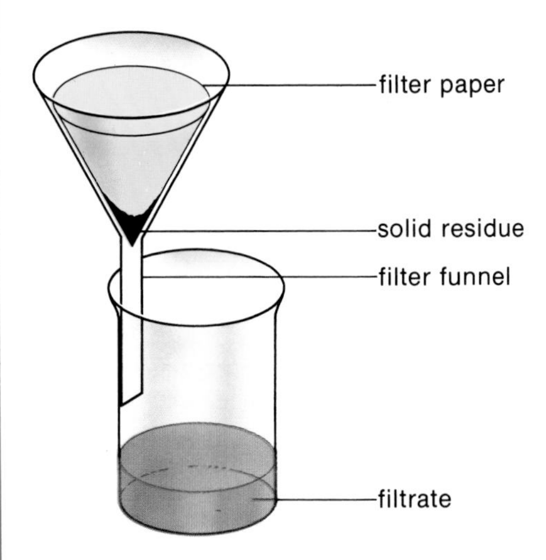

Figure 2

Clean water for our homes

Filtration plays an important part in obtaining clean water for our homes. Figure 3 shows the main stages in the purification of river water.

The water is first stored in reservoirs where most of the solid particles can settle out. As the water is needed, it is filtered through clean sand and gravel which trap smaller particles of mud and suspended solids (figure 4).

After filtering, the water is treated with small amounts of chlorine to kill harmful bacteria in the water. The purified water is then pumped to storage tanks and water towers from which it flows to our homes.

Filtration is also used to separate beer from its sediment (yeast) before bottling. The beer is filtered by forcing it through filter cloths to catch the sediment.

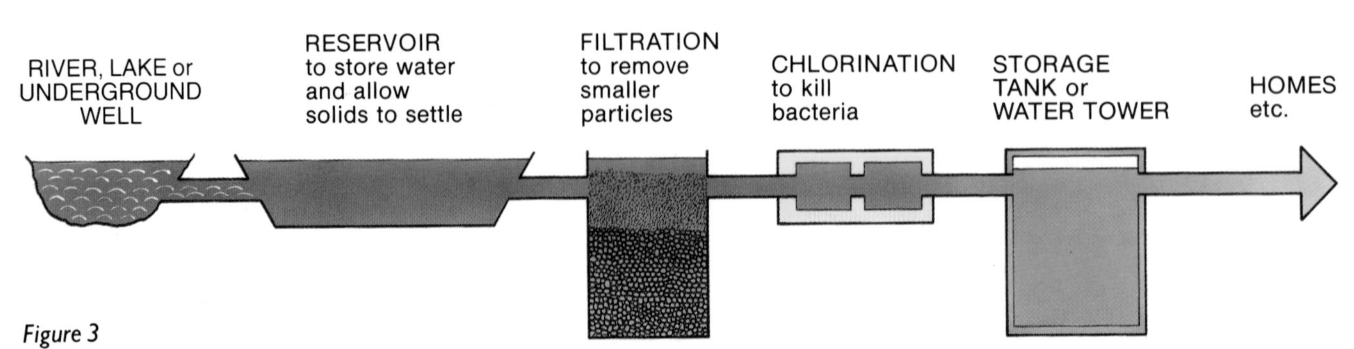

Figure 3

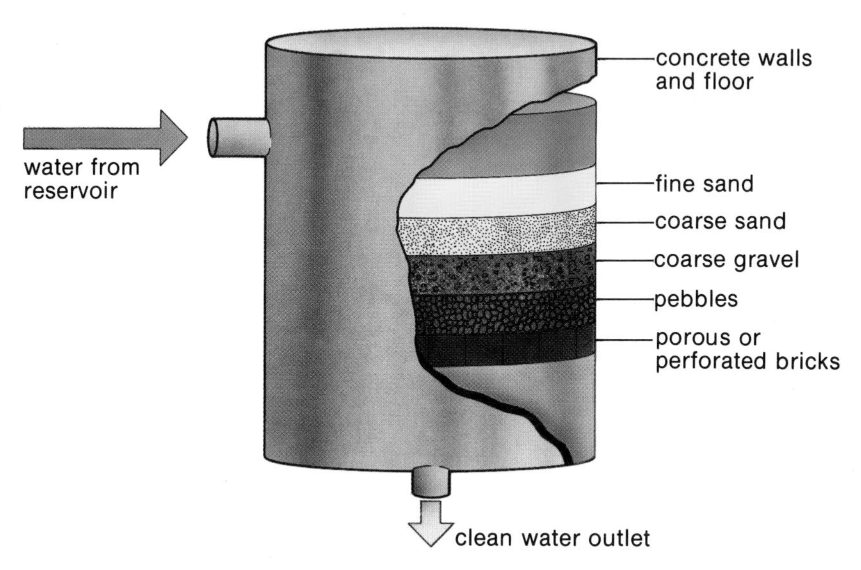

Figure 4

Decanting cream from milk

Separating cream from milk

Another way to separate a solid from a liquid or to separate two immiscible liquids is to use a **centrifuge**. As the centrifuge spins round rapidly, a force pulls outwards and the heavier particles go to the bottom of the mixture. In the same way, a roundabout at the funfair drags you outwards. One of the most important uses of a centrifuge is in separating cream from milk to make skimmed milk. Whole milk is poured into a bank of spinning sloped discs. The heavier milk is forced down and outwards and the cream rises up and inwards (figure 5).

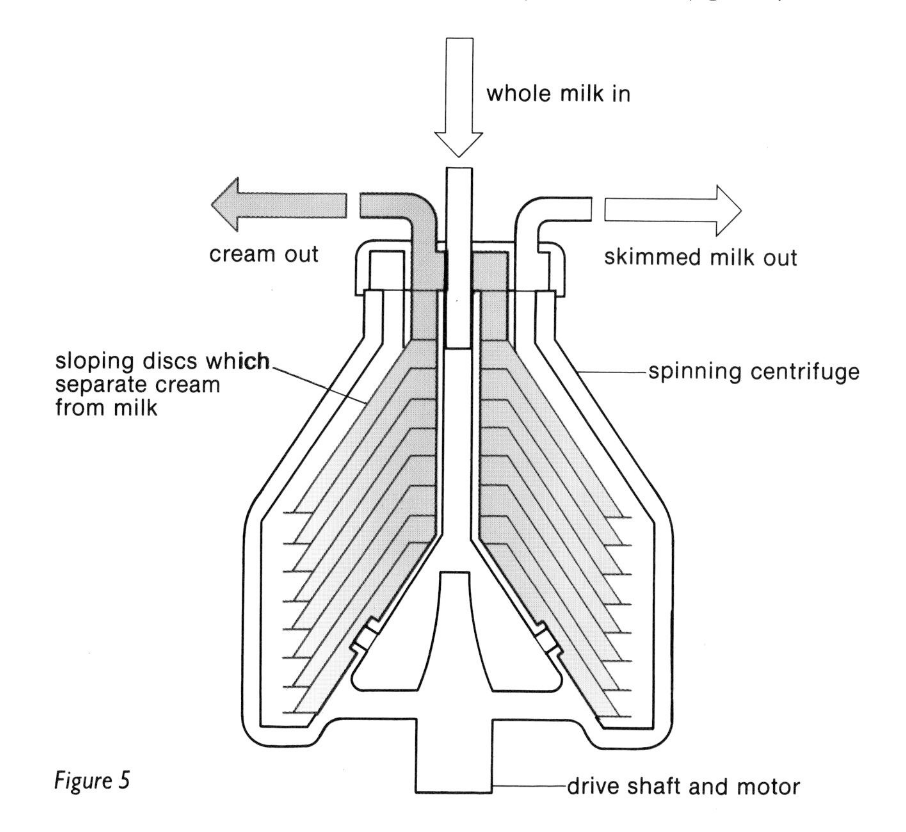

Figure 5

Questions

1 Explain the following:
immiscible liquids; filtrate; residue; suspension.

2 Give an example of separation by
(i) decantation;
(ii) filtration;
(iii) centrifugation.

3 What are the main stages in the purification of river water for use in our homes?

4 Mist is a suspension.
(a) What are the suspended particles?
(b) What are they suspended in?

8 Separating Solutions

Sea-water is not pure—it contains salt and other dissolved substances essential for animals and plants that live in the water

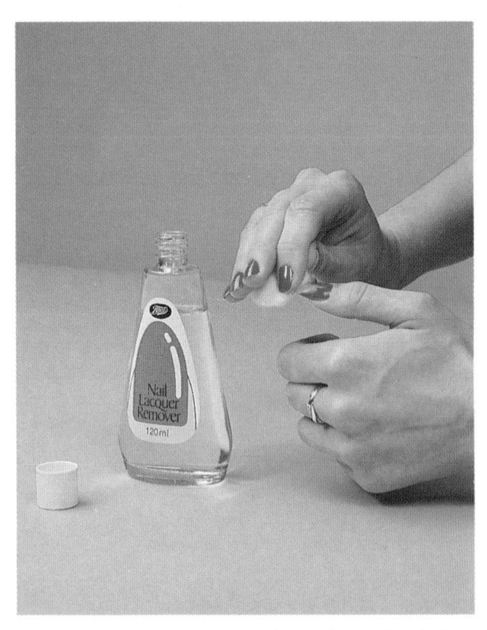

Nail polish remover is a solvent for nail polish. The solvent is propanone (acetone)

Tap water is clean but *not* pure. It contains dissolved gases like oxygen and carbon dioxide and probably dissolved solids which make the water 'hard'. In hard water areas, the dissolved solids are left as scum after washing with soap. Clear sea-water is another example of a liquid containing dissolved solids. You can't see the salt in clear sea-water, but it must be there because you can taste it. The salt has been broken up into tiny particles which are too small to be seen even with a microscope. These particles are so small that they can pass through the holes in filter paper during filtration.

The mixture of dissolved salt and water forms a **solution**. The substance that dissolves is called the **solute**. The liquid in which the solute dissolves is the **solvent**. Solids such as salt and sugar which dissolve are described as **soluble** and solids such as mud which do not dissolve are **insoluble**.

Although water is a very good solvent, many substances are insoluble in it. Oils and greases are insoluble in water but very soluble in other solvents, like trichloroethane. This liquid is used by dry cleaners to remove grease stains from clothes. Nail polish is insoluble in water but it dissolves easily in a liquid called propanone (acetone). Nail polish remover is made of propanone.

Evaporation and crystallization

When sea-water is left to dry in the sun, the water turns into vapour (evaporates) and leaves behind salt. Next time you are at the seaside look for white rings of salt around the edges of rock-pools.

This change of liquid to a gas or vapour is called **evaporation**. Evaporation can be used to separate a dissolved solid from its solvent. If the solvent evaporates slowly, then the dissolved solute is often left behind as well-shaped crystals. We say that it has **crystallized**. In hot countries, salt is obtained from sea-water by allowing the water to evaporate, leaving the crystallized salt. Sugar is also obtained by a process of dissolving and evaporation. The sugar canes are crushed and then mixed with water to dissolve the sugar. The sugar solution is separated from the pulp and then water is evaporated off leaving sugar crystals.

At room temperature, salt solution takes a long time to evaporate and form crystals. We can speed up the process by heating the solution gently (figure 1). If the solution is evaporated quickly, the crystals are usually small. Larger crystals form if the solution is evaporated slowly. As the water evaporates, crystals begin to form at the edges of the solution. When this happens, it is best to stop heating and allow the solution to cool and crystallize slowly from then on. This gives larger, well-shaped crystals. If the crystals form in the solution, they can be dried between filter papers.

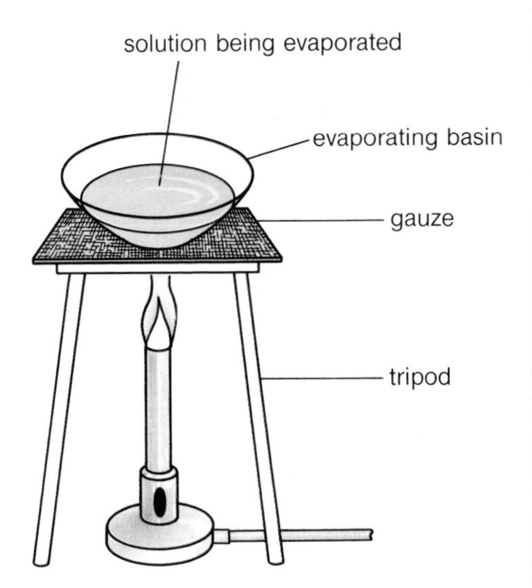

Figure 1

Salt is harvested from sea-water in hot countries

Evaporation is an essential process in drying wet clothes and in producing concentrated evaporated milk. As the solvent evaporates from the solution, the same amount of solid is left in less and less solvent. We say that a solution which contains only a little solute in a given amount of solvent is **dilute**. If a solution contains a lot of solute in a given amount of solvent we say it is **concentrated**. As more and more solvent evaporates, a solution becomes more and more concentrated. Eventually, the solution cannot become any more concentrated. If any more solvent evaporates, some solid solute will come out of solution and form crystals. A solvent that has dissolved all the solute it can is **saturated**. As solvent evaporates from the saturated solution, more and more solid crystallizes out.

Questions

1 Explain what the following mean: *solution; solvent; solute; soluble; insoluble; saturated.*
2 Which of the following are solutions and which are suspensions? *ink; orangeade; steam; tea; exhaust fumes; beer.*
3 Which of the following are soluble in water? *sand; sugar; butter; aspirin; chalk; oil; vinegar.*
4 How would you obtain dry, well-shaped crystals from a solution of salt water?
5 How would you prepare a jug of filter coffee?

9 Separating Liquids

Whisky is made by distilling a liquid called wort in copper vessels. Wort is like weak beer. It is made by fermenting barley

Pure water from sea-water

We can get salt from sea-water by evaporating off the water. But can we get *pure water* from sea-water? When the sea-water is evaporated, water vapour escapes into the atmosphere. If the water vapour is passed into a second container and cooled, it will turn back to water. All the solute (salt) stays in the solution, so the water which evaporates and cools is pure water.

When a vapour changes to a liquid the process is called **condensation**. The apparatus which cools the vapour to liquid is a **condenser**. Figure 1 shows how pure water can be obtained from sea-water by evaporating the water and then condensing the water vapour. This process of evaporating a liquid and condensing the vapour is called **distillation**.

distillation = evaporation + condensation.

When the sea-water in figure 1 is heated, the water vapour which escapes passes through the inner tube of the condenser. This is cooled by the cold water flowing around it, so the water vapour condenses and drips from the lower end of the condenser into a receiver. The pure water which collects is called the **distillate**.

Distillation can be used to separate any solvent from its solution. In parts of the Middle East, where fuel is cheap, distillation is used on a large scale to get pure water from sea-water. Pure water (distilled water) can also be obtained by distilling tap water. Most chemical laboratories have their own distillation apparatus called a *still* which produces distilled water.

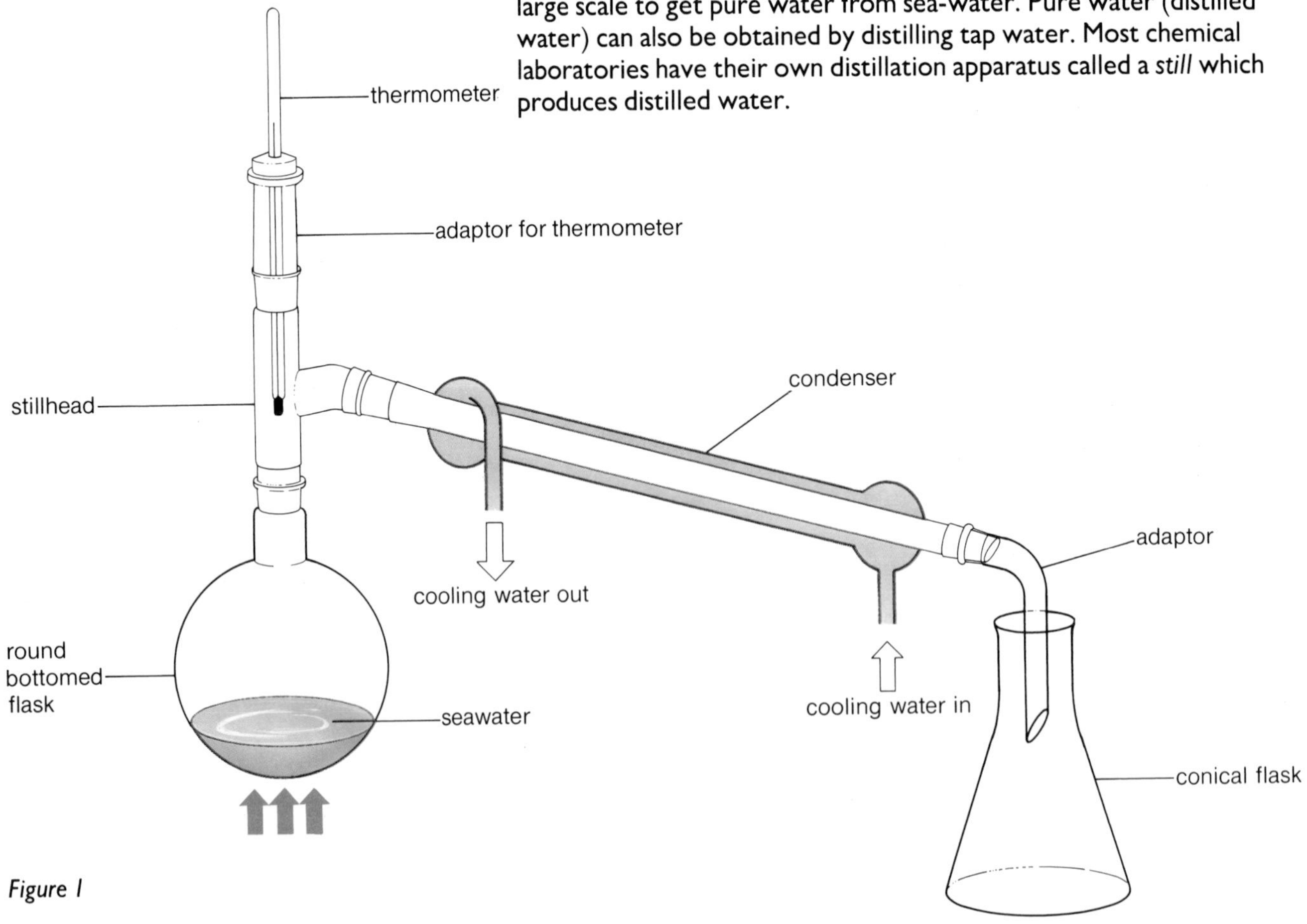

Figure 1

Petrol from crude oil

Crude oil is a mixture of many liquids and dissolved solids. Petrol and paraffin can be separated from it by distillation. When distillation is used to separate a mixture of liquids as in crude oil, it is called **fractional distillation** because it separates the mixture into two or more parts (fractions). Figure 2 shows the temperatures and products at different heights in an industrial fractionating column. Inside the column, there are horizontal trays with raised holes in them. The crude oil is heated by a furnace and the vapours then pass into the lower part of the fractionating column. As the vapours rise up the column through the holes in the trays, the temperature falls. Different vapours condense at different heights in the tower and are tapped off and used for different purposes. Liquids like petrol, which boil at low temperatures, condense high up in the tower. Liquids like lubricating oils, which boil at higher temperatures, condense lower down in the tower.

Fractional distillation is also used in industry to separate oxygen and nitrogen from the other gases in air (section B).

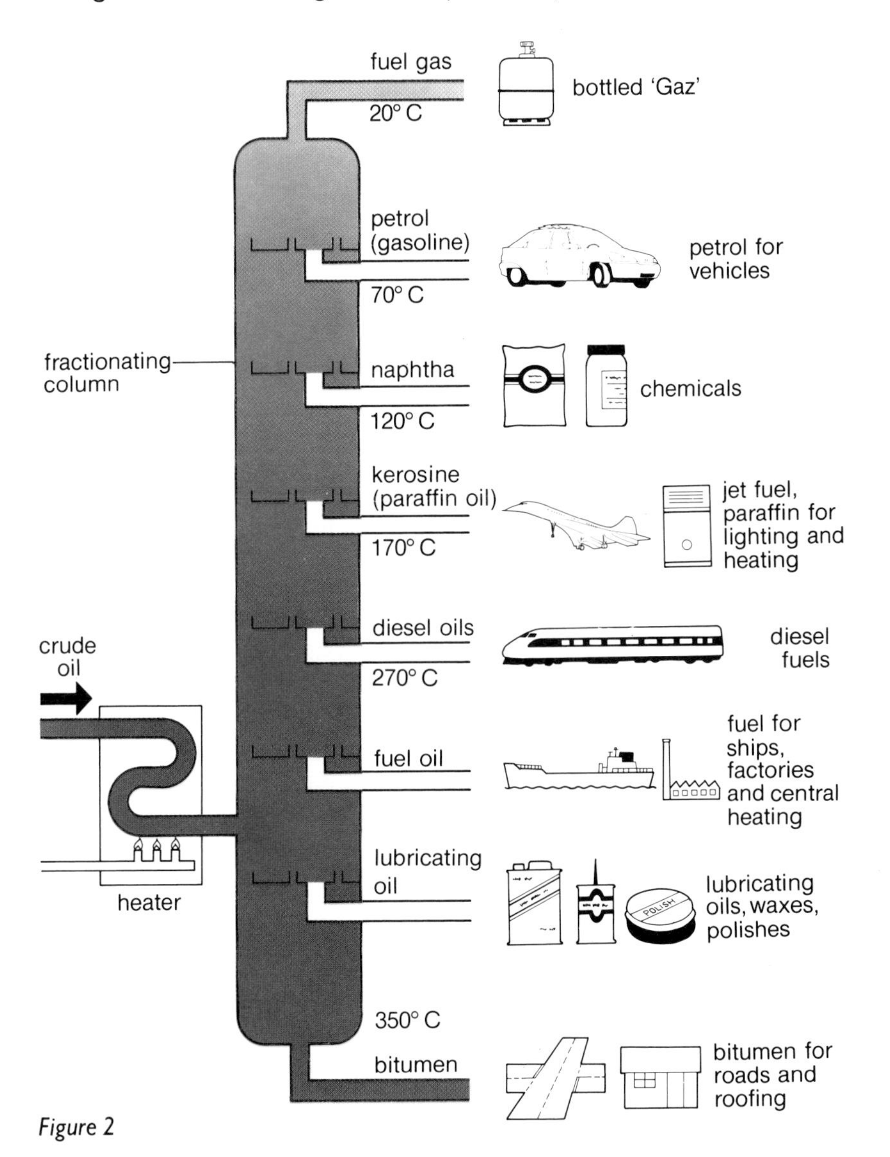

Figure 2

Questions

1 Explain the following:
evaporation; condensation; distillation; distillate.

2 What is the difference between simple distillation and fractional distillation?

3 Two important processes that involve distillation are the manufacture of brandy and whisky. Brandy is made by distilling wine; whisky is made by distilling beer. Both brandy and whisky contain alcohol (boiling point, 78° C) and water (b.pt. 100° C).

(a) Which liquid boils at the lower temperature, alcohol or water?

(b) Which of these two liquids evaporates more easily?

(c) If beer is distilled, will the distillate contain a larger or smaller percentage of alcohol?

(d) Why is whisky more alcoholic than beer?

10 Pure Substances

How pure is 'pure' honey?

The label in the photograph says that the honey is pure. This means that it has had nothing added to it or taken out of it. The so-called 'pure honey' contains many different substances. The proportions of these different substances gives each honey its own particular flavour.

> *When chemists say that something is pure, they mean that it is a single substance and not a mixture of substances.*

The honey is not pure to a chemist because it contains many substances. Most substances that occur naturally are mixtures, *not* pure substances. For example, air is a mixture of gases, mainly nitrogen and oxygen; petrol is a mixture of octane and other liquids and ink is a mixture of water and coloured dyes. If you have spilt ink on blotting paper you may have noticed that as the blot spreads out, it forms a series of different colours. Each of the coloured rings contains one of the dyes in the ink. Figure 1 shows how the dyes in ink can be separated more clearly.

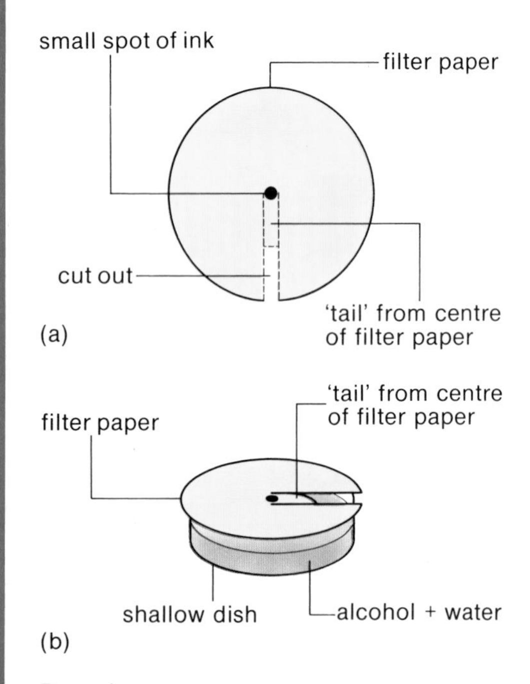

Figure 1

Cut the filter paper as shown in figure 1(a) and fold back the tail from the centre. Put a spot of ink at the centre of the filter paper and leave it to dry. Then, place the filter paper on the shallow dish so that the 'tail' dips into the mixture of equal parts water and alcohol (figure 1(b)). As the liquid soaks up the 'tail' and through the filter paper, the different dyes in the ink spread out in a series of coloured rings (figure 2). The green dye in grass and the various coloured dyes in flower petals can be separated by a similar process. The coloured substances are first removed from the grass or petals by grinding in a pestle and mortar with alcohol.

This method was first used to separate coloured substances. The process was therefore called **chromatography** from the Greek word *khroma* meaning colour. Nowadays, the substances being separated may be colourless. After chromatography has taken place, the paper is sprayed with a liquid which reacts with the colourless substances to produce coloured substances.

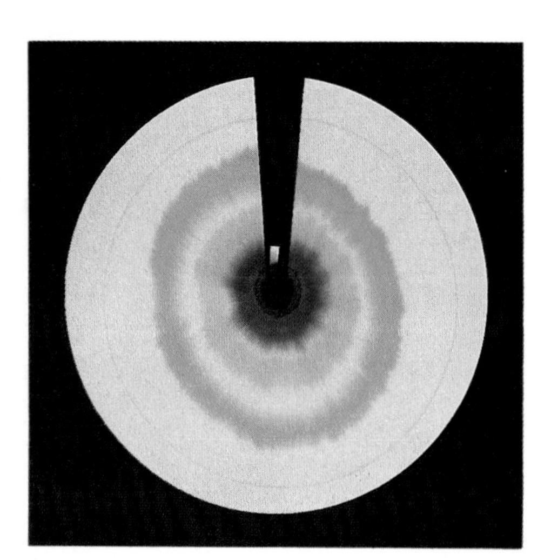

Figure 2

A chromatogram of black ink

The solvent was butan-1-ol: ethanol: 0.88 ammonia solution 3:1:1 by vol.)

Chromatography can be used to separate mixtures and to decide whether a substance is pure. It is very important in medicine because it can be used to separate drugs and other constituents in blood and urine.

The natural colours in fruit, vegetables and other foods can be improved by synthetic dyes. Chemists can detect these dyes using chromatography and check that they are harmless.

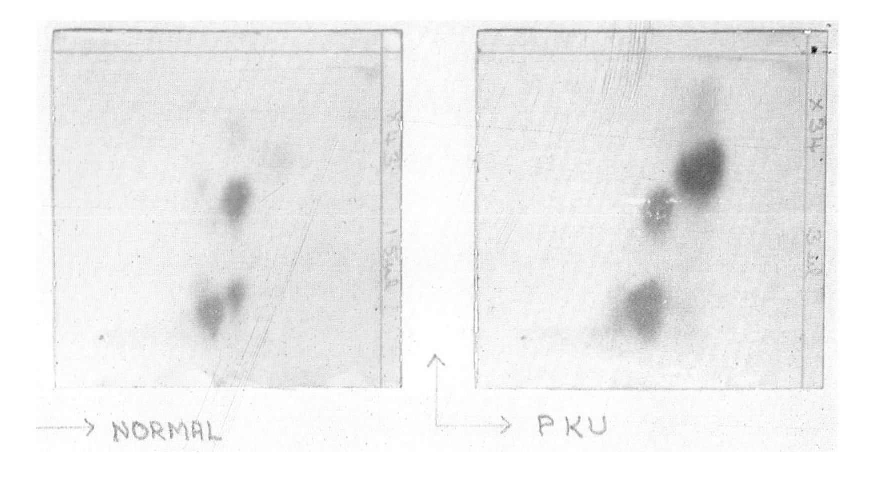

This photograph shows two chromatograms of substances called 'ketones' in the urine. The one on the left is from a normal child. The one on the right labelled 'PKU' is from a child suffering from phenylketonuria. This causes the child to excrete ketones in the urine, which show up as larger spots on the chromatogram

Testing for pure substances

There are three ways of testing that a substance is pure:
- *using chromatography*
- *checking its melting point*
- *checking its boiling point*

If a substance is pure it will contain only one kind of material. So, if a coloured substance is pure, it will give only one coloured ring or band during the chromatography.

The boiling point of a liquid changes as the atmospheric pressure rises and falls. On top of Mount Everest, where the pressure is much lower than at sea level, water boils at about 70° C. But, if the atmospheric pressure stays constant, a pure substance will always boil at the same temperature. Because of this, we must pick a standard pressure and measure all boiling points and melting points at this standard pressure. Scientists have chosen the pressure exerted by a column of mercury 760 mm high as standard atmospheric pressure. This pressure is called 1 atmosphere (atm). In SI units, 1 atm = 101 325 $N\ m^{-2}$.

Pure water always boils at 100° C and pure ice always melts at 0° C at 1 atm pressure. The measurement of the melting point or the boiling point of a substance is the best way to decide if it is pure. If the substance is impure, its melting point and boiling point will differ from the values for the pure substance. For example, salty water boils above 100° C. As the water boils away and the solution gets more concentrated, the boiling point slowly rises.

Questions

1 What is a pure substance?
2 Describe how you would use chromatography to test whether some red ink contains a single pure dye or a mixture of dyes.
3 How would you check whether a colourless liquid is pure water? Draw a diagram of the apparatus you would use.
4 Which of the following are pure substances and which are mixtures?
sea water; petrol; iron; steel; steam; soil; milk; salt.

Section A: Study Questions

1 This question concerns the following practical methods: *A* Chromatography; *B* Crystallization: *C* Distillation; *D* Electrolysis; *E* Filtration.

Choose, from *A* to *E*, the method that would be used to
(i) isolate nitrogen from liquid air;
(ii) separate coloured substances in a sample of a soft drink;
(iii) separate petrol from crude oil;
(iv) separate a precipitate from a solution [4]

LEAG

2 The table shows the melting points and boiling points of substances *A*, *B*, *C* and *D*

Substance	Melting point /°C	Boiling point /°C
A	645	1300
B	−7	59
C	−165	−92
D	27	98

(a) Which substance is a liquid at room temperature? (1)
(b) Which substance is a gas at room temperature? (1)
(c) Which substance cannot be a metal? (1)
(d) Which substance is likely to be an ionic compound? (1)

SEG

3 A company that separates cream from milk knows that there is an increased demand for cream during the summer. Assuming that the company can produce enough cream to satisfy the increased demand, what extra equipment will be needed during the summer?

4 Some properties of six elements labelled *A* to *F* are given in the table.

Element	Density /g cm^{-3}	Boiling point /°C	Electrical conductivity
A	3.12	58	Poor
B	8.65	765	Good
C	19.3	2970	Good
D	3.4×10^{-3}	−152	Poor
E	0.53	1330	Good
F	2.07	445	Poor

(a) Which of the six elements are metals?
(b) Which element will float on water as a solid at 20°C?
(c) Which element is a gas at room temperature?
(d) Which non-metal may be a solid at 60°C
(e) Which metal is probably a liquid at 2000°C?
(f) What is the boiling point of the least dense metal?
(g) The elements in the list are cadmium, gold, lithium, bromine, krypton and sulphur. Which element is which?

5 What elements do the following contain? octane (in petrol); marble (calcium carbonate); anhydrous copper sulphate; sodium chloride; sugar; bread; cheese; wood; a penny; alcohol; polythene; water.

6 Some of the processes used in chemistry to separate mixtures are: filtration; distillation; fractional distillation; solution followed by filtration; chromatography; crystallisation; evaporation to dryness.

For each part choose the most suitable process
(i) to obtain copper (II) sulphate crystals from copper (II) sulphate solution;
(ii) to obtain pure water from sodium chloride solution;
(iii) to find out how many coloured substances are present in a purple dye;
(iv) to obtain calcium carbonate from a mixture of calcium carbonate and water;
(v) to obtain sand from a mixture of salt and sand;
(vi) to separate a mixture of ethanol and water. [6]

MEG

7 The following chromatogram was obtained in an experiment to analyse two mixtures.

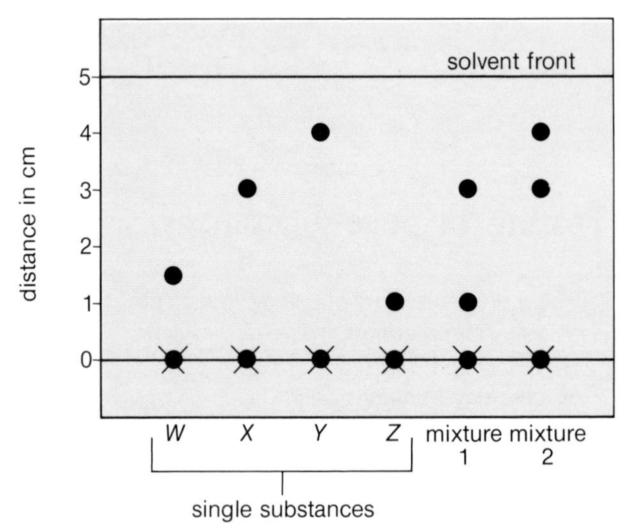

(a) (i) A pencil was used to mark the chromatogram. Why was a pencil used in preference to a ball point or ink pen?
(ii) Why must the base line, on which a small drop of each sample was placed, be above the level of the solvent at the start? (2)
(b) From the chromatogram, which of the single substances were
(i) present in mixture 1;
(ii) present in mixture 2;
(iii) not found in either mixture? (3)
(c) The R_f value of a substance in chromatography is defined as:

$$R_f = \frac{\text{distance moved by substance}}{\text{distance moved by solvent front in the same time}}$$

Calculate the R_f value of substance Y (2)

NEA

(Total: 7 marks)

SECTION B

The Environment—Air and Water

Smoke pollution over Manchester and river pollution in the Grand Union Canal

1 The Air

The air is all around us. We need it to live. We need it to burn fuels and keep warm. Air is a mixture, but the substance in it that we need for both burning and breathing is oxygen.

What percentage of the air is oxygen?

Figure 1 shows an apparatus to find the amount of air used up when copper reacts with the oxygen in it.

copper + oxygen → copper oxide

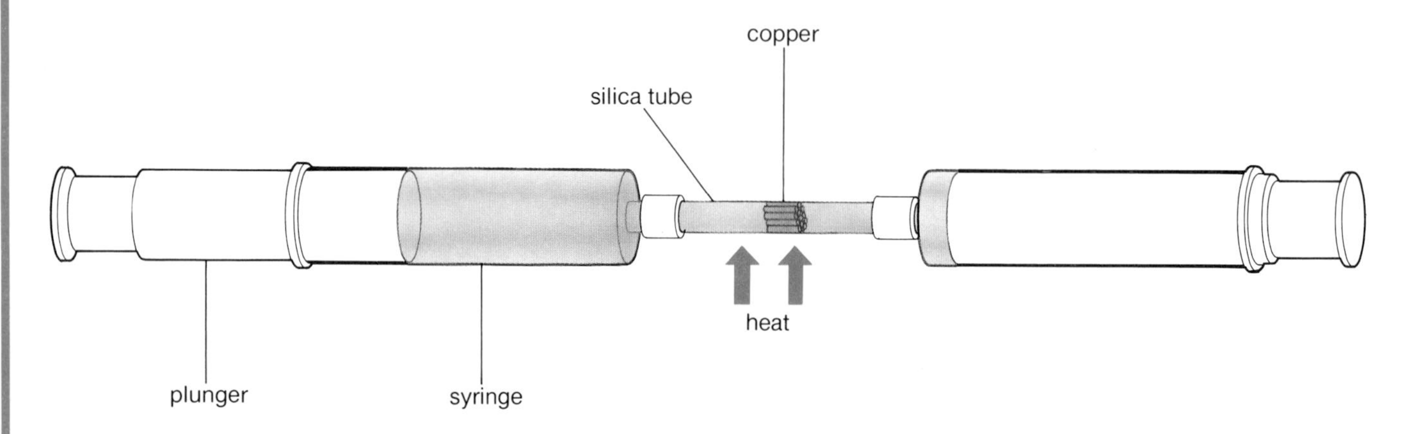

Figure 1

Two syringes are joined through a hard-glass silica tube containing copper. At the beginning, one syringe is empty and the other is filled with 100 cm^3 of air. The silica tube is heated strongly and the air is pushed to and fro several times so that all the oxygen in the air reacts with the copper. The tube is then allowed to cool and the volume of air in the syringe is measured. The heating and cooling are repeated until the volume of air which remains in the syringe is constant. The following table shows the results from one experiment.

Volume of air in syringe before heating	100 cm^3
Volume of gas after first heating and cooling	82 cm^3
Volume of gas after second heating and cooling	79 cm^3
Volume of gas after third heating and cooling	79 cm^3

The results of an experiment to find the percentage of oxygen in air.

1 Has all the oxygen been used up after the first heating?
2 Has all the oxygen been used up after the second heating?
3 Why is the heating and cooling repeated three times?
4 How much oxygen did the copper remove?
5 What is the percentage of oxygen in the air?

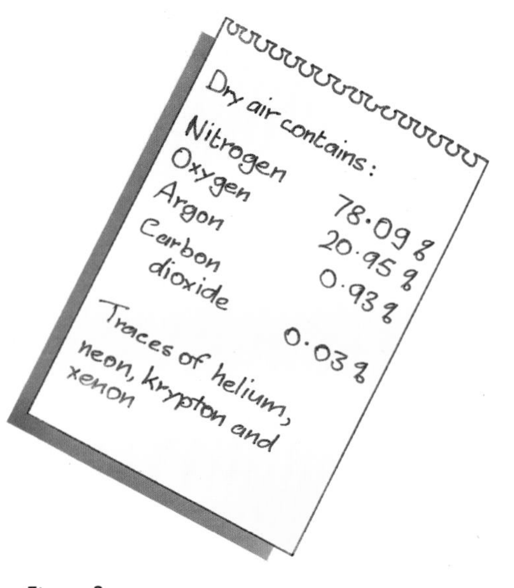

Figure 2

Other gases in the air

Figure 2 shows the percentages of the gases in dry air.

- **Nitrogen** is a very unreactive gas. Its most important industrial use is in making ammonia for nitric acid and fertilizers (section K unit 9).

- **Argon** is even less reactive than nitrogen. Electric light bulbs are filled with argon. The argon will not react with the hot filament when the light is switched on.
- **Carbon dioxide.** Fresh air always contains about 0.03% carbon dioxide. This is because plants and animals release carbon dioxide (in a process called respiration), and it is also formed when fuels burn. Carbon dioxide is used up when plants *photosynthesize* (see section I).

Although *dry* fresh air contains only those gases in figure 2, ordinary air also contains water vapour and waste materials from industry such as sulphur dioxide, nitrogen dioxide, soot and smoke.

Humans have no gills. They must carry a supply of air for underwater swimming

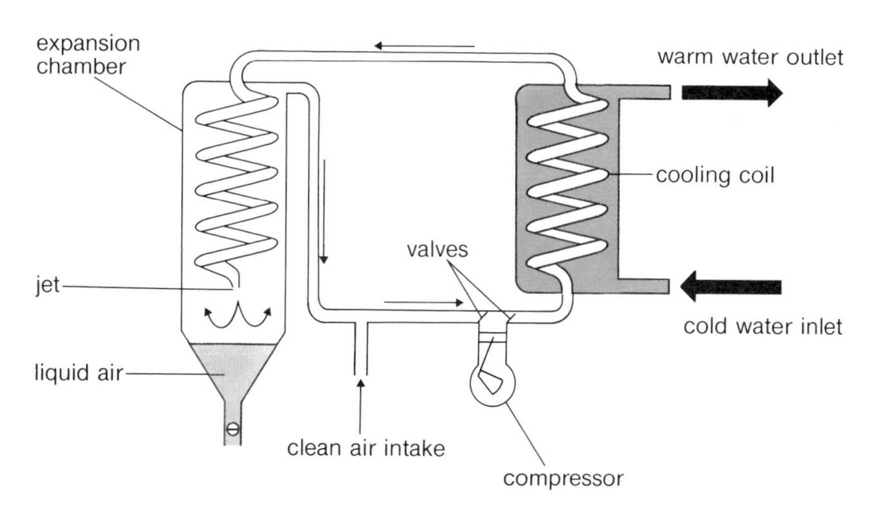

Figure 3

Obtaining gases from the air

The air is an important source of oxygen, nitrogen and argon. The gases are separated by making liquid air (figure 3) and then fractionally distilling this. The clean dust-free air is first compressed, which causes it to warm up. To cool it down again, it is passed through a water-cooled coil.

The air is then allowed to escape rapidly from a jet in the expansion chamber. This sudden expansion causes cooling. The same cooling happens when compressed carbon dioxide escapes from a fire extinguisher. The colder, expanded air further cools the air approaching the jet. As the cooled air circulates round the apparatus, its temperature gets lower and lower. In time the temperature gets so low that the air starts to liquefy and collect below the jet. The liquid air is then removed from the chamber and separated into pure gases by fractional distillation. When liquid air is warmed slowly, liquid nitrogen (b.pt. −196°C) becomes a gas at a lower temperature than liquid oxygen (b.pt. −182°C). The nitrogen boils off and can be stored under pressure in steel cylinders. When all the nitrogen has boiled off, the argon and oxygen can be separated and stored in the same way.

Questions

1 In the syringe experiment to find the percentage of oxygen in the air, the hard-glass tube containing copper was weighed before and after the experiment.

Vol. of oxygen in air	21 cm^3
Wt. of tube + contents at start	11.365 g
Wt. of tube + contents finally	11.393 g

(a) Why does the tube + contents increase in mass?
(b) What volume of oxygen does the copper react with?
(c) What mass of oxygen does the copper react with?
(d) What is the density of this oxygen (i.e. the mass of 1 cm^3 of it)?

2 Describe an experiment which you could do to find the percentage of oxygen, by volume, in the air you breathe out. Draw a labelled diagram of the apparatus you would use, describe the measurements you would make and explain how you would calculate a result.

3 Is air a mixture or a compound? Give three reasons to support your answer.

2 Oxygen

Metal sculptors use oxy-acetylene welders to shape and cut their designs

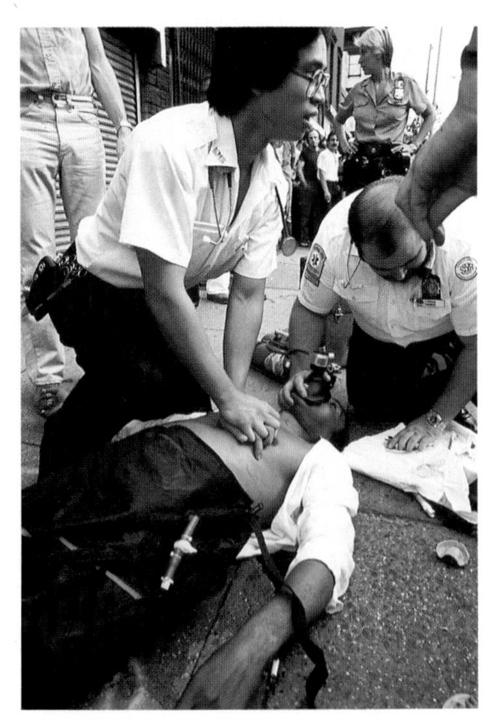

Oxygen being used to revive someone

The uses of oxygen

Oxygen is essential for burning and breathing, but the oxygen does not have to be pure. Pure oxygen is, however, required in large quantities for industrial and medical uses.

- **Manufacture of steel**. One tonne of pure oxygen is needed to produce every 10 tonnes of steel from impure iron ore. Oxygen is blown through the molten iron to remove carbon and sulphur impurities. The carbon and sulphur are converted to carbon dioxide and sulphur dioxide and then escape as gases.
- **Welding and cutting**. Pure oxygen is used in oxy-acetylene welding and cutting. When acetylene burns in oxygen, the temperature reaches 3200° C. This is hot enough to melt most metals which can then be cut or welded together.
- **Breathing apparatus**. Pure oxygen is used in life-support machines in hospitals. Oxygen helps the breathing of patients with lung diseases such as pneumonia. It is also mixed with anaesthetizing gases during surgical operations.
  Mountaineers and deep sea divers also use oxygen mixed with other gases to breathe when supplies of air are not available.

Element	Reaction with oxygen	Product	Add water to product, then universal indicator
Carbon	Glows red hot, reacts slowly	Colourless gas (carbon dioxide)	Dissolves pH = 5, acidic
Sulphur	Burns readily with a blue flame	Colourless gas (sulphur dioxide)	Dissolves pH = 3, acidic
Sodium	Bright yellow flame—white smoke and powder	White solid (sodium oxide)	Dissolves pH = 11, alkaline
Magnesium	Dazzling white flame—white clouds and powder	White solid (magnesium oxide)	Dissolves slightly pH = 8, alkaline
Iron	Glows red hot and burns with sparks	Black-brown solid (iron oxide)	Insoluble
Copper	Does not burn, but the surface turns black	Black solid (copper oxide)	Insoluble

Comparing the reactions of some elements with oxygen

How does oxygen react with other elements?

Things made of iron are slowly covered with rust (iron oxide) on exposure to the air. Similarly, shiny aluminium surfaces get covered with a layer of white aluminium oxide as the aluminium reacts with oxygen in the air. The table shows the results obtained when various elements are heated strongly in pure oxygen.

What substances are produced?

All the elements burn better in oxygen than in air. The substances produced are **oxides**.

carbon + oxygen → carbon dioxide
sodium + oxygen → sodium oxide

- *The metal oxides (sodium oxide, magnesium oxide, iron oxide and copper oxide) are all solids.*
- *The non-metal oxides (carbon dioxide and sulphur dioxide) are both gases.*

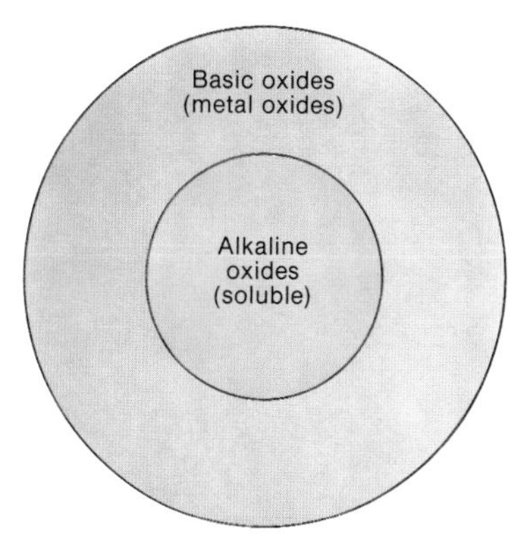

Figure 1

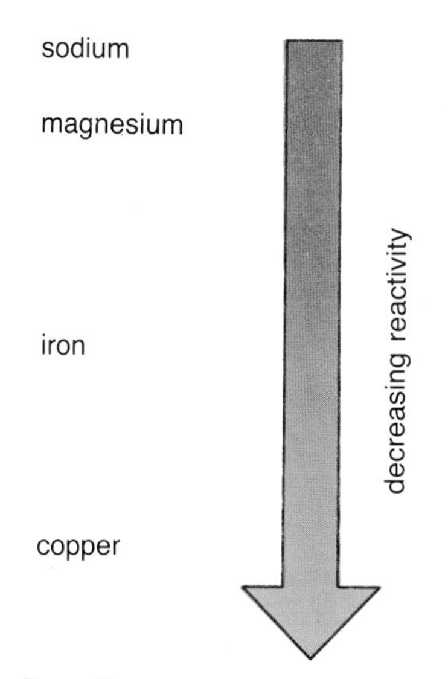

Figure 2

Do oxides react with water?

Each of the oxides was shaken with water and the solution produced was then tested with universal indicator. The results are given in the table.

Sodium and magnesium oxides react with water to form alkaline solutions with a pH greater than 7.

sodium oxide + water → sodium hydroxide
magnesium oxide + water → magnesium hydroxide

Iron oxide and copper oxide do not react with water or dissolve in it. So, they do not change the colour of the indicator.

The oxides of metals are called **basic oxides**. Most metal oxides are insoluble in water but a few, like sodium oxide and magnesium oxide, react with it to form alkaline solutions. These oxides are called **alkaline oxides**. Figure 1 shows the relationship between basic oxides and alkaline oxides in a Venn diagram.

The non-metal oxides react with water to form acidic solutions with a pH less than 7. *These oxides of non-metals which give acids in water are called* **acidic oxides**.

carbon dioxide + water → carbonic acid
sulphur dioxide + water → sulphurous acid

What is the order of reactivity?

Some metals, such as sodium and magnesium, react very vigorously with oxygen producing a bright flame. Others, like copper, react very slowly with no flame. These metals only form a thin layer of oxide. We can arrange the metals in order of reactivity with the most reactive metal at the top and the least reactive at the bottom. This is called a **reactivity series** for the metals. Figure 2 shows a reactivity series for the four metals that we used.

Questions

1 (a) What are the main uses of oxygen?
(b) Why do rockets carry liquid oxygen?

2 What is meant by the following: *basic oxide; acidic oxide; alkaline oxide?*

3 Barium burns in air to form a solid oxide.
This oxide reacts with water to give an alkaline solution.
(a) Is barium a metal or a non-metal?
(b) Write word equations for the reactions described above.
(c) Which of the following elements would react like barium?
copper; sulphur; sodium; iron; nitrogen; calcium; carbon; potassium.

3 Burning and Fuels

The Greeks believed that Prometheus took pity on humans and stole fire from the gods to give warmth to men and women

Some fuels produce sparks and flames when they burn

Why is burning so important?

Burning was the first chemical process used by humans. Our ancestors burnt fuels to keep warm, to cook food and to produce new materials like metals and clay pots. Burning is just as important to us. We burn fuels to keep warm, cook food, drive motor cars and generate electricity. Any substance which burns in air to produce heat is a **fuel**.

The most important fuels are:

- *coal* }
- *oil* } *fossil fuels which come from the decay of dead animals and plants*
- *natural gas* }
- *uranium—nuclear fuel* *(see section L)*

Lavoisier showed that part of the air is used up when things burn. He said this part of the air was oxygen

Fossil fuels are mostly compounds containing carbon and hydrogen. During burning, these elements combine with oxygen in the air to produce carbon dioxide, water and heat.

fuel + oxygen → carbon dioxide + water + heat

Reactions like this which give out heat are called **exothermic reactions**.

Sometimes, when a fuel burns, so much heat is produced that the products burst into flames. This happens in a Bunsen burner or a gas cooker when gases react with oxygen in the air to produce the flame.

Some fuels, like hydrogen, and explosives, like TNT and dynamite, react so fast that they cause explosions. Because of this *great care is needed in handling and using fuels.*

Is there an ideal fuel?

Firelighters are ideal for starting a fire. They can be set alight easily and they keep on burning steadily. But firelighters are too expensive to use all the time. Charcoal is excellent for a barbecue because it smoulders slowly and doesn't produce any smoke, but it would be useless on an open fire. Coal and oil are used as fuels for our homes and industries because they are relatively cheap, they produce a lot of heat and they occur naturally. But, you wouldn't use either of these to cook in a tent.

These points will make you see that there is no single ideal fuel. Different fuels are suitable for different purposes. What is more, chemists have found ways of making more useful and more suitable fuels. These include firelighters, coke, petrol, GAZ, lighter fuel and diesel oil.

The important questions to ask in choosing a fuel for a particular purpose are:

1. does it produce a lot of heat?
2. does it produce a lot of ash and/or smoke?
3. how cheap is it?
4. does it occur naturally?
5. does it light easily?
6. does it burn steadily?
7. is it easy to store?
8. is it easy to transport?

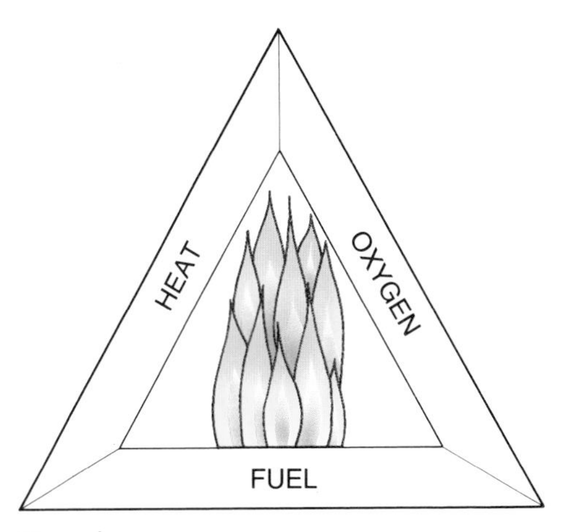

Figure 1
The fire triangle. To start a fire you need a fuel, oxygen and usually heat

What fuel is used to get a spaceship into orbit?

Questions

1 Figure 1 shows a diagram of the *fire triangle*.
(a) Why are heat, oxygen and fuel shown along the sides?
(b) What happens if one of these is missing?
(c) What must you do to stop a fire?
(d) Have you ever needed to stop a fire? If so, what did you do and why?
(e) How would the fire service deal with
(i) *a petrol fire*; (ii) *a forest fire*;
(iii) *a chip-pan fire*?
Say what they would do and why.

2 (a) What are the fossil fuels?
(b) Why are fossil fuels running into short supply?
(c) How did fossil fuels form originally?

3 (a) What fuels are used for each of the following?
barbecueing; cooking in a tent; heating a school; getting a spaceship into orbit; lighting a housefire; running a car.
(b) For each of the uses in part (a) say why the particular fuel is chosen.

4 Air Pollution

Air pollution is worst in heavily industrialized areas.

The two photographs show the Public Record Building, London before and after cleaning

Air pollution causes windows, curtains and clothes to become dirty very quickly. It also attacks the stonework of buildings and damages our health.

Most air pollution is caused by burning fuels. The problems are therefore worst in industrial areas. When a fuel burns, it reacts with oxygen to form oxides. If the fuel burns completely, then all the carbon in it is turned into carbon dioxide which is only slightly acidic. If there is not much air available, the carbon may form **soot and smoke** or it may be turned into **carbon monoxide**, which is a very poisonous gas. Carbon monoxide reacts with a substance in the blood (*haemoglobin*) and stops it carrying oxygen to the brain and other parts of the body. It is dangerous to run a car engine in a garage with the doors closed because the lack of air may lead to the production of carbon monoxide. Some fuels, like coal and coke, contain small amounts of sulphur. When these fuels burn, **sulphur dioxide** is produced. This is a colourless, choking gas which irritates our eyes and lungs. The sulphur dioxide dissolves in water to form an acidic solution containing sulphurous acid.

$$\text{sulphur dioxide} + \text{water} \rightarrow \text{sulphurous acid}$$

When sulphur dioxide gets into rain, the rain water become acidic. This **acid rain** harms plants and attacks the stonework of buildings. Acid rain has been blamed for the poor growth of trees in Scandinavia. It is thought that the sulphur dioxide is produced in the industrial areas of Northern England and Southern Scotland and then carried by the prevailing winds across the North Sea to Scandinavia.

In a car engine, petrol burns in the cylinders. A lead compound is added to the petrol to help it burn smoothly, but waste lead compounds pass out with the exhaust gases. These **lead compounds** are poisonous because they affect the brain. In some countries there are strict laws to control the amounts of lead compounds in exhaust gases, and lead compounds cannot be added to petrol. This makes the petrol a little more expensive.

Car engines need air to burn the petrol. When the mixture is sparked, nitrogen and oxygen in the air combine to produce **nitrogen dioxide** which is an acidic gas. Nitrogen dioxide causes further pollution of the air. It irritates our eyes and lungs like sulphur dioxide.

When solid fuels like coal and coke burn, they leave ash. The ash contains solid metal oxides, called basic oxides. If the ash contains oxides of reactive metals (e.g. sodium oxide and calcium oxide), it will form an alkaline solution with water. Oxides of less reactive metals (like aluminium oxide, iron oxide and copper oxide) are insoluble and do not react with water. Notice how the products and pollutants from burning are mostly oxides. The properties of these oxides are shown in the table.

Acid rain has caused damage to this statue on Wells Cathedral, Somerset

The Tokyo policeman is checking the exhaust gases of the car for pollutants

Air pollution shows that advances in modern technology have both advantages and disadvantages. We have the benefits of more fuels and more vehicles, but we also have the problems of more pollution. Fortunately, several countries now have strict laws to control air pollution. In 1956, the Clean Air Act in Britain made it an offence to pollute the air with soot and smoke from factories and homes. Local authorities were given the power to set up smoke control areas ('smokeless zones') to prevent air pollution. Only smokeless fuels, such as coke and anthracite, can be used in smokeless zones.

	Non-metal oxides	**Metal oxides**
Examples	Sulphur dioxide, carbon dioxide, nitrogen dioxide	Calcium oxide, sodium oxide, iron oxide, lead oxide
State	*Gases*	*Solids*
Nature	*Acidic*	*Basic*
Reaction with water	React to form *acids* e.g. carbon dioxide + water → carbonic acid	Most are insoluble, but oxides of reactive metals form *alkalis*. sodium oxide + water → sodium hydroxide

The properties of metal oxides and non-metal oxides

Questions

1 (a) What are the main substances that cause air pollution?
(b) How do they get into the air?
(c) Suggest 3 ways in which air pollution causes damage.

2 (a) What is acid rain and how is it caused?
(b) How does acid rain affect (i) lakes; (ii) forests?

3 What is (i) the Clean Air Act (ii) a 'smokeless zone'?

4 (a) What further steps could be taken to reduce air pollution in heavily industrialised areas?
(b) What problems are there with stricter controls over air pollution?

5 Water Supplies

Water controls our lives. It affects our crops, our food, our health and the weather. 60% of the people in Asia and Africa cannot obtain clean water easily

Water is the commonest liquid. We drink it, wash in it, swim in it and complain when it rains. In many ways, water controls our lives. It determines where we can live, it determines whether we can grow crops and produce enough food, it determines which sports we can enjoy and it determines the weather we have. All living things need water—people, animals and plants. About two thirds of your body is water. Everyday you need about 2 litres (3½ pints) of water. This water may be part of your food or drinks. It replaces the water that you lose in urine, in sweat and when you breathe. Water is more important than food—most people can survive for 50 to 60 days without food, but only 5 to 10 days without water.

Using water

Most of the time, we take water for granted. It is only during a very dry summer that we need to use it carefully. In the UK, each person uses about 180 litres of water every day. In some parts of Africa, each person must survive on less than 10 litres per day. Every time you flush the toilet about 10 litres of water are used and taking a bath may use 100 litres. Figure 1 shows the amounts of water that we might use for different purposes each day.

Large amounts of water are also used in industry. Most of the water is used for cooling. A large power station uses about 5 million litres of water per day. Because of this, power stations are often built near the coast or near rivers so that they have a large source of water for the cooling towers. The other large industrial uses of water are for washing and as a solvent. When all these uses are added together, it takes:

10 litres of water to make 1 litre of lemonade or beer;
200 litres of water to make 1 newspaper;
50 000 litres of water to make 1 average-sized car.

Where does water come from?

We expect to get as much clean water as we need at the turn of a tap. But where does this water come from? In section A, we learnt how water is purified by the water authorities. Where does this treated water come from originally? If we trace the flow of water backwards, we find that it comes from:
surface water in rivers and lakes or
ground water in underground wells.

We can trace the water supply back even further and see that all surface water and underground water comes from rain. When rain falls on the ground, some of it soaks deep into the earth as ground water. The rest runs off the land into rivers and lakes as surface water. But where does the rain come from?

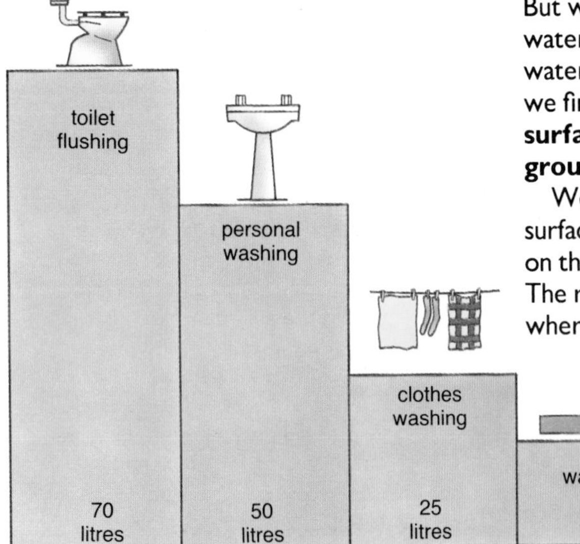

Figure 1
Amounts of water used by one person each day in the UK

A large power station uses 5 million litres of water every day for cooling

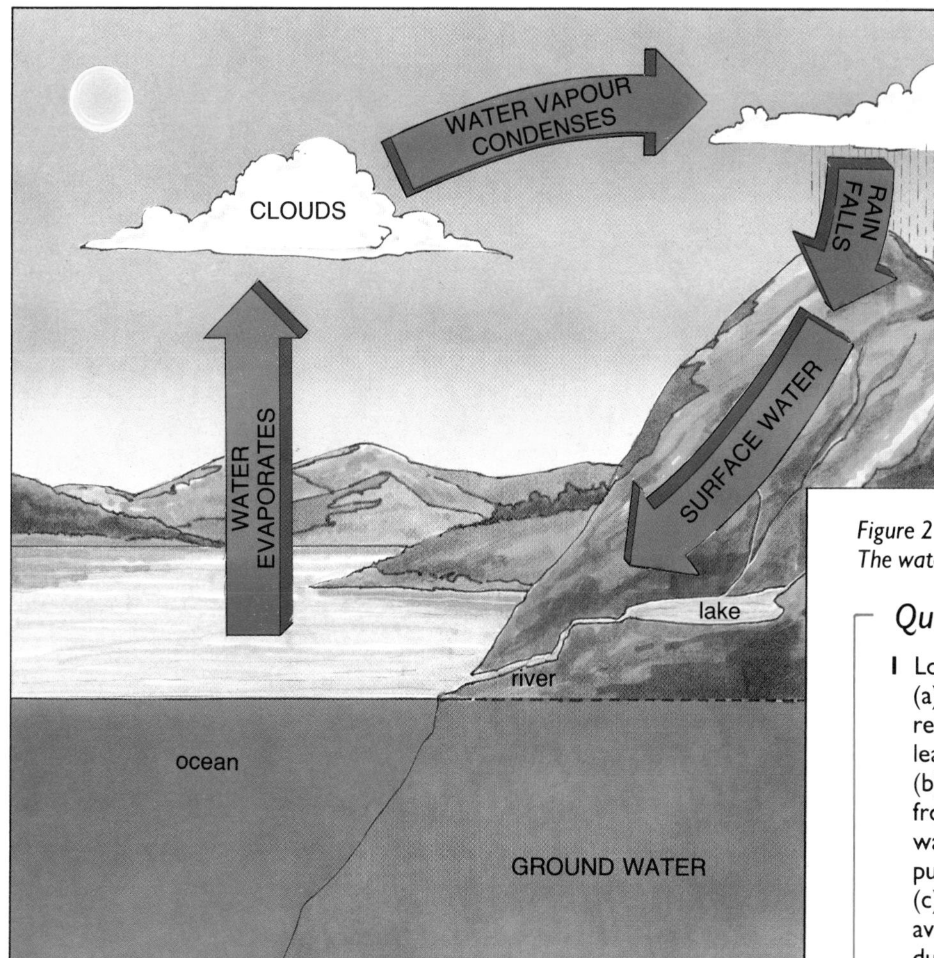

Figure 2
The water cycle

Heat from the sun causes water to evaporate into the air from rivers, lakes and oceans. This water vapour then collects into clouds. As the clouds rise, they cool down and the water vapour condenses to form drops of water. These fall back to the earth as rain which soaks into the soil or joins rivers and oceans. The whole cycle then begins again. This continuous movement of water from the earth's surface to clouds and then back to the earth as rain is called the **water cycle** (figure 2).

Questions

1 Look at figure 1.
(a) Which of the uses of water requires (i) the purest water; (ii) the least pure water?
(b) Write down the uses, in order, from those needing the purest water to those needing the least pure water.
(c) Which of the uses could be avoided in order to save water during a drought?
(d) Suppose you saved all the water from personal washing. To which of the other uses could this water then be put?

2 What do you understand by *surface water; ground water; water cycle?*

3 Where does water come from?

6 Water Pollution

This aerial photograph shows sea-water being badly polluted as a result of material being dumped at sea

Main sources of water pollution
Sewage
Fertilizers
Industrial chemicals
Pesticides
Oil
Detergents

Table 1: the main sources of water pollution

There are vast amounts of water in oceans, lakes and rivers, but it is easily polluted. The main sources of water pollution are listed in table 1.

Sewage is the main cause of water pollution. At one time, all sewage was pumped into rivers and the sea. This caused health hazards and led to diseases such as cholera. Sewage also upsets the balance of life in the water. If the amount of sewage is small, then bacteria in the water can break it down to harmless materials like carbon dioxide, nitrates and water. But, if the amount of sewage is large, then the bacteria use up all the oxygen dissolved in the water as they feed on the sewage. Once the oxygen concentration gets too low, most of the living organisms in the water (including the bacteria) die and the water becomes cloudy and smelly. In order to avoid this pollution, sewage plants treat the waste before it is returned to rivers and the sea. The sewage is pumped into large tanks and mixed with air (aerated) so that it can be decomposed more rapidly by bacteria.

During the 1950s, the River Thames was almost lifeless between London Bridge and the sea, mainly because of the low oxygen concentration in the water. Since then, there have been improvements in pollution control. In 1974, the first salmon was caught since 1835 and now more than 70 species of fish have returned.

Table 2 summarises the sources and effects of other substances which cause water pollution.

Some people believe that heat is also a serious water pollutant. It is a particular problem near the cooling water oulets from large nuclear power stations where the water temperatures may be 10°C higher than normal. The higher water temperatures increase the corrosion of steel structures in the water and also reduce the amount of oxygen dissolved in the water. The reduced oxygen content can be harmful to some fish such as salmon and trout, although other species thrive in the warmer water.

Polluting substance	Source	Effect
Fertilizers	Rain washes fertilizers into rivers and lakes	Bacteria and algae grow faster, use up all the dissolved oxygen and then die
Industrial chemicals	Oils, metal compounds, acids, alkalis, dyes, etc. from factories	Poisonous to animals, plants and bacteria in the water
Pesticides	Spraying of crops with chemicals	Poisonous chemicals accumulate in the bodies of larger animals
Oil	Oil from refineries and from ship-wrecked tankers	Covers sea birds with oil; pollutes beaches
Detergents	Factories, offices, homes	Causes water to foam; poisonous to organisms in the water

Table 2: the sources and effects of various water pollutants

This guillemot and these razorbills have been killed by oil pollution at sea

This beach in the Gulf of Aqaba has been badly polluted from a factory which processes phosphate rock (calcium phosphate)

It is easy to adopt a very simple view of pollution and say, 'All pollution is wrong'. But, if we did this and passed laws to protect our rivers and lakes, then industries would have to spend more money on avoiding pollution and getting rid of waste in some other way. These additional costs would be passed on to the customers and the goods from our factories would be more expensive. As a society we have to decide between more pollution control and higher prices or less pollution control and lower prices. Remember, too, that as members of society and consumers of goods and materials that require oil, fertilizers, pesticides and detergents, we are all responsible for pollution.

Questions

1 What are the main sources of water pollution?

2 (a) Why does sewage cause fresh water to become murky and smelly?
(b) Why are bacteria mixed with sewage at some treatment plants?
(c) Why is solid sewage sometimes heated to 30° C and aerated regularly at the sewage plant?

3 (a) Suppose that each person in your family uses 150 litres of water every day.
(i) How much water is used by your family in one day?
(ii) How much water is used by your family in one year?
(b) Suppose your family pays £50 in water rates for the supply and treatment of water each year.
(i) What is the cost of water per litre for your family?
(ii) Compare this with the cost of one litre of petrol or milk.

7 Properties of Water

Living things, like this seal, can survive in water even in the Antarctic because ice floats on water

Water covers more than 70% of the Earth's surface in oceans, lakes and rivers. It freezes at 0° C and boils at 100° C so it is a liquid at most places on the Earth. This is unusual. Can you think of another substance which occurs naturally as a liquid?

Water is also a very good solvent. It will dissolve substances as different as salt, sugar, alcohol and oxygen. Some substances are more soluble than others as you can see in the table.

When the temperature falls below 0° C, water turns to ice. If the water is trapped in pipes as it freezes, then the pipes may split open and cause a burst. The pipes split because water expands when it freezes and the ice takes up more space than the water. This is another unusual property of water. Almost all other substances contract when they freeze. The expansion of water as it changes to ice means that ice is less dense than water. It causes rocks to crack and potholes to appear in the roads.

When ice forms on the surface of a pond, it acts as an insulator and it prevents the water below from freezing. Because of this, living things can survive in water even in Arctic conditions. If the water contracted as it froze, ice would be denser than water and it would sink. Lakes and ponds would then freeze from the bottom upwards and in a really cold spell the whole pond would freeze solid. All the fish and plants and other water creatures would die if this happened.

Water expands as it freezes. This can cause serious damage such as burst pipes

Notice that water has some unusual properties:
- *it occurs naturally on the Earth as a liquid;*
- *it is a solvent for many different substances;*
- *it expands when it freezes.*

Water as a solvent

When you stir a spoonful of instant coffee into a cup full of *hot* water, it dissolves very quickly. If you tried to make a cup of coffee with *cold* water, the coffee would not dissolve. You may have noticed that sugar dissolves better in hot coffee than in cold coffee. These everyday examples show that solids dissolve better in water as the temperature increases.

When sugar is stirred into coffee, it dissolves until a certain amount has been added. If more sugar is then added, it remains undissolved, provided the volume of the liquid and its temperature do not change. The coffee is saturated with sugar. The extent to which a solute like sugar can saturate a solvent like coffee is expressed in terms of **solubility**.

Substance	Mass which dissolves in 100 g of water at room temp. (20° C)
Sand	insoluble
Salt	36.0 g
Sugar	204 g
Alcohol	infinite
Oxygen	0.004 g
Carbon dioxide	0.014 g

The solubility of some substances in water at 20°C

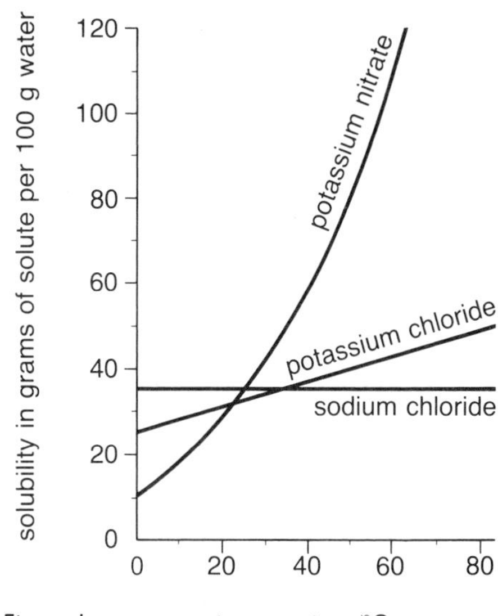

Figure 1

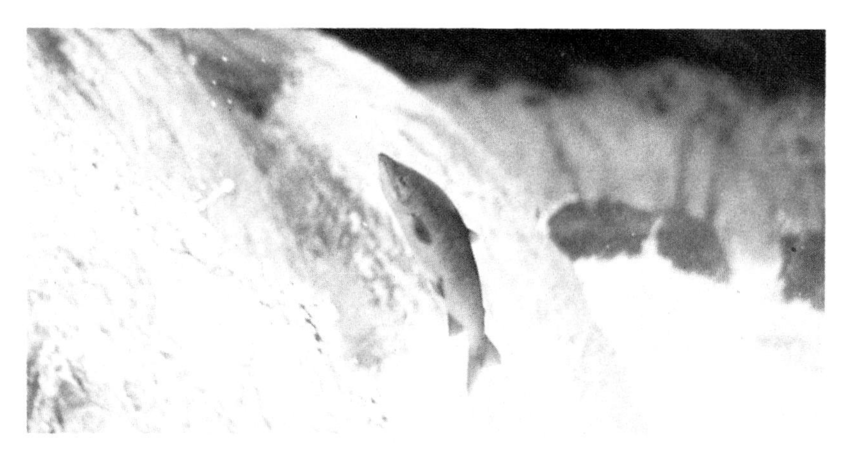

Some fish species require more dissolved oxygen in the water than others. Because of this they cannot survive at higher temperatures. Salmon require higher concentrations than most fish. They cannot survive if the water temperature is above 15°C

The solubility of a solute in a particular solvent is the mass of solute that saturates 100 g of solvent at a given temperature. The table shows the solubilities of various substances in water at 20° C. The solubility of most solids increases with temperature. Figure 1 shows how the solubilities of salt (sodium chloride), potassium chloride and potassium nitrate change as the temperature rises.

Most solids become *more soluble* in water as the temperature rises, but gases become *less soluble* as the temperature rises. Natural water in lakes and rivers contains dissolved gases from the air such as oxygen and carbon dioxide. More gas dissolves if the water is colder or if the water flows over waterfalls where it mixes more freely with air.

Water animals depend on oxygen dissolved in the water. Fish die if there is less than 0.0004 g of dissolved oxygen per 100 g of water. This explains why fish caught in rivers cannot survive in fish tanks indoors where the water is warmer and cannot dissolve as much oxygen.

Notice in the table that carbon dioxide is more soluble than oxygen in water. This is because it reacts with water to form carbonic acid:

$$\text{carbon dioxide} + \text{water} \rightarrow \text{carbonic acid}$$

The carbonic acid which forms in natural waters reacts with chalk and limestone to form hard water. We shall return to this in a later unit.

Questions

1 Why is water an unusual substance?

2 (a) Explain the word *solubility*.
(b) How does the solubility of solids differ from that of gases?

3 Look at figure 1.
(a) Which of the three substances has the largest solubility at
(i) 20° C; (ii) 60° C?
(b) At what temperature do sodium chloride and potassium chloride have the same solubility?
(c) Estimate the solubility of potassium nitrate at
(i) 20° C; (ii) 60° C?
(d) What mass of potassium nitrate crystallizes out of solution when 100 g of water saturated with potassium nitrate at 60° C cools to 20° C?

4 (a) Why is it necessary to give the temperature at which a solubility is measured?
(b) How would you make sure that a solution is saturated at a particular temperature?

5 Design an experiment to show that water expands when it freezes.

8 Reactions of Water with Metals

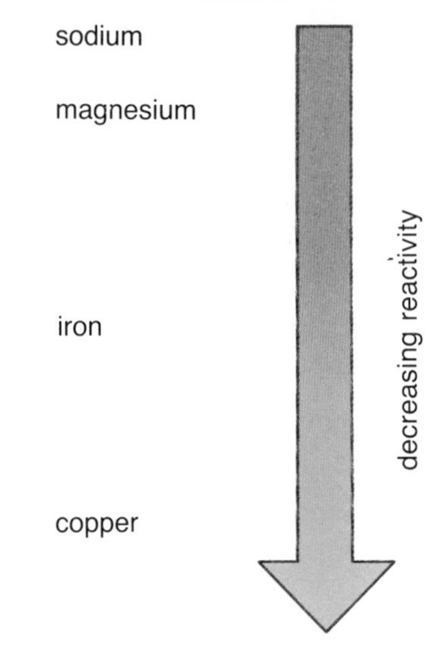

Figure 1

Metals vary in their reactivity with air and oxygen. Some metals, such as sodium and magnesium, react very fast and burn with bright flames to form their oxides:

$$\text{metal} + \text{oxygen} \rightarrow \text{metal oxide}$$

Others, like iron, produce sparks when they react, whilst metals like copper react very slowly and only form a layer of oxide.

> *As a result of these reactions, we can arrange the metals in a rough order of reactivity called an* **activity series** *or* **reactivity series**.

The most reactive metals, sodium and magnesium, are at the top of the series (figure 1) and the least reactive (copper) is at the bottom.

We can test our idea of an activity series using the reaction of metals with water. Sodium reacts with oxygen far more vigorously than copper. Thus, sodium should remove oxygen from water much more readily than copper.

$$\text{sodium} + \underset{\text{(hydrogen oxide)}}{\text{water}} \longrightarrow \text{sodium oxide} + \text{hydrogen}$$

When a small piece of sodium is added to water, it reacts very vigorously (figure 2). The sodium melts to a silvery bead which skates about the water surface. The bead gets smaller as it reacts with water and gives off hydrogen gas. The other product is sodium oxide. This reacts with the water to form sodium hydroxide solution which is alkaline.

$$\text{sodium oxide} + \text{water} \rightarrow \text{sodium hydroxide}$$

Figure 2
Sodium reacting with water

The table summarises the reactions of various metals with water. From these reactions, we can write a reactivity series for the six metals: potassium, sodium, calcium, magnesium, iron and copper. This order of reactivity of metals with water is very similar to the order of

Metal	Observation
Calcium (figure 3)	Sinks in the water and gives off a steady stream of hydrogen. The solution becomes alkaline and cloudy due to the formation of calcium hydroxide.
Copper	No reaction.
Iron	Iron rusts very slowly in water provided oxygen is present. The product is hydrated iron oxide.
Magnesium	Tiny bubbles of gas appear on the surface of the magnesium after a few minutes. The solution slowly becomes alkaline.
Potassium	A violent reaction occurs. Hydrogen and potassium hydroxide form. The potassium gets so hot, that it burns with a lilac flame.

The reactions of various metals with water

reactivity with oxygen (figure 1). This is not surprising since in both cases the metal is reacting to form its oxide.

metal + water (hydrogen oxide) ⟶ metal oxide + hydrogen

metal + oxygen → metal oxide

Notice that copper is the least reactive metal in the activity series in figure 4. Copper does not even react with steam, never mind water! Because of this, it is used to make pipes and storage tanks for hot water.

Another way of comparing the reactivity of metals is to see how they compete with each other for oxygen. The competition of metals for oxygen is used to obtain chromium, iron and titanium from their oxides. The oxides of these metals are mixed with powdered aluminium or another reactive metal. The mixture is then made to react using a fuse of burning magnesium.

A mixture of powdered aluminium and iron oxide (known as 'thermit') is sometimes used for welding. Aluminium is more reactive than iron so it removes oxygen from iron oxide to form aluminium oxide and iron.

aluminium + iron oxide → aluminium oxide + iron

This reaction is very exothermic and the iron is produced in a molten state. As the iron solidifies, it will weld two pieces of steel together. Welding together Inter-City railway lines using this method has increased the average 'life' of the track from 23 to 30 years (figure 5).

Figure 5
Molten iron from the reaction runs into a mould around the rails to be joined. When the iron has cooled the mould is removed and excess metal trimmed off

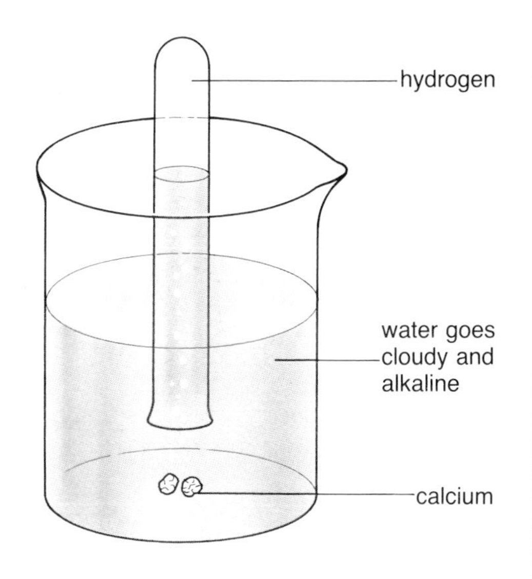

Figure 3
Calcium reacting with water

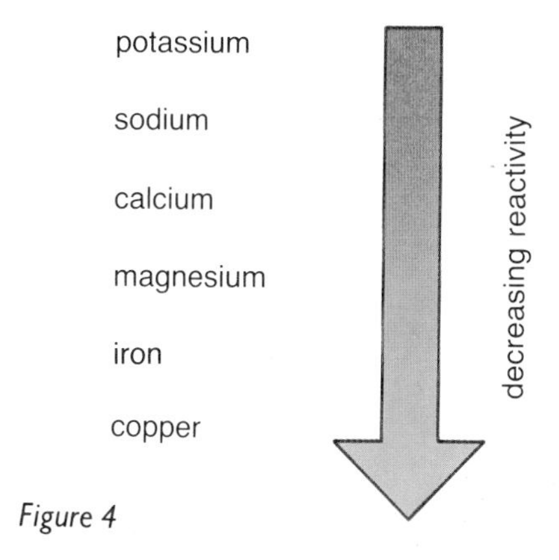

Figure 4

Questions

1 What do you understand by the following:
reactivity series; thermit reaction?

2 (a) Which metals in figure 4 can remove oxygen from water?
(b) Which metals in figure 4 are more reactive than hydrogen?
(c) Where would you place hydrogen in figure 4?

3 (a) Will copper oxide and magnesium react on heating? Explain your answer.
(b) Will magnesium oxide and copper react on heating? Explain your answer.

4 Element *W* reacts with the oxide of element *X* but not with the oxide of element *Y*. Write *W*, *X* and *Y* in order of reactivity (most reactive first).

9 Rusting

Rusted old cars

Articles made of iron and steel rust much faster if they are left outside in wet weather. A garden fork rusts more quickly than a pair of coal tongs kept at the fireside. This suggests that water plays a part in rusting. But what other substances are involved?

When aluminium is exposed to the air, it becomes coated with a layer of oxide. The metal has reacted with oxygen in the air. If rusting is similar to the reaction of aluminium with air, it is likely that oxygen also takes part in rusting.

The apparatus in figure 1 can be used to decide whether water and oxygen are involved in rusting. The test tubes are set up and left for several days. Tube 1 is the *control experiment*. It is the standard which we use to compare the results in the other tubes. It contains iron nails in moist air. Tube 2 contains iron nails and anhydrous calcium chloride to absorb water vapour and keep the air dry. Tube 3 contains nails covered with boiled distilled water. The water has been boiled to remove any dissolved air. The layer of olive oil prevents air dissolving in the water.

The nails in tube 1 and tube 4 rust, but those in tubes 2 and 3 do not.

1 Does iron rust if there is
(i) no water; (ii) no air; (iii) no oxygen?
2 Will iron rust in (i) pure water; (ii) pure oxygen?
3 What conditions are necessary for rusting?

What is rust?

> *Iron will only rust if both*
> - *oxygen*
> - *and water are present.*

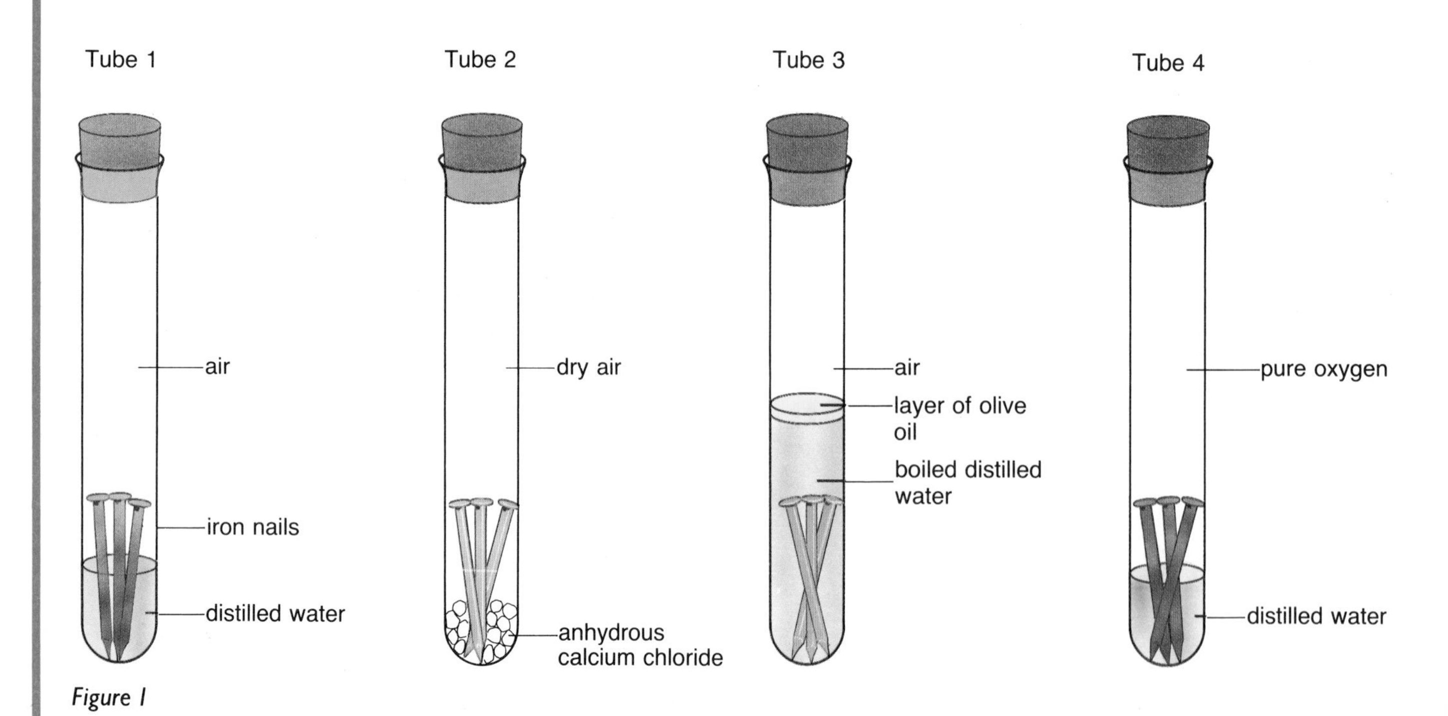

Figure 1

During rusting, iron reacts with oxygen to form brown iron oxide:

iron + oxygen → brown iron oxide

At the same time, the iron oxide combines with water to form hydrated brown iron oxide—rust:

iron oxide + water → hydrated iron oxide (rust)

Substances, like rust, that have water as part of their structure, are described as **hydrated** and we call them **hydrates**.

Blue copper sulphate crystals are also hydrated. They form when water is added to white anhydrous copper sulphate:

anhydrous copper sulphate (white) + water → hydrated copper sulphate (blue)

The water present in hydrates is called **water of crystallization**.

Rusting costs millions of pounds every year because of

- the need to protect iron and steel objects;
- the replacement of rusted articles.

Although iron and steel rust more easily than several other metals, steel is used for ships, cars, bridges and other structures because it is cheaper and stronger than other building materials.

These car valves are being dipped in a solvent to prevent them from rusting

Large structures such as the Forth Rail Bridge are being painted all the time. As soon as the painters have finished they begin again at the other end. It would take one man 72 years to paint the Forth Rail Bridge

How is rusting prevented?

In order to stop iron and steel rusting, we must protect them from water and oxygen. The most important ways of doing this are:

- **Painting.** This is the usual method of preventing rusting in ships, vehicles and bridges.
- **Oiling.** The moving parts of machines cannot be protected by paint which would get scratched off. Instead they are oiled or greased (and this also helps lubrication).
- **Alloying.** Iron and steel can be mixed with other metals to form alloys. Stainless steel contains chromium, nickel and manganese mixed with iron.
- **Covering with a non-rusting metal.** Buckets and dustbins are coated ('galvanized') with a layer of zinc. Other articles, like car bumpers, taps and kettles, are chromium plated.

Questions

1 (a) What is rust?
(b) How does rust form?

2 What methods are used to stop rusting?

3 Explain the following:
control experiment; hydrates; anhydrous; water of crystallization.

4 Explain the following statements.
(a) Iron objects rust away completely in time.
(b) Iron railings rust more quickly at the bottom.
(c) Iron on shipwrecks in deep sea-water rusts very slowly.

5 What experiments would you do to find out whether:
(i) iron rusts more quickly in sea-water or in distilled water;
(ii) steel rusts more quickly than iron?
Draw a labelled diagram of the apparatus you would use in each case. What results would you expect?

10 Redox

An unusual redox process. What substances are being oxidized and reduced?

Many reactions which occur in everyday life involve substances combining with oxygen to form oxides. Burning, breathing and rusting are three important examples.

During burning, fuels containing carbon and hydrogen react with oxygen to form carbon dioxide and water (hydrogen oxide):

fuel + oxygen → carbon dioxide + water

During respiration, foods containing carbon and hydrogen react with oxygen to form carbon dioxide and water:

food + oxygen → carbon dioxide + water

During rusting, iron reacts with oxygen and water to form hydrated iron oxide:

iron + oxygen + water → hydrated iron oxide

Chemists use a special word for reactions in which substances combine with oxygen. They call the reaction **oxidation** and the substance is said to be **oxidized**. But if one substance combines with oxygen, another substance (possibly oxygen itself) must lose oxygen. We say that substances which lose oxygen in chemical reactions are **reduced** and we call the process **reduction**.

Oxidation and reduction always happen together. If one substance combines with oxygen and is oxidized, another substance must lose oxygen and be reduced. We call the combined process **redox** (*RED*uction + *OX*idation). Figure 1 shows what happens when iron oxide is heated with aluminium.

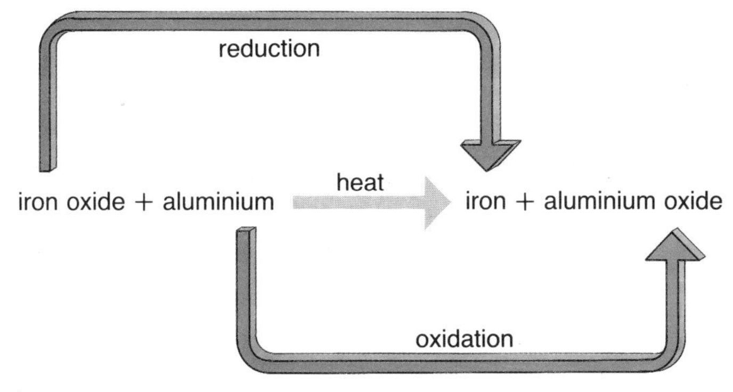

Figure 1

Aluminium gains oxygen and is oxidized, forming aluminium oxide. Iron oxide loses oxygen and is reduced to iron. The whole process involves both oxidation and reduction. It is an example of redox. Figure 2 shows the redox processes during burning.

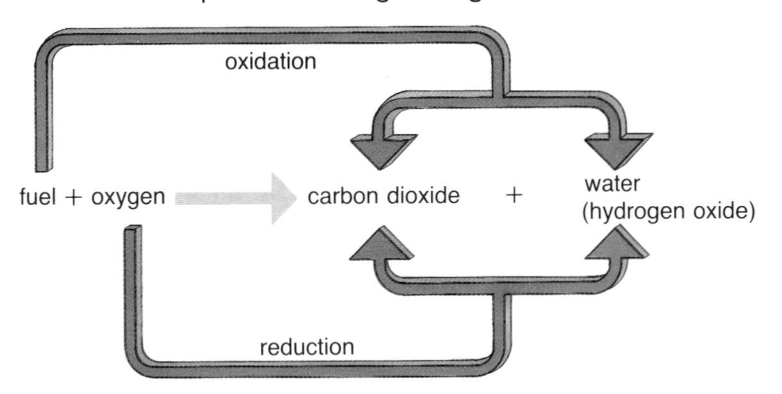

Figure 2

Iron ore is impure iron(III) oxide. This is reduced to iron in industry

Reducing metal oxides

Hydrogen can be used to reduce some metal oxides. If hydrogen has a stronger 'liking' for oxygen than the metal, then it will remove oxygen from the metal oxide and form water:

$$\text{metal oxide} + \text{hydrogen} \rightarrow \text{metal} + \underset{\text{(water)}}{\text{hydrogen oxide}}$$

Figure 3 shows how to find out which metal oxides can be reduced by hydrogen. Pass a stream of hydrogen over the heated metal oxide. Allow a few seconds for hydrogen to push the air out of the silica tube before lighting the unused hydrogen at the jet.

Copper oxide and lead oxide are reduced easily by hydrogen. Black copper oxide changes to red-brown copper. Yellow lead oxide gives a silvery globule of lead.

$$\text{copper oxide} + \text{hydrogen} \xrightarrow{heat} \text{copper} + \text{water}$$

Copper and lead have a weaker attraction for oxygen than hydrogen. Hydrogen can therefore remove oxygen from their oxides to form copper and lead respectively. When the experiment is tried with iron oxide and calcium oxide, the metals are not produced. These metals have a stronger 'liking' for oxygen than hydrogen.

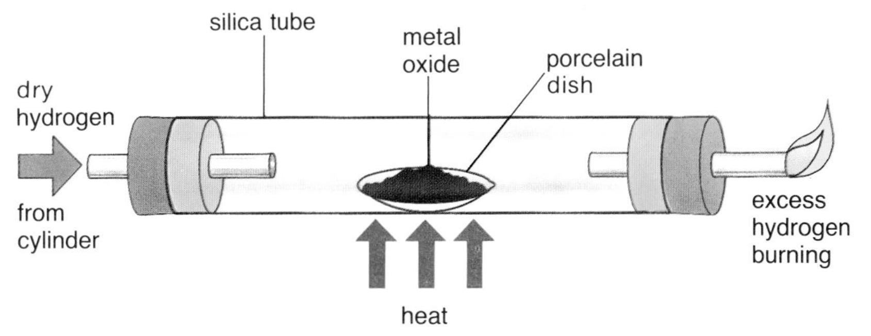

Figure 3

Questions

1 Explain the following words: *oxidation; reduction; redox.*

2 Which substance is oxidized and which is reduced in each of the redox reactions below?

(a) aluminium + water → aluminium oxide + hydrogen

(b) hydrogen + oxygen → water

(c) copper oxide + hydrogen → copper + water

3 Give three pieces of evidence to show that iron is a more reactive metal than copper. Write word equations for the reactions you mention.

4 This question concerns *magnesium oxide, sodium oxide, copper oxide, iron oxide.*

(a) Which are basic oxides?

(b) Which are alkaline oxides?

(c) Which give a green or blue colour with universal indicator?

(d) Which can be reduced to the metal by heating in hydrogen?

(e) Which can be reduced to the metal by heating with calcium?

Section B: Study Questions

1 (a) Which gas in the air do you use up in order to keep you alive?
(b) Approximately how much of this gas would there be in 100 cm^3 of air?
(c) What would you use to test for this gas in the laboratory? [3]
LEAG

2 The diagram shows apparatus which can be used to find the composition of the air. 100 cm^3 of air were placed in syringe *A* with syringe *B* empty. The copper was heated strongly and the air was passed to and fro between syringes *A* and *B* over the hot copper, and finally returned to syringe *A*.

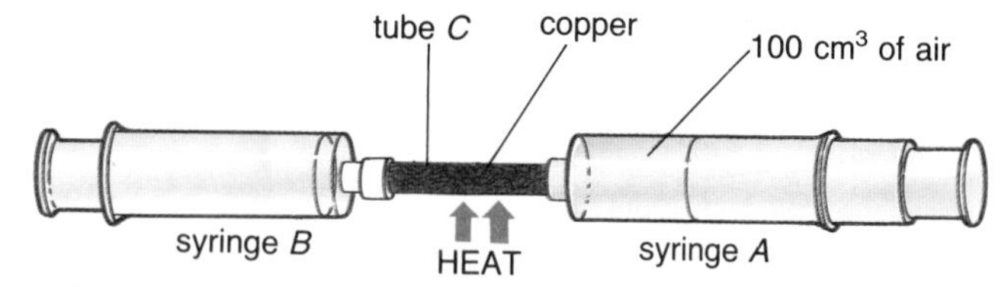

(a) Name the compound formed in tube *C*.
(b) Which gas does the copper remove from the air?
(c) What volume of gas would be left in syringe *A* at the end of the experiment?
(d) Which would be the most abundant gas remaining in syringe *A* at the end of the experiment? [4]
LEAG

3 Name:
(a) *one* source of air pollution. (1)
(b) *one* compound responsible for air pollution. (1)
(c) *one* substance that can cause excessive growth in water plants. (1)
SEG

4 Many fuels contain some sulphur impurity which forms sulphur dioxide when the fuel is burned.
(a) Write an equation to show the formation of sulphur dioxide. (1)
(b) How does sulphur dioxide cause rain water to become acidic? (2)
(c) Why is acidic rain water a nuisance? (2)
SEG

5 The Clean Air Act of 1956 forbade the emission of dark smoke from a chimney. A definition of dark smoke was made in terms of a white card with a scale composed of areas with differing degrees of black shading. New industrial plants burning solid fuel at rates of 10 tonnes per hour or more must install efficient devices to trap the grit and dust produced.
(a) How would you construct a shaded card for use in testing the darkness of smoke emitted from a chimney?
(b) How could you represent the extent of darkness of the smoke using this shaded card?
(c) What devices could industrial plants use to trap grit and dust?
(d) How would you design and locate:
(i) industrial furnaces
(ii) industrial chimneys
to reduce smoke pollution?

6 (a) What processes cause smoke pollution in the atmosphere?
(b) Why do industrial plants get rid of their smoke through tall chimneys?
(c) What effect does smoke pollution of the atmosphere have on health? Which diseases are most linked to smoke pollution?
(d) Devise experiments:
(i) to show that city air is polluted with smoke particles.
(ii) to obtain clean air.

7 (a) 'Atmospheric' nitrogen is prepared from air by removing oxygen, carbon dioxide and water vapour. Name a substance which could be used to remove
(i) oxygen
(ii) carbon dioxide
(iii) water vapour. (3)
(b) Nitrogen can be prepared by passing ammonia over heated copper(II) oxide.
$3CuO(s) + 2NH_3(g) \rightarrow 3Cu(s) + N_2(g) + 3H_2O(l)$
(i) What colour change would you see? (1)
(ii) What type of reaction is this? (1)
(iii) Ammonia is a base. Name a substance which could be used to remove any unreacted ammonia from the nitrogen. (1)
(c) 'Atmospheric' nitrogen has a slightly greater density than pure nitrogen. Explain this fact. (2)
(d) Since the year 1900 the percentages of carbon dioxide and lead compounds in the atmosphere have increased. Give the reason for the increase in
(i) carbon dioxide
(ii) lead compounds. (3)
NEA

8 Steam is passed over red-hot coke and a mixture of two gases (*A* and *B*) is produced. This mixture is then burned in pure oxygen and, when the products of the reaction are cooled to room temperature, a colourless liquid, *C*, and a colourless gas, *D*, remain. *D* reacts with a dilute solution of calcium hydroxide (lime water) to form a white precipitate, *E*. *D* allows magnesium to burn in it forming two solids, *F* and *G*. When *D* is passed over red-hot coke the gas, *A*, is formed.
(a) Identify the substances *A*–*G*.
(b) Write word equations for all the reactions involved.

9 Some rust was dried carefully for several days in a desiccator. It was then heated strongly in a vacuum, and water condensed on the cooler parts of the test tube. The solid residue was heated strongly in a stream of hydrogen. Water condensed on the cooler parts of the apparatus and a dark grey magnetic solid remained which conducted electricity.
(a) Why was the rust dried so carefully?
(b) Where did the water come from when the rust was heated in a vacuum?
(c) Write a word equation for the process taking place when rust is heated in a vacuum.
(d) How is the water produced when the residue is heated in a stream of hydrogen?
(e) What do you think the residue is? How does the residue differ from the initial rust?
(f) Describe what you think rust is.

SECTION C

Particles

A painting of John Dalton collecting marsh gas which is mainly methane. John Dalton was the first scientist to use the word 'atom'

1 Evidence for Particles

Only a small amount of curry powder gave this meal its hot taste

Anyone who cooks knows that a small amount of pepper, ginger or curry powder will give something a really strong taste. Too much spice can spoil the whole meal. This suggests that tiny particles in the spice can spread throughout the whole dish. Figure 1 shows an experiment which provides evidence that matter contains tiny particles.

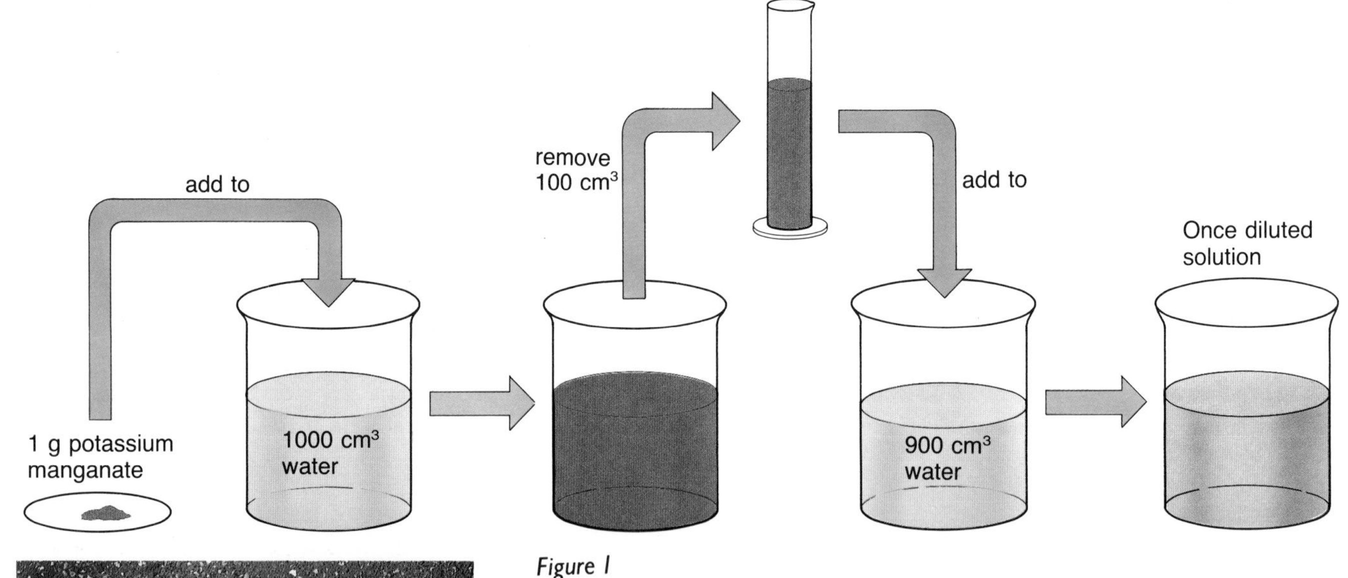

Figure 1

When oil is spilt on water or on the road, it covers the surface in a thin, brightly coloured film

Dissolve about 1 g of dark purple crystals of potassium manganate (VII) in 1000 cm^3 of water. Take 100 cm^3 of this solution and dilute it to 1000 cm^3 with water. Now take 100 cm^3 of the once diluted solution and dilute this to 1000 cm^3 with water. Repeat the dilution again and again until you get a solution in which you can just see the pink colour. It is possible to make 5 or 6 dilutions before the pink colour is so faint that it is only just noticeable. When the potassium manganate (VII) dissolves, its particles spread throughout the water making a dark purple solution. When this solution is diluted, the particles spread further apart.

This experiment shows that the tiny particles in only 1 g of potassium manganate (VII) can colour about 1 000 000 000 cm^3 of water—good evidence that there must be millions and millions of tiny particles in only 1 g of potassium manganate (VII).

Similar experiments show that the particles in all substances are extremely small. For example, there are more air particles in a thimble than there are grains of sand on a large beach.

How large are the particles?

When oil (or petrol) is spilt onto water, it covers the surface in a thin, brightly-coloured film. If the puddle is large and flat, the oil will spread out until it is only one particle thick. We can use the idea that it is one particle thick to find the size of oil particles. It is better to use olive oil rather than engine oil or petrol. Unfortunately, one drop of olive oil will cover the whole surface of a swimming pool, so we must use a solution containing 1 cm^3 of olive oil in 1000 cm^3 of alcohol. When one drop of this solution is put on the surface of the water, the alcohol dissolves in the water and the olive oil spreads out as a thin film.

Using a burette with a fine jet, count the number of drops in 1 cm^3 of the olive oil solution. Now fill a shallow tray with water and sprinkle fine powder (e.g. talcum or lycopodium) on the surface. Let 1 drop of the olive oil solution fall gently onto the water surface. The powder is pushed aside as the oil spreads out forming a circular patch (figure 2). Quickly measure the diameter of the oil film.

Some results for this experiment are shown in the table below.

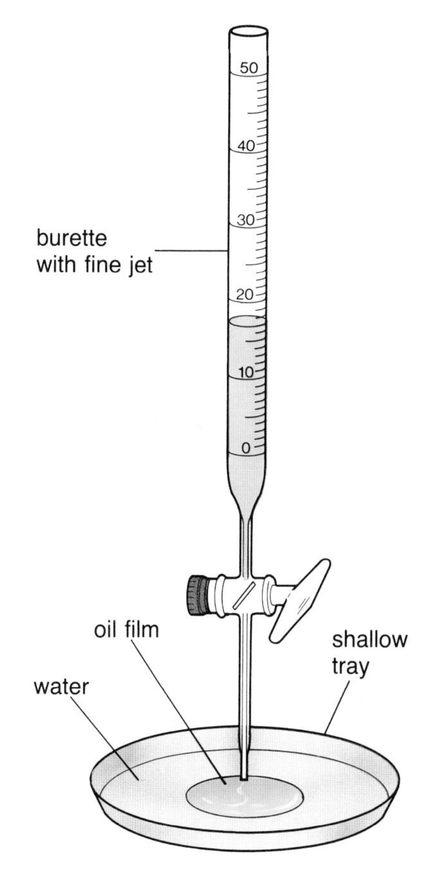

Figure 2
The oil drop experiment

Number of drops in 1 cm^3 of olive oil solution	= 50
Diameter of oil film	= 16 cm
∴ Area of oil film	$= \pi r^2$ $= \pi \times 8^2$ $= 200\ cm^2$
If the thickness of the oil film is t cm, volume of oil film	= area × thickness $= 200 \times t\ cm^3$

Finding the size of oil particles

1 50 drops of olive oil solution occupy 1 cm^3. What is the volume of 1 drop?
2 What is the volume of olive oil in 1 drop of solution? (Remember only 1/1000 of the drop is oil, the rest is alcohol.)
3 The volume of olive oil from the last question equals 200 × the thickness of the olive oil layer, t, cm^3. What is t?

Your answer to question 3 should be 1/10 000 000 cm. This can be written as 10^{-7} cm or 0.000 000 1 cm. Assuming that the oil film is only one particle thick, we can conclude that the olive oil particles are 10^{-7} cm thick. The oil particles are so small that 10 000 000 of them laid end-to-end would take up only 1 cm.

Questions

1 Look back at the oil drop experiment described above.
(a) What errors may occur if too much powder is sprinkled on the water surface?
(b) We assumed that the alcohol in the olive oil solution dissolves in the water. How would you test this?
(c) Why is the drop added to the water surface gently?
(d) If the shape of the oil film is only roughly circular, how would you estimate its area?

2 A goldsmith used 1.93 g of gold (density 19.3 g/cm^3) to make an extremely thin sheet of gold of 100 cm^2 in area.
(a) What is the volume of the gold foil? (Use the density to obtain this.)
(b) What is the thickness of the gold foil?
(c) What is the largest possible size of gold particles?

2 Particles in Motion

Why is it possible to smell the perfume that someone is wearing from several metres away?

Evidence for moving particles

The best evidence that particles of matter are constantly moving comes from studies of *diffusion* and *Brownian motion*.

- **Diffusion.** Fish and chips have a delicious smell. You can smell them a long way from where they are being fried. How does the smell get from the fish and chips to your nose?

Particles of gas are released from the fish and chips. The particles mix with air particles and move away from the chip pan. *This movement and mixing of particles is called* **diffusion**.

Gases diffuse to fill all the space available to them—even heavy gases like bromine (figure 1). How does the bromine vapour get to the top of the gas jar? Scientists believe that gases consist of tiny particles, moving around haphazardly and colliding with each other and the walls of their container. The gas particles don't care where they go, so sooner or later they will spread into all the space available.

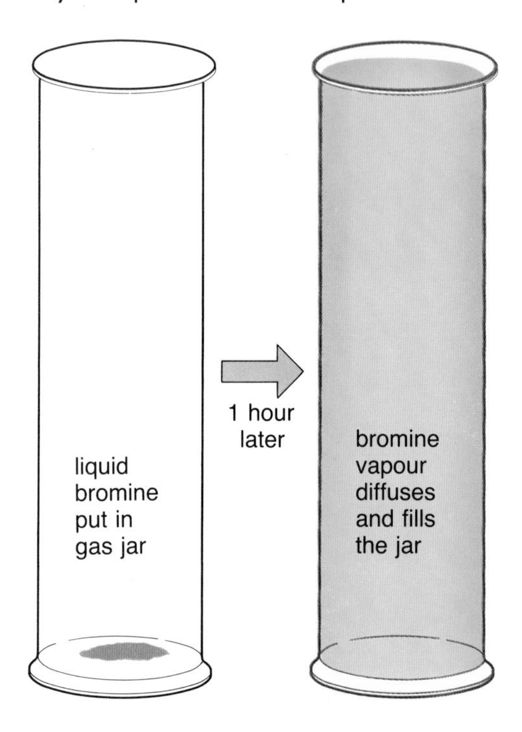

Figure 1

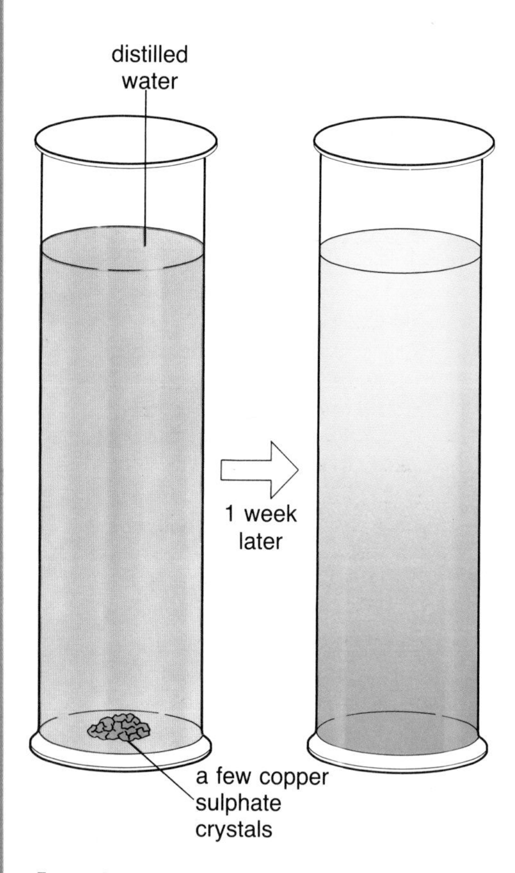

Figure 2

Diffusion also occurs in liquids, although it takes place more slowly than in gases (figure 2). This means that liquid particles move around more slowly than gas particles. Diffusion does not happen in solids.

1 Why is the blue colour darker near the bottom of the gas jar?
2 Will all the solution eventually be the same blue colour? Explain.

Diffusion is very important in living things. It explains how the food you eat gets to different parts of your body. After a meal, food passes into the stomach and through the intestines. Large particles are broken down into smaller ones. The smaller particles can then diffuse through the walls of the intestines into the bloodstream.

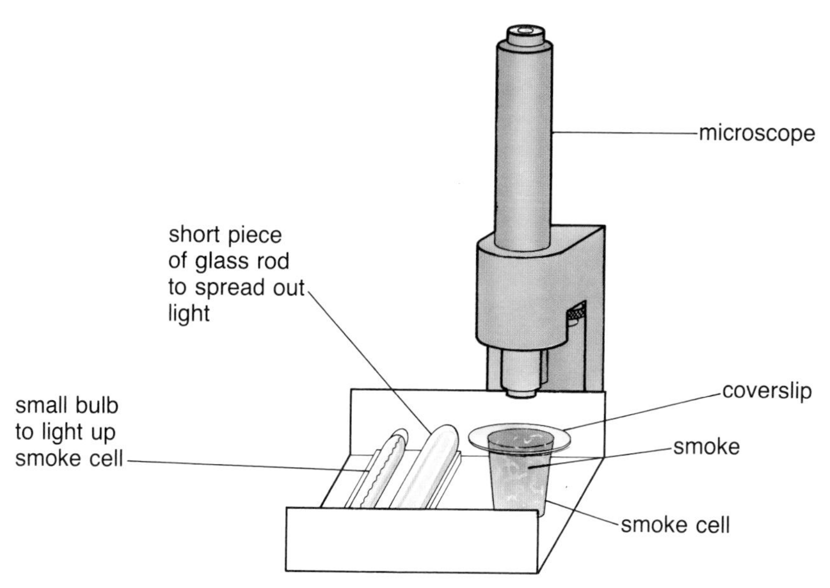

Figure 3

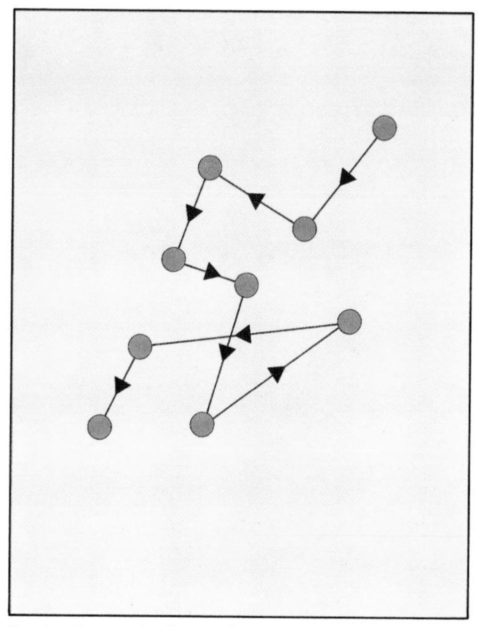

Path of particle during Brownian motion

Figure 4

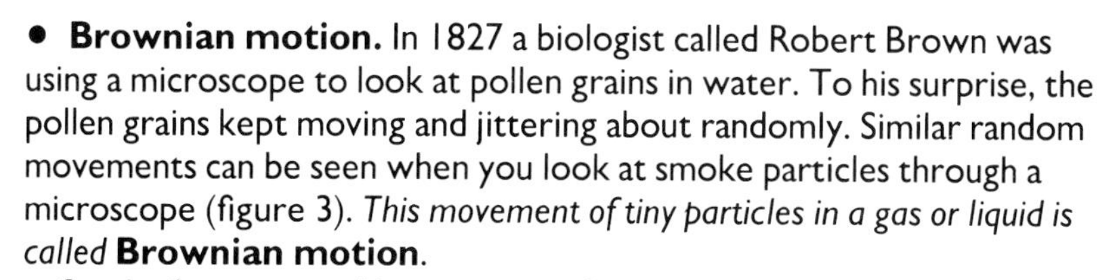

• **Brownian motion.** In 1827 a biologist called Robert Brown was using a microscope to look at pollen grains in water. To his surprise, the pollen grains kept moving and jittering about randomly. Similar random movements can be seen when you look at smoke particles through a microscope (figure 3). *This movement of tiny particles in a gas or liquid is called* **Brownian motion**.

Smoke from a smouldering piece of string is injected into the smoke cell using a teat pipette. Under the microscope, the smoke particles look like tiny pinpoints of light which jitter about (figure 4).

The movement of the smoke particles is caused by the random motion of oxygen and nitrogen particles in the air around them. The particles of smoke are small, but they are much larger than air particles. Through the microscope we can see smoke particles, but air particles are much too small to be seen. These air particles move very fast and hit the smoke particles at random. The smoke particles are therefore knocked first this way and then that way so they appear to jitter about (figure 5). As the temperature rises, particles have more energy and they move about faster. This means that gases and liquids diffuse faster when the temperature rises. Particles undergoing Brownian motion also jitter about faster as the temperature rises because they are being bumped more often by the small particles in the gas or the liquid.

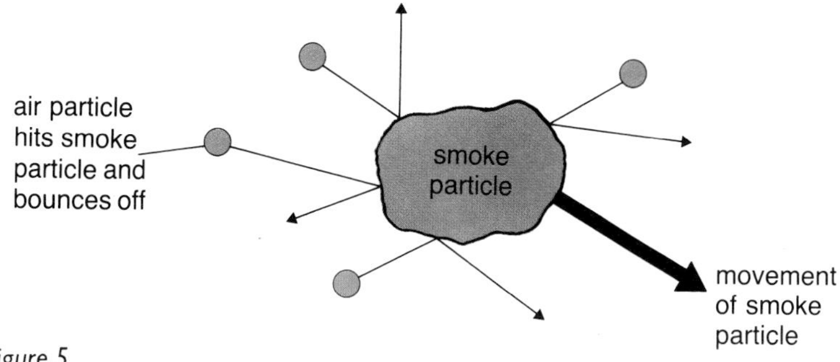

Figure 5

Questions

1 What is (i) diffusion, (ii) Brownian motion?

2 (a) Why is it important to switch off the engine when filling a car's petrol tank?
(b) Why is it possible to smell some cheeses even when they are wrapped in cling film?

3 You can smell hot, sizzling bacon several metres away, yet you have to be near cold bacon to smell it. Explain this difference in terms of particles.

4 (a) Why does bromine vapour eventually fill the gas jar in figure 1?
(b) How would the bromine diffuse if all the air was removed from the gas jar?
(c) Draw a labelled diagram to show how you would compare the rates of diffusion of bromine and chlorine (a pale green gas).
(d) Bromine particles are about twice as heavy as chlorine particles. Which gas will diffuse faster? Explain your answer.

3 Change of State

Figure 1

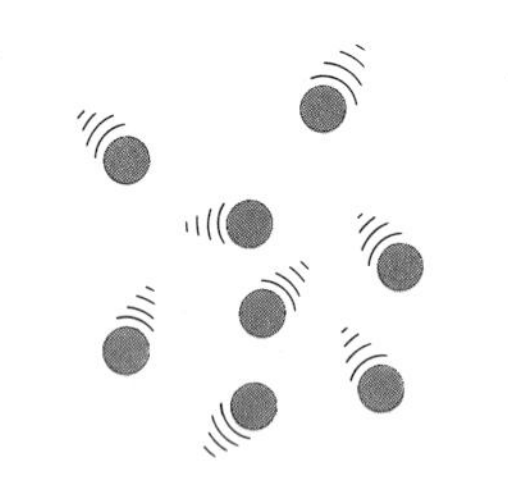

Figure 2

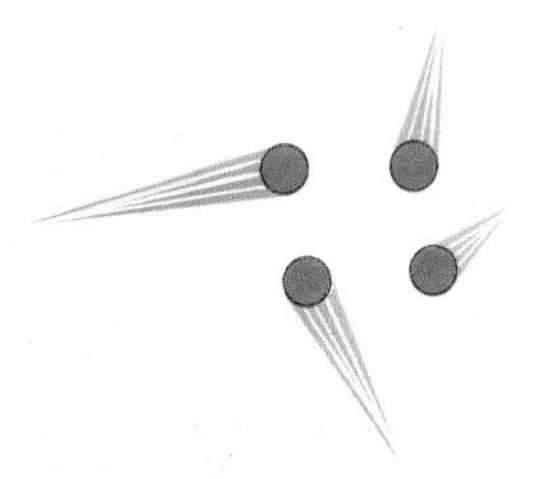

Figure 3

The kinetic theory of matter

The idea that matter is made of moving particles is called the **kinetic theory**. (The word 'kinetic' comes from a Greek word *kineo* which means 'I move'.) The main points of the theory are as follows:

1 All matter is made up of tiny, invisible, moving particles.

2 Particles of different substances have different sizes.

3 Small particles move faster than heavier ones at a given temperature.

4 As the temperature rises, the particles move faster.

5 In a solid, the particles are very close and they can only vibrate about fixed positions (figure 1).

6 In a liquid, the particles are a little further apart. They have more energy and they can move around each other (figure 2).

7 In a gas, the particles are far apart. They move rapidly and randomly in all the space they can find (figure 3).

The movement of particles in solids, liquids and gases can be compared to the movement of children in school. During lessons, the pupils stay in their seats but they are not still. They write notes, answer questions and wriggle (*vibrate*) about in their seats. They are behaving like the particles of a *solid*. When the lesson ends, the pupils pack up their things and move in a group to another part of the school for the next lesson. The pupils stay together, but *move around each other*. They resemble the particles in a *liquid*. When the bell rings at break, the children spill out into the playground where they *move around freely and randomly*. The pupils now resemble a *gas*.

During lessons pupils resemble solid particles

Between lessons pupils resemble liquid particles

In the playground pupils resemble gas particles

Changes of state

Solids, liquids and gases are sometimes called the three **states of matter**. The kinetic theory can be used to explain how a substance changes from one state to another. A summary of the different changes of state is shown in figure 4. These changes are usually caused by heating or cooling a substance.

- **Melting and freezing.** When a solid is heated, the particles gain energy and vibrate faster and faster. Eventually, they break free from their fixed position and begin to move round each other. The solid melts to form a liquid. *The temperature at which the solid melts is the melting point.*

The temperature at which a solid melts tells us how strongly its particles are held together. Substances with high melting points have strong forces between their particles; substances with low melting points have weak forces between their particles. Metals and alloys, like iron and steel, have high melting points. This suggests that there are strong forces between their particles. This is why metals can be used as girders and supports.

The forces between metal particles in steel are so strong that steel can be used in thin cables to lift heavy loads

- **Evaporating and boiling.** When a liquid is heated, the particles gain energy and move around each other faster and faster. Some particles near the surface of the liquid have enough energy to escape from those around them into the air, and some of the liquid evaporates to form a gas.

Eventually, a temperature is reached at which the particles are trying to escape from the liquid so rapidly that bubbles of gas actually start to form inside the liquid. *The temperature at which this evaporation begins to occur within the bulk of the liquid is the boiling point.* Liquids which evaporate at low temperatures are described as *volatile*.

The temperature at which a liquid boils tells us how strongly the particles are held together in the liquid. Liquids with high boiling points have stronger forces between their particles than liquids with low boiling points.

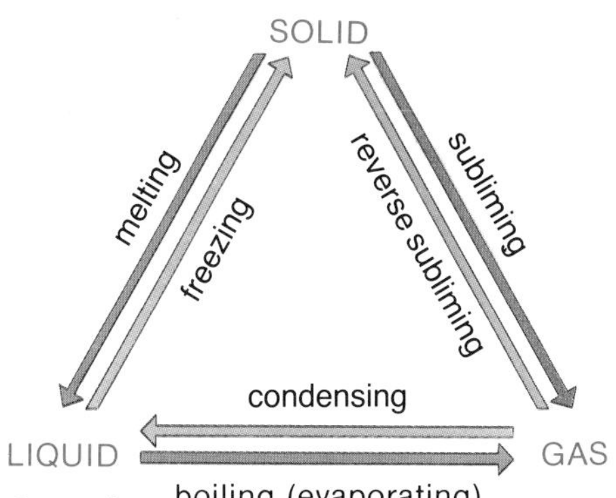

Figure 4

Gases fill their container completely

Questions

1 How do the particles move in (i) a solid; (ii) a liquid; (iii) a gas?

2 What do you understand by the following:
kinetic theory; states of matter; melting point; boiling point?

3 What happens to the particles of a liquid (i) as it cools down; (ii) as it freezes?

4 Use the kinetic theory to explain why (i) gases exert a pressure on the walls of their container; (ii) solids have a fixed size and a fixed shape;
(iii) liquids have a fixed size but not a fixed shape; (iv) solid blocks of air freshener used in toilets can disappear without leaving any solid.

4 Atoms—The Smallest Particles

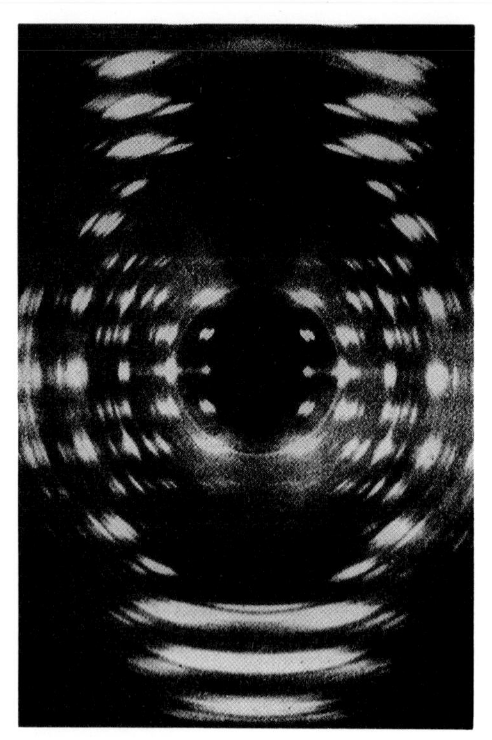

This X-ray photograph of DNA has bright spots caused by X-rays being reflected by particles in the crystal

John Dalton—the first scientist to use the name 'atom' for the smallest particle of an element

Dalton's theory of atoms

Experiments involving diffusion and Brownian motion (unit 2 of this section) show that substances are made up of small particles. Further evidence for particles has been obtained from photographs taken by X-rays and electron microscopes. X-ray photographs of crystals, such as the one here, show tiny white spots on a black background. The tiny white spots are caused by X-rays being reflected by regularly-spaced particles in the crystal. Electron microscopes can magnify objects more than a million times. In 1958, scientists in the USSR observed individual particles of barium and oxygen using an electron microscope. Figure 1 shows an electron microscope photo of a complex manganese compound. The magnification is 127 000 times. The dark lines are rows of manganese particles in the crystal. These particles are the smallest possible particles of manganese —they are manganese *atoms*.

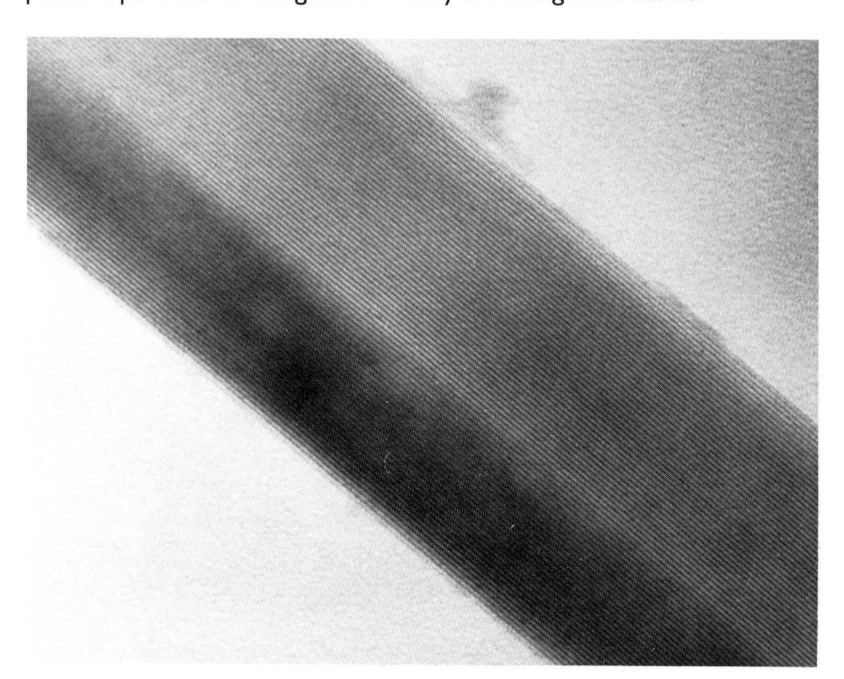

Figure 1
An electron microscope photograph of a manganese compound

The smallest particle of any element is called an **atom**. The word 'atom' was first used by John Dalton in 1807, when he put forward his **Atomic Theory of Matter**. The main points in Dalton's theory are:

1 All matter is made up of tiny particles called atoms.

2 Atoms cannot be made or destroyed.

3 Atoms of the same element are exactly alike, with the same mass, colour, etc.

4 Atoms of different elements have different masses, colours, etc.

5 Atoms can join together to form larger particles in compounds.

Nowadays, we call these larger particles **molecules**. Dalton's theory still works very well.

Representing atoms with symbols

Dalton also suggested a method of representing atoms with symbols. Figure 2 shows how he represented an atom of carbon, an atom of oxygen and a molecule of carbon dioxide. The modern symbols which we use for different elements are based on his suggestions.

The table gives a list of the symbols of some of the common elements. (A longer list of symbols appears on page 283.) Notice that most elements have two letters in their symbol; the first letter is a capital, the second letter is *always* small. These symbols come from either the English name (O for oxygen, C for carbon) or the Latin name (Au for gold—Latin: *aurum*; and Cu for copper—Latin: *cuprum*).

carbon –

oxygen –

carbon dioxide –

Figure 2

Element	Symbol	Element	Symbol	Element	Symbol
Aluminium	Al	Helium	He	Oxygen	O
Argon	Ar	Hydrogen	H	Phosphorus	P
Bromine	Br	Iodine	I	Potassium	K
Calcium	Ca	Iron	Fe	Silicon	Si
Carbon	C	Lead	Pb	Silver	Ag
Chlorine	Cl	Magnesium	Mg	Sodium	Na
Chromium	Cr	Mercury	Hg	Sulphur	S
Cobalt	Co	Neon	Ne	Tin	Sn
Copper	Cu	Nickel	Ni	Uranium	U
Gold	Au	Nitrogen	N	Zinc	Zn

The symbols for some elements

Using symbols, we can represent compounds as well as elements. For example, water is represented as H_2O because the smallest particle of water (a molecule) contains two hydrogen atoms and one oxygen atom. Carbon dioxide is written as CO_2—one carbon and two oxygen atoms. H_2O and CO_2 are called **formulas**. Formulas show the relative numbers of atoms of each element in a compound.

Chemists use symbols in the same way that typists use shorthand. Using symbols, they can represent substances and chemical changes quickly. For example, the formula for potassium manganate (VII) is $KMnO_4$. If you have to name this substance many times, how would you prefer to write it—using symbols or using the long name!

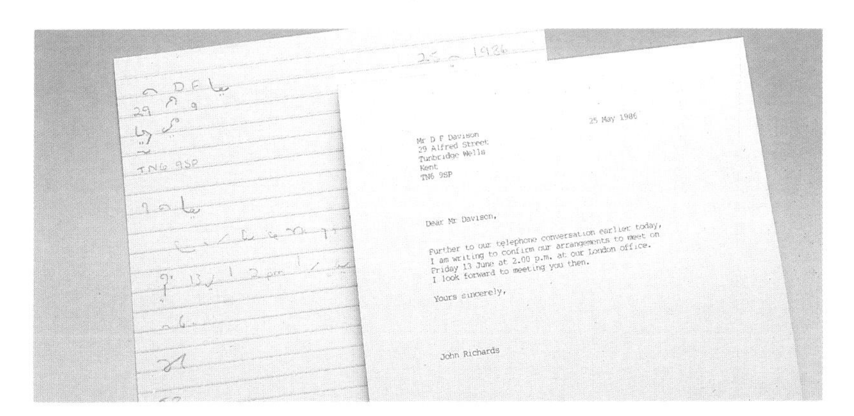

25 May 1986

Mr D F Davison
29 Alfred Street
Tunbridge Wells
Kent
TN6 9SP

Dear Mr Davison,

Further to our telephone conversation earlier today, I am writing to confirm our arrangements to meet on Friday 13 June at 2.00 p.m. at our London office. I look forward to meeting you then.

Yours sincerely,

John Richards

A secretary can write a letter in a much shorter fashion using shorthand. Chemists use chemical symbols as their shorthand

Questions

1 Write down the symbols for the following elements: (i) *copper, cobalt, calcium, carbon*; (ii) *nickel, neon, nitrogen*; (iii) *chlorine, chromium, curium.*
2 Name the elements with the following symbols:
Mn; Mg; Na; K; P; Si; Ag; Ar.
3 What is (i) an atom, (ii) a molecule?
4 Explain the words: *symbol; formula.*
5 How many atoms of the different elements are there in one molecule of
(i) methane (natural gas) CH_4;
(ii) sulphuric acid, H_2SO_4; (iii) sugar, $C_{12}H_{22}O_{11}$?

5 Measuring Atoms

How large are atoms?

Experiments with thin films of oil on water show that olive oil particles are about $1/10\,000\,000$ cm thick. But olive oil particles are large molecules containing more than 50 atoms. If we estimate that one molecule of olive oil is about 10 atoms thick, how big is a single atom?

Atoms are about $1/100\,000\,000$ *of a centimetre across*. This means that if you put 100 million of them side-by-side, they will only measure 1 cm. It is very difficult to imagine anything as small as this. You will get some idea of the size of atoms from figure 1. If atoms were magnified to the size of marbles then, on the same scale, marbles would have a diameter of 1500 km—one third of the distance between New York and London.

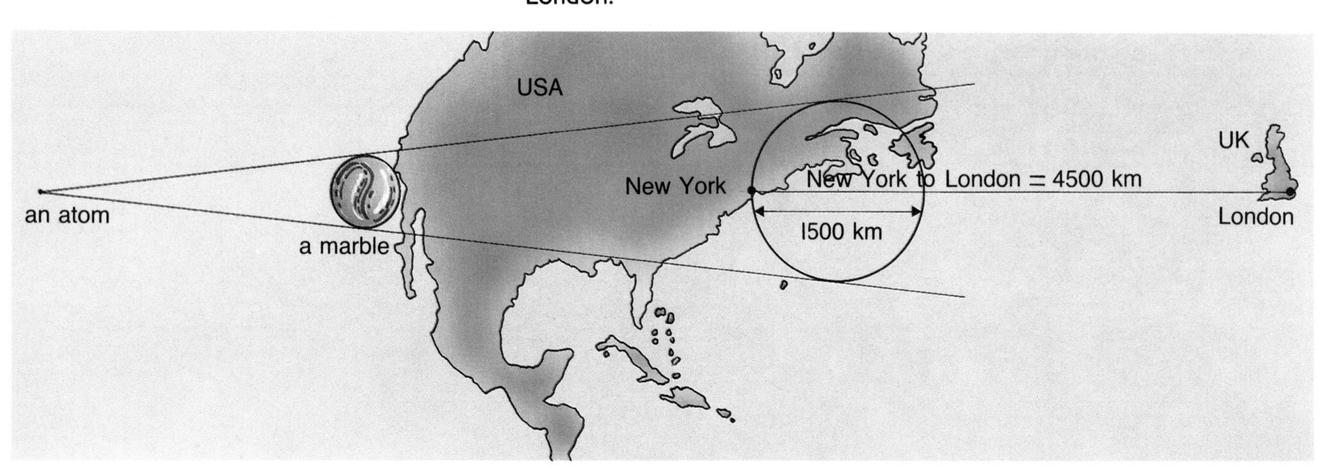

Figure 1
If atoms are magnified to the size of marbles, on the same scale a marble would have a diameter of 1500 kilometres

Figure 2 will also help you to realize just how small atoms are. It shows a step-by-step decrease in size from 1 cm to $1/100\,000\,000$ cm. Each object is one hundred times smaller than the one before it. The dice on the left is about 1 cm wide. In the next square, the grain of sand is about $1/100$ cm across. The bacterium in the middle is 100 times smaller again—about $1/10\,000$ (10^{-4}) cm from end to end. In the next square, the molecule of haemoglobin is 100 times smaller than this—about $1/1\,000\,000$ (10^{-6}) cm in diameter. Finally, in the right-hand square, the atom is one hundredth of the size of the haemoglobin molecule—about $1/100\,000\,000$ (10^{-8}) cm in diameter.

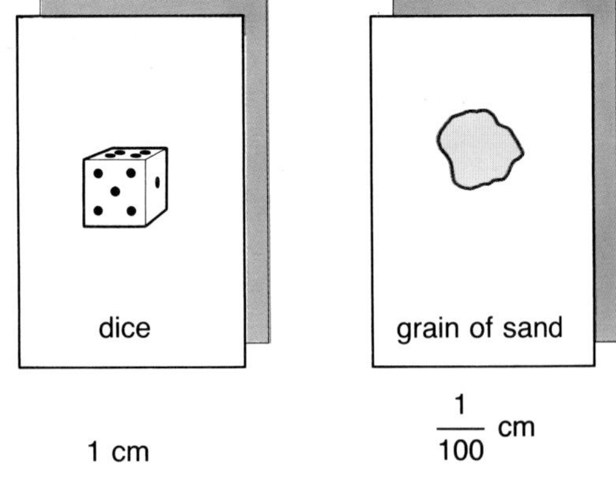

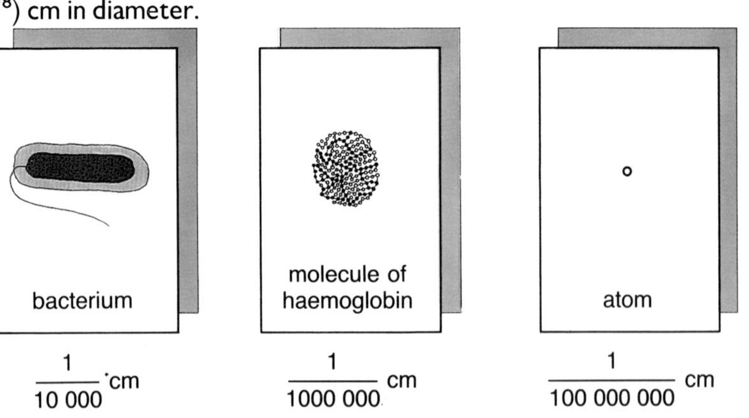

Figure 2
Step-by-step to the size of atoms

How heavy are atoms?

A single atom is so small that it cannot be weighed on a balance. However, *the mass of one atom can be compared with that of another atom using an instrument called* **a mass spectrometer** (figure 3).

In a mass spectrometer, atoms are passed along a tube and focused into a thin beam. This beam of particles passes through an electric field (which speeds them up) and then through a magnetic field where they are deflected. The extent to which an atom is deflected depends on its mass—the greater the mass, the smaller the deflection. Using the amount of deflection, it is possible to compare the masses of different atoms and make a list of the relative masses.

Element	Symbol	Relative atomic mass
Carbon	C	12.0
Hydrogen	H	1.0
Oxygen	O	16.0
Iron	Fe	55.8
Copper	Cu	63.5
Gold	Au	197.0

The relative atomic masses of a few elements

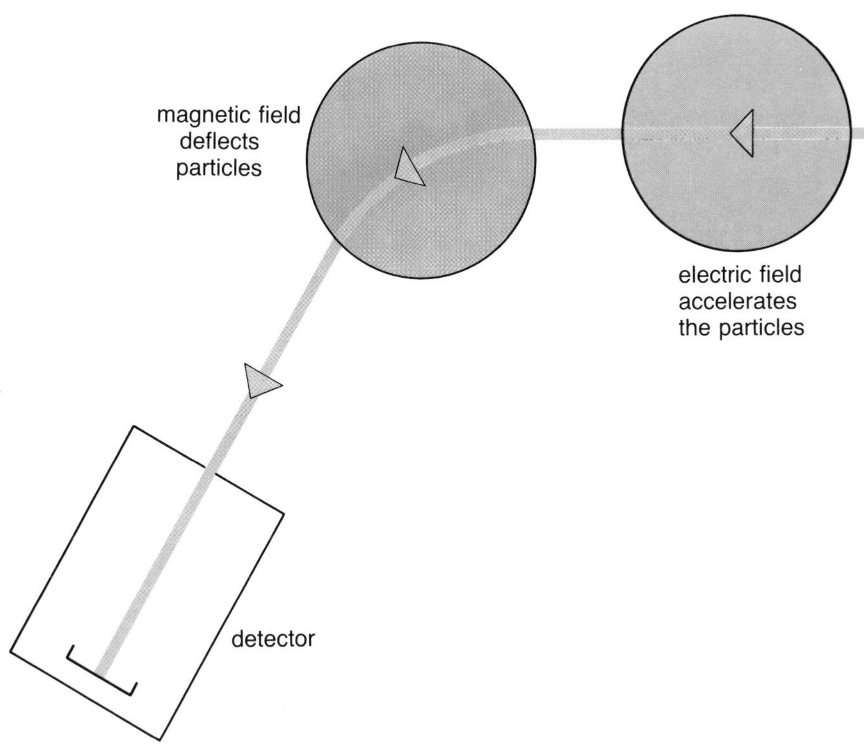

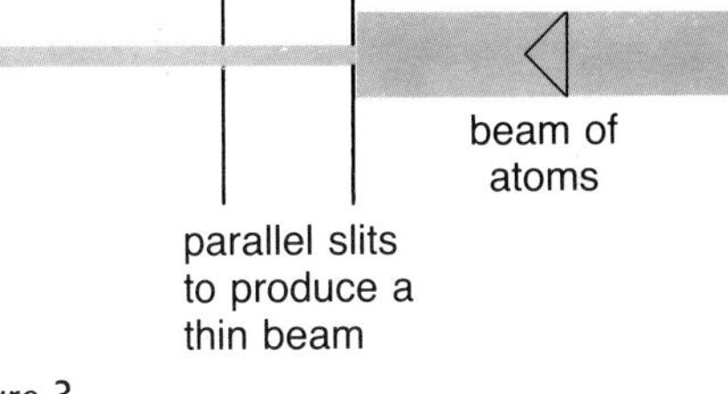

Figure 3
A simple diagram of a mass spectrometer

Questions

1 The radius of a potassium atom is $\frac{2}{100\,000\,000}$ cm. How many potassium atoms can be arranged next to each other to make a line 1 cm long?

2 Use the relative atomic masses on page 283 to answer the following questions.

(a) Which element has the lightest atoms?
(b) Which element has the next lightest atoms?
(c) How many times heavier are carbon atoms than hydrogen atoms?
(d) Which element has atoms four times as heavy as oxygen?

3 Put the following in order of size from the largest to the smallest:
a bacterium; the thickness of a human hair; a molecule of sugar (which contains about 50 atoms); a smoke particle; a copper atom; a fine dust particle.

4 Write the following elements in order of decreasing deflection in a mass spectrometer (put the element which is deflected the most first):
copper; calcium; carbon; cobalt; chlorine.

Relative atomic masses

Chemists compare the masses of different atoms using **relative atomic masses** (RAMs). The symbol A_r is sometimes used for relative atomic mass. The element carbon is chosen as the standard against which the masses of other atoms are compared. Carbon atoms are given a relative atomic mass of 12 and the relative masses of other atoms are obtained by comparison with the mass of a carbon atom. A few relative atomic masses are listed in the table. Other relative atomic masses are given on page 283. From the table you will see that carbon atoms are 12 times as heavy as hydrogen atoms; iron atoms are 55.8 times as heavy as hydrogen atoms.

6 Using Relative Atomic Masses

The photograph shows an aluminium colander (mass 108g), a copper bracelet (mass 16g), some iron nails (mass 5.58g) and some barbecue charcoal (mass 150g). How many moles of each element do the objects contain? (Al=27, Cu=64, Fe=55.8, C=12)

The bank clerk is counting money by weighing the coins. In the same way, chemists count atoms by weighing

Counting atoms

Relative atomic masses show that one atom of carbon is 12 times as heavy as one atom of hydrogen. Therefore, 12 g of carbon will contain the same number of atoms as 1 g of hydrogen. An atom of oxygen is 16 times as heavy as an atom of hydrogen, so 16 g of oxygen will also contain the same number of atoms as 1 g of hydrogen. In fact, *the relative atomic mass (in grams) of every element* (1 g of hydrogen, 12 g carbon, 16 g oxygen, etc.) *will contain the same number of atoms.* This number is called **Avogadro's constant** in honour of the Italian scientist Amedeo Avogadro. *The relative atomic mass in grams is known as one* **mole** *of the element.* So, 12 g of carbon is 1 mole of carbon and 1 g of hydrogen is also 1 mole. 24 g of carbon is 2 moles and 240 g of carbon is 20 moles.

Notice that the number of moles $= \frac{\text{mass}}{\text{RAM}}$.

To work out the number of moles, divide the mass by the RAM.

Experiments show that Avogadro's constant is 6×10^{23}. Written out in full this is 600 000 000 000 000 000 000 000. Thus *1 mole of an element always contains 6×10^{23} atoms.*

We can use the mole idea to count the number of atoms in a sample of element. For example:

12 g (1 mole) of carbon contains 6×10^{23} atoms
so 1 g ($\frac{1}{12}$ mole) of carbon contains $\frac{1}{12} \times 6 \times 10^{23}$ atoms
$\Rightarrow$ 10 g ($\frac{10}{12}$ mole) of carbon contains $\frac{10}{12} \times 6 \times 10^{23} = 5 \times 10^{23}$ atoms
i.e. number of atoms = number of moles $\times 6 \times 10^{23}$

Chemists are not the only people who 'count by weighing'. Bank clerks use the same idea when they count coins by weighing them. For example, 100 1p coins weigh 356 g; so it is quicker to take 100 1p coins by weighing out 356 g of them than by counting.

Chemists often need to count atoms. In industry, nitrogen is reacted with hydrogen to form ammonia, NH_3, which is then used to make fertilizers. In a molecule of ammonia, there is one nitrogen atom and three hydrogen atoms. In order to make ammonia, chemists must therefore react:

1 mole of nitrogen + 3 moles of hydrogen
(14 g of nitrogen) (3×1 g = 3 g of hydrogen)

not 1 g of nitrogen and 3 g of hydrogen.

To get the right quantities, chemists must measure in moles *not* in grams. Thus the mole is the chemist's counting unit.

Finding formulas

We have used some formulas already, but how are they obtained? How do we know that the formula of water is H_2O?

All formulas are obtained by doing experiments. When water is decomposed into hydrogen and oxygen, results show that

18 g of water give 2 g of hydrogen + 16 g of oxygen
= 2 moles of hydrogen + 1 mole of oxygen
= $2 \times 6 \times 10^{23}$ atoms of hydrogen + 6×10^{23} atoms of oxygen

Since 12 × 10^{23} hydrogen atoms combine with 6 × 10^{23} atoms of oxygen, it means that 2 hydrogen atoms combine with 1 oxygen atom. Therefore, the formula must be H_2O. These results are set out in table 1. By finding the masses of the elements which form a compound, we can use relative atomic masses to calculate the number of moles of atoms that are present and this gives us the formula.

	H	O
Masses present	2 g	16 g
Mass of 1 mole	1 g	16 g
∴ moles present	2	1
Ratio of atoms	2	1
⇒ Formula	H_2O	

Table 1: finding the formula of water

Finding the formula of magnesium oxide

When magnesium ribbon is heated, it burns with a very bright flame to form white, powdery magnesium oxide:

$$\text{magnesium} + \text{oxygen} \rightarrow \text{magnesium oxide}$$

Weigh 0.24 g of clean magnesium ribbon. Heat this strongly in a crucible until all of it forms magnesium oxide (figure 1). Put a lid on the crucible to stop magnesium oxide escaping, but keep a small gap so that air can enter. When the magnesium has finished reacting, reweigh the crucible + lid + magnesium oxide. 0.40 g of magnesium oxide should have formed.

0.24 g magnesium gives 0.40 g magnesium oxide
⇒ mass of oxygen reacting = 0.40 − 0.24 = 0.16 g

Table 2 shows how to obtain the formula of magnesium oxide from these results.

	Mg	O
Masses reacting	0.24 g	0.16 g
Mass of 1 mole	24 g	16 g
∴ moles present	0.01	0.01
Ratio of moles	1	1
∴ ratio of atoms	1	1
⇒ Formula	MgO	

Table 2: finding the formula of magnesium oxide

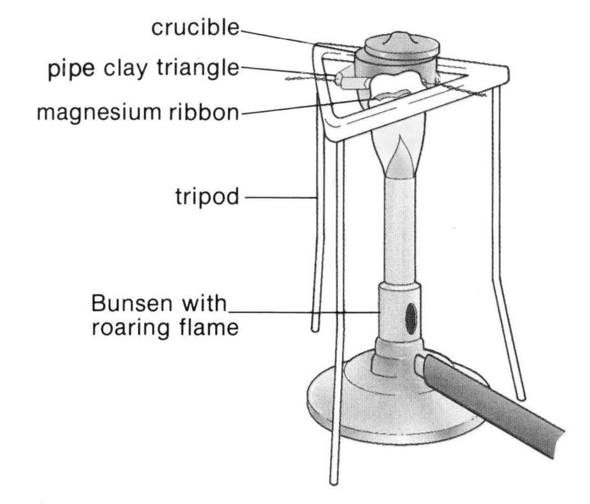

Figure 1

Questions

1 (a) What is the relative atomic mass of hydrogen?
(b) How many atoms are there in 1 g of hydrogen?
(c) How many atoms are there in 20 moles of hydrogen atoms?
(d) How heavy are 20 moles of hydrogen atoms?
(e) How many moles are 24 × 10^{23} hydrogen atoms?

2 How many atoms are there in (i) 52 g chromium (Cr = 52) (ii) 2.8 g nitrogen (N = 14) (iii) 0.36 g carbon (C = 12) (iv) 20 g bromine (Br = 80)?

3 What is the mass of (i) 3 moles of bromine (Br = 80) (ii) ¼ mole of calcium (Ca = 40) (iii) 0.1 mole of sodium (Na = 23)?

4 Methane in natural gas is found to contain 75% carbon and 25% hydrogen. Calculate the formula of methane using a method like that in table 1.

5 What is the formula of the following compounds?
(a) A compound in which 10.4 g chromium (Cr = 52) combines with 48 g bromine (Br = 80).
(b) A nitride of chromium in which 0.26 g chromium forms 0.33 g of chromium nitride (N = 14).

7 Particles in Elements

Look at the properties of iron and oxygen in table 1. These properties show the difference between metals, like iron, and non-metals, like oxygen. What are the particles in iron and oxygen? How do these particles explain the properties of the two elements?

Element	Iron	Oxygen
State	solid	gas
Melting point (M.Pt.)	high	v. low
Boiling point (B.Pt.)	high	v. low
Density	high	v. low
Electrical conductivity	good	poor

Table 1: the properties of iron and oxygen

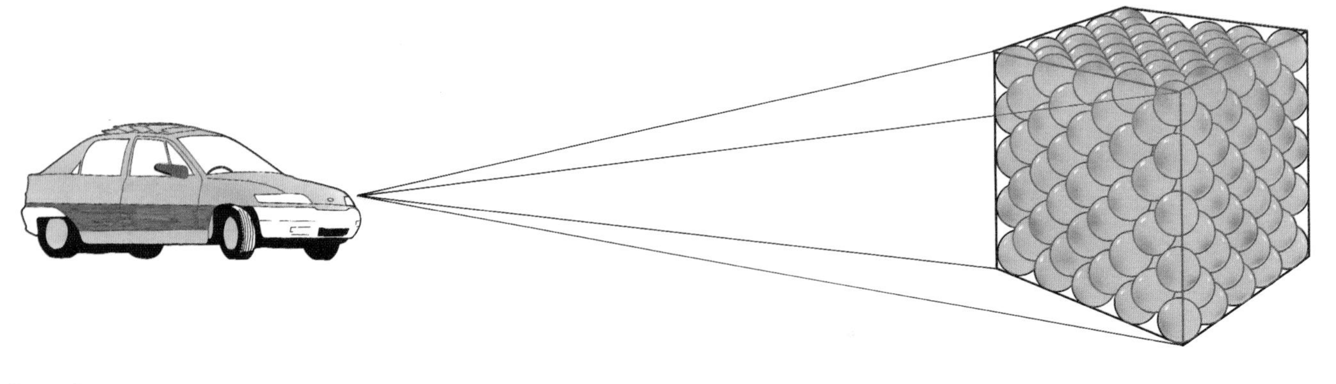

Figure 1

Iron and metals

In iron and other metals, the particles are single atoms. These atoms are packed together very closely and in a regular fashion (figure 1). The regular arrangement of particles builds up to give a solid with a regular shape called a **crystal**. The atoms are packed so tightly that there are strong forces between them. This gives the metal a high density and makes it hard and strong. The strong forces between the atoms hold them together and the melting point and boiling point are high.

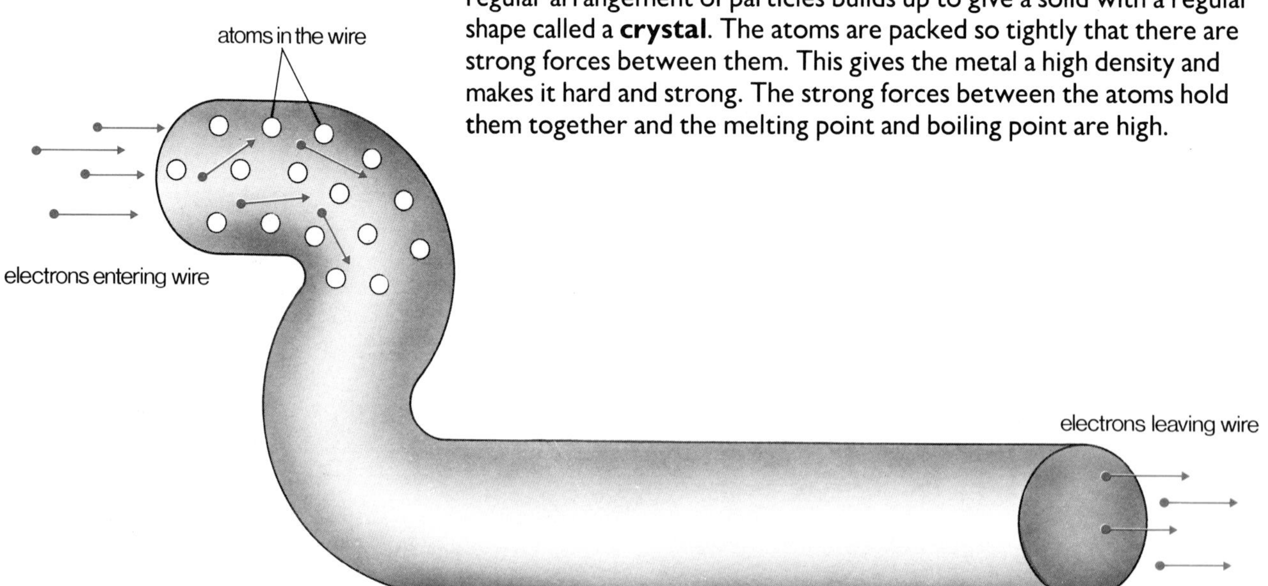

Figure 2

Metals will conduct electricity because there are tiny negatively charged particles called **electrons** near the outside of the atoms. These electrons are attracted to the positive terminal of a battery and form an electric current when the metal conducts (figure 2). Metals are studied further in section F.

Oxygen and non-metals

The relative atomic mass of oxygen is 16, but experiments show that particles of oxygen in the air have a relative mass of 32 (figure 3). This shows that oxygen consists of molecules containing two atoms joined together (figure 4) and we say it has a **relative molecular mass** (RMM) of 32. The formula of oxygen gas is therefore O_2, not O.

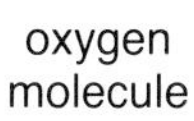
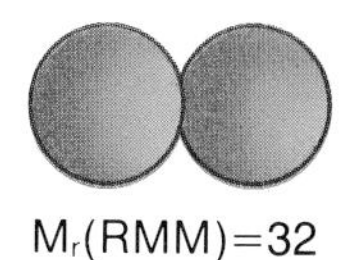

Figure 3

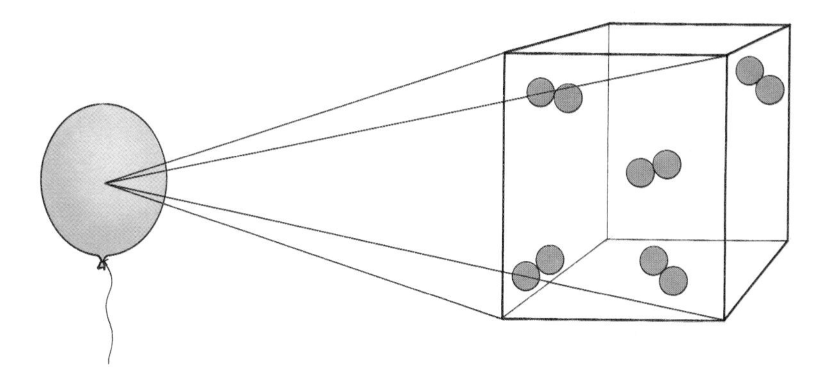

Figure 4

Most non-metals are like oxygen—they have a small number of atoms joined tightly together to form a molecule (table 2). All the common gaseous elements have molecules containing two atoms. They are often described as **diatomic molecules**. There are strong forces between the two atoms in one molecule of oxygen, but very weak forces between one molecule and another. Therefore, the O_2 molecules are easy to separate and oxygen has a very low melting point and boiling point. The particles are widely-spaced as a gas at room temperature so the density is very low. Since the particles are widely spaced, it is possible to squeeze them into a much smaller volume. This explains why gases are so compressible.

Element	Formula
Oxygen	O_2
Hydrogen	H_2
Nitrogen	N_2
Chlorine	Cl_2
Sulphur	S_8
Argon	Ar

Table 2: the formulas of some non-metals

Questions

1 The formula for nitrogen is N_2. What does this mean?
2 How many atoms are there in one molecule of sulphur?
3 Why do metals have a high density?
4 Why does sulphur melt at a much lower temperature than copper?
5 What is the difference between an atom and a molecule?
6 Explain the following:
crystal; relative molecular mass; diatomic molecule.

8 Particles in Reactions

The original balance with which Dalton studied the amounts of substances which react together. Dalton suggested that atoms were simply rearranged in chemical reactions

Writing equations

So far, we have used word equations to show what happens when substances react. When magnesium reacts with oxygen, the product is magnesium oxide. We can summarise this in a word equation as:

$$\text{magnesium} + \text{oxygen} \rightarrow \text{magnesium oxide}$$

It would be easier to use formulas instead of names when we write equations. We could write magnesium as Mg, oxygen as O_2 and magnesium oxide as MgO. This gives

$$Mg + O_2 \rightarrow MgO$$

But, notice that this does not balance. There are two oxygen atoms in the O_2 molecule on the left and only one oxygen atom in MgO on the right. So, MgO on the right must be doubled to give

$$Mg + O_2 \rightarrow 2MgO$$

Unfortunately, the equation still does not balance. There are two Mg atoms on the right in 2MgO, but only one on the left in Mg. This can easily be corrected by writing 2Mg on the left, i.e.

$$2Mg + O_2 \rightarrow 2MgO$$

This example shows the three stages in writing an equation:

1 *Write a word equation for the reaction,*

e.g. hydrogen + oxygen → water

2 *Write formulas for the reactants and products,*

e.g. $H_2 + O_2 \rightarrow H_2O$

3 *Balance the equation by making the number of atoms of each element the same on both sides.*

e.g. $2H_2 + O_2 \rightarrow 2H_2O$

An equation is a summary of the starting substances and the products in a chemical reaction. Equations with formulas also help us to see how the atoms are rearranged in a chemical reaction. We can see this even better using models to represent the equation, as in figure 1.

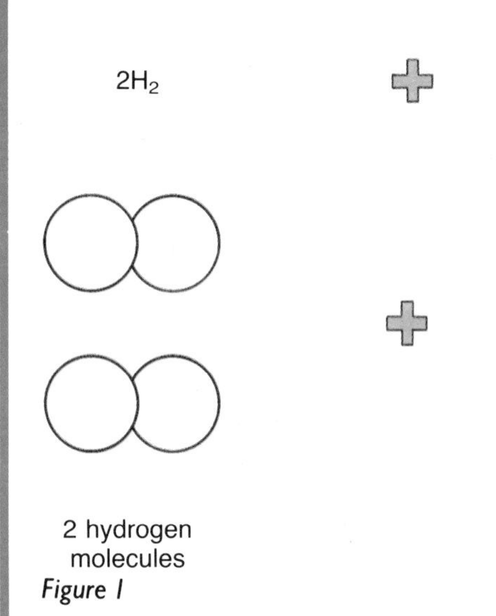

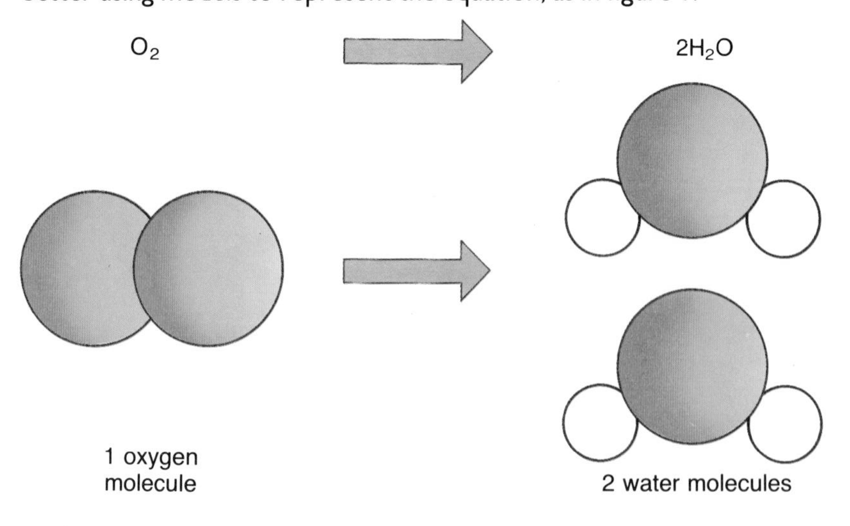

Figure 1

The elements hydrogen, oxygen, nitrogen, chlorine, bromine and iodine exist as diatomic molecules. Thus, they are written in equations as H_2, O_2, N_2, Cl_2, Br_2 and I_2. All other elements are shown as single atoms and represented by a monatomic symbol in equations; i.e. Mg for magnesium, Cu for copper, S for sulphur, etc. There is one other important point to remember in writing equations. *Never change a formula* to make an equation balance. For example, the formula of magnesium oxide is always MgO. Mg_2O and MgO_2 do not exist. Equations can only be balanced by putting a number *in front* of the whole formula, i.e. 2MgO or 3MgO.

The copper dome on Brighton Pavilion is green because the copper has reacted with oxygen in the air to form copper oxide and then this has reacted very slowly with water to form green copper hydroxide. Write equations for these two reactions

Extra information in equations

- **State symbols** show the state of a substance. The symbol (s) after a formula indicates the substance is a solid; (l) is used for liquid; (g) for gas and (aq) for an aqueous solution (i.e. a substance dissolved in water). For example,

$$\text{zinc} + \text{sulphuric acid} \rightarrow \text{zinc sulphate} + \text{hydrogen}$$
$$Zn(s) + H_2SO_4(aq) \rightarrow ZnSO_4(aq) + H_2(g)$$

- **Reaction conditions.** The conditions needed for a reaction can be shown above the arrow in the equation. For example,

$$\text{hydrogen peroxide} \xrightarrow{MnO_2\ \text{catalyst}} \text{water} + \text{oxygen}$$
$$2H_2O_2 \xrightarrow{MnO_2\ \text{catalyst}} 2H_2O + O_2$$
$$\underset{\text{(limestone)}}{\text{calcium carbonate}} \xrightarrow{\text{heat}} \underset{\text{(lime)}}{\text{calcium oxide}} + \text{carbon dioxide}$$
$$CaCO_3 \xrightarrow{\text{heat}} CaO + CO_2$$

When natural gas burns, methane (CH_4) reacts with oxygen in the air to form carbon dioxide and water. Write an equation for this

Questions

1 What is an equation?
2 What are the important rules to follow in writing an equation?
3 Why is oxygen written as O_2 and not O in equations?
4 Write equations for the reaction of
(i) copper with oxygen to give copper oxide (CuO);
(ii) copper oxide with water to give copper hydroxide ($Cu(OH)_2$);
(iii) aluminium with oxygen to give aluminium oxide (Al_2O_3);
(iv) copper oxide with sulphuric acid to give copper sulphate ($CuSO_4$) and water;
(v) nitrogen with hydrogen to give ammonia (NH_3);
(vi) charcoal (carbon) burning in oxygen to give carbon dioxide;
(vii) natural gas (methane, CH_4) burning in oxygen to give carbon dioxide and water;
(viii) iron with chlorine to give iron chloride ($FeCl_3$).

9 Formulas and Equations

	C	H
Masses reacting	0.18 g	0.06 g
Mass of 1 mole	12 g	1 g
∴ Moles reacting	0.015	0.060
Ratio of moles	1	4
∴ Ratio of atoms	1	4
= Formula	CH_4	

Finding the formula of methane

Using formulas

In unit 6, we found that the formulas of water and magnesium oxide were H_2O and MgO. The table shows the results of an experiment to find the formula of methane, the main constituent of natural gas. First, we must find the masses of carbon and hydrogen which react to form a sample of methane. Then, using the relative atomic masses as the mass of one mole, we find the number of moles which react. Finally, we calculate the simplest ratio of the moles. This is the same as the ratio of atoms in the formula. Figure 1 shows models of the structures of water, methane and carbon dioxide. Notice that hydrogen atoms are smaller than carbon atoms and carbon atoms are smaller than oxygen atoms.

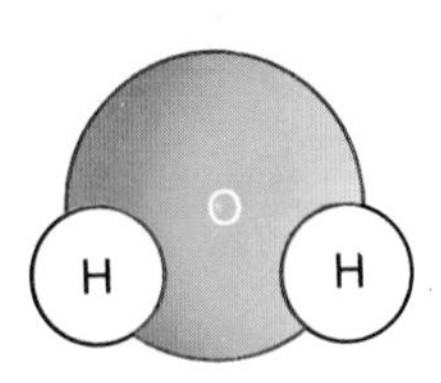

H_2O
water

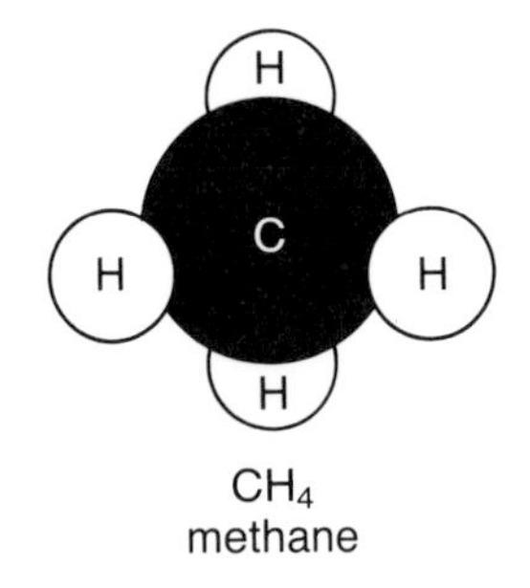

CH_4
methane

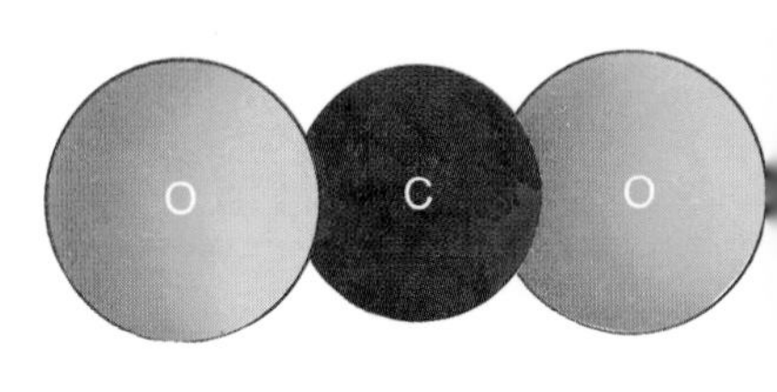

CO_2
carbon dioxide

Figure 1

Formulas are useful in several ways.

1 They provide a shorthand way of writing a substance (see section C, unit 4).

2 They enable us to compare the relative masses of different molecules. For example, using relative atomic masses for hydrogen, carbon and oxygen of 1, 12 and 16 respectively,

relative molecular mass of water, H_2O,
$= 1 + 1 + 16 = 18$

relative molecular mass of carbon dioxide, CO_2,
$= 12 + 16 + 16 = 44$

relative molecular mass of methane, CH_4,
$= 12 + 1 + 1 + 1 + 1 = 16$

3 They help us to find the mass of 1 mole of different substances. For example:

the mass of 1 mole of water, H_2O, $= 18$ g;

the mass of 1 mole of sodium chloride, NaCl,
$= 23.0 + 35.5 = 58.5$ g;

the mass of 1 mole of calcium carbonate, $CaCO_3$,
$= 40.0 + 12.0 + 16.0 + 16.0 + 16.0 = 100$ g.

One mole of sodium chloride, NaCl (common salt) has a mass of 58.5 g. The salt cellar gives you some idea of the amount of sodium chloride in one mole

4 They help us to calculate the masses of elements which have combined to form a compound. For example, when aluminium articles are left in the air, they slowly form a thin white coating of aluminium oxide. The formula of this is Al_2O_3.
Thus, 2 moles of Al (Al = 27) combine with 3 moles of O
$\Rightarrow 2 \times 27$ g of aluminium combine with 3×16 g of oxygen
i.e. 54 g of aluminium combine with 48 g of oxygen.

The yellow substance in yellow 'no parking' lines is lead chromate, $PbCrO_4$. What does this formula tell us about lead chromate?

What do equations tell us?

Look at the equation for the reaction of sodium with chlorine to form sodium chloride:

$$2Na + Cl_2 \rightarrow 2NaCl$$

This equation tells us that:

1 sodium reacts with chlorine to form sodium chloride;

2 the formula of chlorine is Cl_2 and that of sodium chloride is NaCl;

3 2 atoms of Na react with 1 molecule of Cl_2 to give 2 particles of NaCl;

4 2 moles of Na react with 1 mole of Cl_2 to give 2 moles of NaCl;

5 Using relative atomic masses (Na = 23.0 and Cl = 35.5);
2×23 g sodium react with 2×35.5 g chlorine to give
2×58.5 g sodium chloride.

This example shows that equations can tell us
- *the reactants and products*
- *the formulas of these substances*
- *the numbers of particles (or moles) of the reactants and products*
- *the masses of reactants and products.*

It is useful in industry to know the amounts of reactants and products. Industrial chemists need to know how much product they can make from a given amount of starting material. For example, using an equation and relative atomic masses, we can calculate how much lime (calcium oxide) can be obtained from 1 kg of pure limestone (calcium carbonate).

calcium carbonate → calcium oxide + carbon dioxide

$$CaCO_3 \rightarrow CaO + CO_2$$

$$\therefore 1 \text{ mole } CaCO_3 \rightarrow 1 \text{ mole } CaO$$

$$(40 + 12 + 16 + 16 + 16) \text{ g } CaCO_3 \rightarrow (40 + 16) \text{ g } CaO$$

$$100 \text{ g } CaCO_3 \rightarrow 56 \text{ g } CaO$$

$$\therefore 1000 \text{ g } CaCO_3 \rightarrow 560 \text{ g } CaO$$

i.e. 1 kg $CaCO_3$ produces 0.56 kg CaO.

This lime kiln converts limestone into lime

Questions

1 What are the main uses of formulas?
2 What are the main uses of equations?
3 60 g of a metal M (M = 60) combine with 24 g of oxygen (O = 16).
(a) How many moles of O react with one mole of M?
(b) What is the formula of the oxide of M?
4 The fertilizer 'nitram' (ammonium nitrate) has the formula NH_4NO_3.
(a) What are the masses of nitrogen, hydrogen and oxygen in 1 mole of nitram?
(N = 14, H = 1, O = 16)
(b) What are the percentages of nitrogen, hydrogen and oxygen in nitram?
5 Nitram is made from ammonia and nitric acid.

$$\underset{\text{ammonia}}{NH_3} + \underset{\text{nitric acid}}{HNO_3} \rightarrow \underset{\text{ammonium nitrate}}{NH_4NO_3}$$

(a) What does this equation tell you?
(b) How much NH_3 reacts with HNO_3 to give 1 mole of nitram?
(N = 14, H = 1, O = 16)
(c) How much NH_3 reacts with HNO_3 to give 1 kg of nitram?

Section C: Study Questions

1 (a) The kinetic theory of matter states that all substances contain particles which are moving. Use this theory to explain the following:
(i) A solid has its own shape, but a liquid takes up the shape of its container. (2)
(ii) The pressure exerted by a gas in a sealed container increases with the temperature of the gas. (2)
(iii) A gas will move to fill any container. (2)
(b) Pollen grains suspended in water are observed using a microscope. Smoke particles in air are also observed.
(i) What would you see in both cases? (1)
(ii) How would the behaviour of the smoke particles differ from that of the pollen grains? (1)
(iii) What causes the particles to behave in the way described in b(i)? (2)
Total [10]
NEA

2 (a) What is the name given to the smallest particle of a compound which can freely exist?
(b) What is the name given to the smallest particle of an element which has the chemical properties of that element?
(c) Which particles within an atom are used in chemical bonding?
(d) Sometimes we talk about 'splitting the atom' to create a new element. What part of the atom is actually 'split'?
(e) If the formula for water is H_2O, how many atoms are there in one molecule of water?
(f) Write the formula for 'ten molecules of water'.
LEAG

3 In the boxes below, ○ and ● represent two different kinds of atoms. Use the information in the diagram to answer the questions which follow.

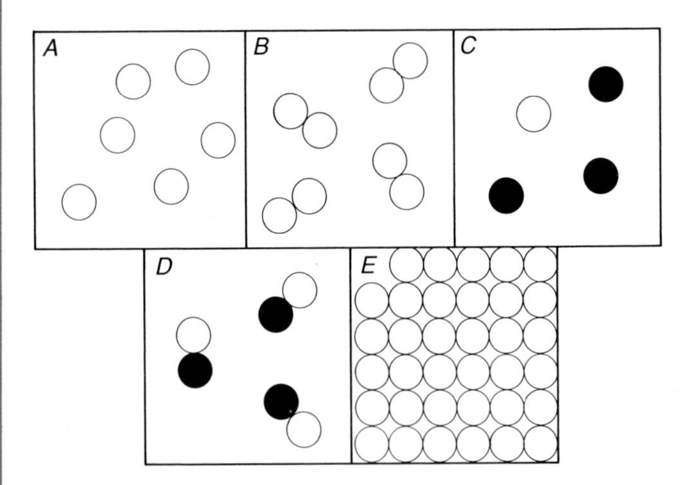

(a) Which box is most likely to contain a mixture of gases?
(b) Which box is most likely to contain a compound?
(c) Which box is most likely to contain a solid metal?
(d) Which box is most likely to contain hydrogen gas?
LEAG

4 What atoms and molecules do the following contain?
Air, vinegar, sulphuric acid (H_2SO_4), limestone ($CaCO_3$).

5 Zinc carbonate decomposes, when it is heated, into zinc oxide and carbon dioxide according to the equation

$$ZnCO_3 \rightarrow ZnO + CO_2 \quad .$$

An empty test tube was weighed, some zinc carbonate was put in it, and it was weighed again. The diagrams show the scale of the top-pan balance used for each of these weighings.

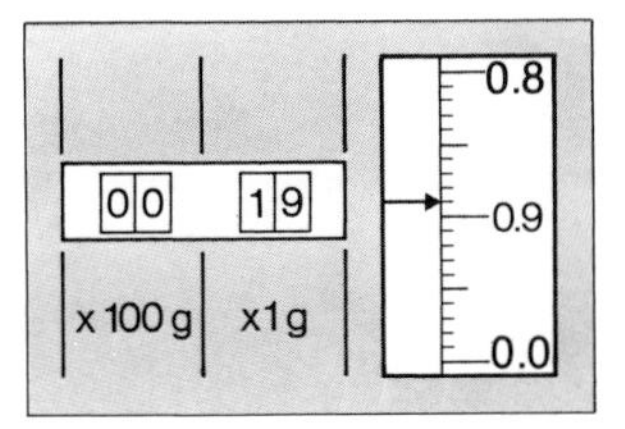

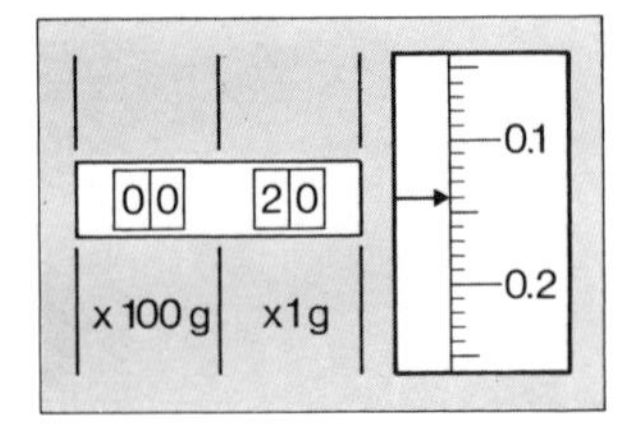

(a) Complete the results table below:
Mass of empty test tube ______ g
Mass of test tube + zinc carbonate ______ g
Mass of zinc carbonate taken ______ g (3)
(b) Calculate the mass of 1 mole of:
(i) zinc carbonate
(ii) zinc oxide (2)
(c) Calculate the expected mass of zinc oxide remaining in the experiment in which the test tube was heated. (2)
(d) The test tube was allowed to cool before it was reweighed. The mass remaining was found to be 0.19 g. Compare this result with the expected result obtained in part (c) and account for any difference. (2)
(e) Why was the test tube allowed to cool before it was weighed? (1)
(f) What would the experimenter do to ensure that the reaction has been completed? (1)
MEG

6 (a) Rust may be considered as a hydrated form of an oxide of iron $Fe_xO_y.zH_2O$. How would you determine experimentally the values of x, y and z? (7)
(b) It was found that a sample of rust contained 28.0 g of iron, 12.0 g of oxygen and 13.5 g of water. What is the formula of rust? (3)
MEG

7 A solid compound contains 39% potassium, 1% hydrogen, 12% carbon and 48% oxygen, by weight.
(a) What is the ratio of the weights of K: H: C: O in the compound?
(b) What is the ratio of moles of K: H: C: O in the compound?
(c) What is the formula of the solid?

8 Tinstone, SnO_2, is an important mineral from which tin can be obtained.
(a) How much does 1 mole of tinstone weigh?
(b) How much tin does 1 mole of tinstone contain?
(c) What is the percentage of tin in tinstone?
(d) How much tin is there in 5000 kg of tinstone?
(e) Where are the world's major deposits of tinstone found? (You will need to refer to an encyclopaedia.)
(f) When tinstone is converted into tin, the tinstone is:
A oxidized, B reduced, C synthesized, D decomposed?

SECTION D

Electricity and Electrolysis

1 Electricity in Everyday Life
2 Electric Currents
3 Which Substances Conduct Electricity?
4 Investigating Electrolysis
5 The Faraday Constant
6 Charges on Ions
7 Ionic Compounds
8 Molecular Compounds
9 Electrolysis in Industry
10 Electroplating
11 Electricity from Chemicals
12 Cells and Batteries

Electricity is transmitted across the country from power stations to our homes, schools and factories. Sometimes the pylons and cables spoil the countryside. Why are the cables not buried below ground?

1 Electricity in Everyday Life

A city at night. How could we survive without electricity?

Look at all the electrical appliances in this kitchen. It is difficult to imagine life without electrical appliances

It is hard to imagine life without electricity. Every day we depend on electricity for cooking, for lighting and for heating. At the flick of a switch, we want to use electric fires, electric kettles and dozens of other electrical gadgets. All these electrical appliances use mains electricity. The electricity is generated in power stations from coal, oil or nuclear fuel. Heat from the fuel is used to boil water. The steam produced drives turbines and generates electricity. In this way, energy in the fuel is converted into electrical energy.

In addition to the many appliances which use mains electricity, there are many others like torches, radios and calculators that use electrical energy from cells and batteries. Electricity can also be used to manufacture some important chemicals. For example, salt (sodium chloride) cannot be decomposed into sodium and chlorine using heat, but it can be decomposed using electricity. Sodium and chlorine are manufactured by passing electricity through a mixture of sodium chloride and calcium chloride at 600° C.

$$\text{sodium chloride} \xrightarrow{\text{electricity}} \text{sodium} + \text{chlorine}$$

These uses of electricity show why it is so useful to our society.

- *It can be used to transfer energy easily from one place to another.*
- *It can be converted into other forms of energy and used to warm a room, light a torch or cook a meal.*

Electric charges

Comb your hair quickly with a plastic comb and then use the comb to pick up tiny bits of paper. Why does this happen?

Atoms are composed of three particles—**protons**, **neutrons** and **electrons**. The centre of the atom, called the **nucleus**, contains protons and neutrons. Protons are positive but neutrons have no charge. Electrons occupy the outer parts of the atom and move around the nucleus. The negative charge on one electron just balances the positive charge on one proton so atoms have equal numbers of protons and electrons. For example, hydrogen atoms have 1 proton and 1 electron. Carbon atoms have 6 protons, 6 electrons and 6 neutrons, whereas aluminium atoms have 13 protons, 13 electrons and 14 neutrons (figure 1).

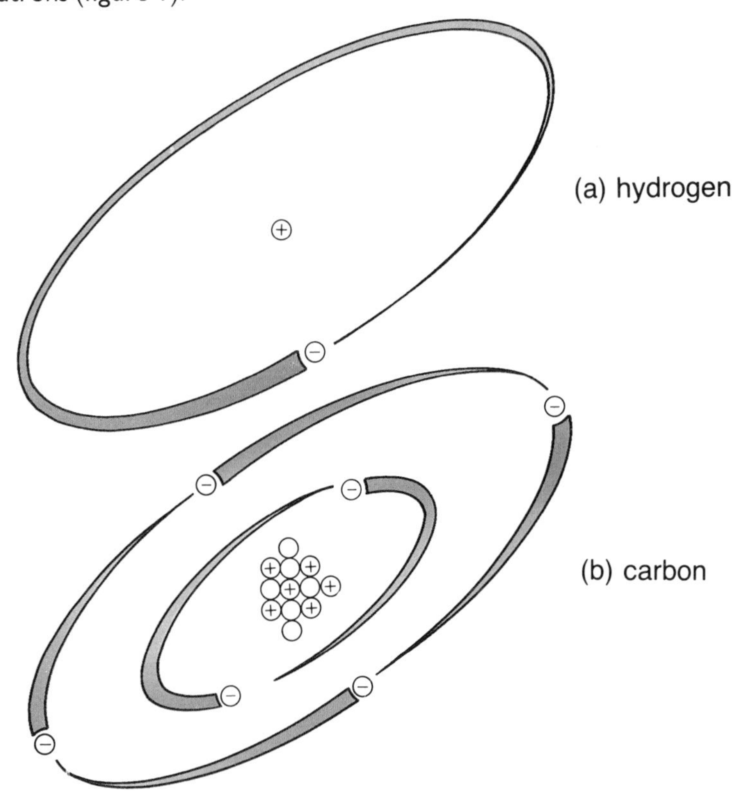

Figure 1
Protons, neutrons and electrons in a hydrogen atom and a carbon atom
(⊕ = proton, ○ = neutron, ⊖ = electron)

When you comb your hair quickly with a plastic comb, the comb pulls electrons off the atoms in your hair. The comb will then have more electrons than protons so it has a negative charge overall. Your hair will have fewer electrons than protons and so it has a positive charge overall.

When you bring the charged comb close to tiny pieces of paper, the negative charge on the comb repels electrons from the area of the paper nearest to it (figure 2(a)). This part of the paper therefore becomes positive and it is attracted to the comb because *unlike charges attract*. If the paper is light enough, it can be picked up (figure 2(b)).

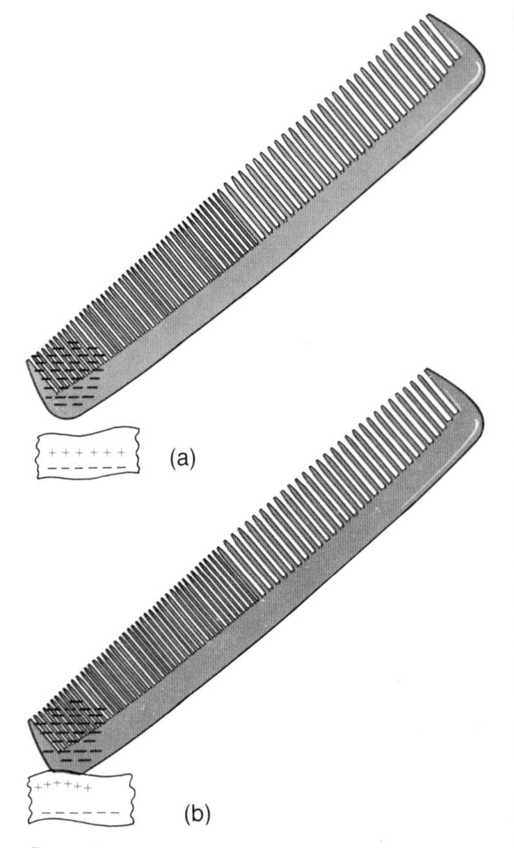

Figure 2

The plastic comb has just been rubbed on a woollen jumper. What is making the water bend?

Questions

1 Electricity is used more widely than gas for our energy supplies. Why is this?

2 (a) List 4 important uses which electricity has in your home.
(b) How would the jobs in part (a) be done without electricity?

3 Lithium atoms have 3 protons, 3 electrons and 4 neutrons. Draw a picture of a lithium atom similar to that for a carbon atom in figure 1.

2 Electric Currents

Plastic and glass insulators are used to hold the cables to pylons in the Grid System

Materials, such as plastic combs, which hold their charge on rubbing and do not allow electrons to pass through them are called **insulators**. Plastics like polythene, cellulose acetate and PVC are used to insulate electrical wires and cables. Materials, and in particular metals, that allow electrons to pass through them are called **conductors** (see the table).

Conductors			**Insulators**
Good	**Moderate**	**Poor**	
metals, e.g. copper, aluminium, iron	carbon, silicon, germanium	water	rubber, air, plastics, e.g polythene, PVC

Some conductors and insulators

Thus, copper, iron and aluminium, which are easily made into wire, are used for fuses, wires and cables in electrical machinery. Copper is used more than any other metal in electrical wires and cables because electrons move through it easily—it is a good conductor. When an electric current flows through the wire, the outermost electrons in the copper atoms are attracted towards the positive terminal of the battery, while extra electrons are repelled into the copper wire from the negative terminal (figure 1). *An electric current is simply a flow of electrons.* Electrons flow through the metal rather like water flows through a pipe or traffic moves along a road.

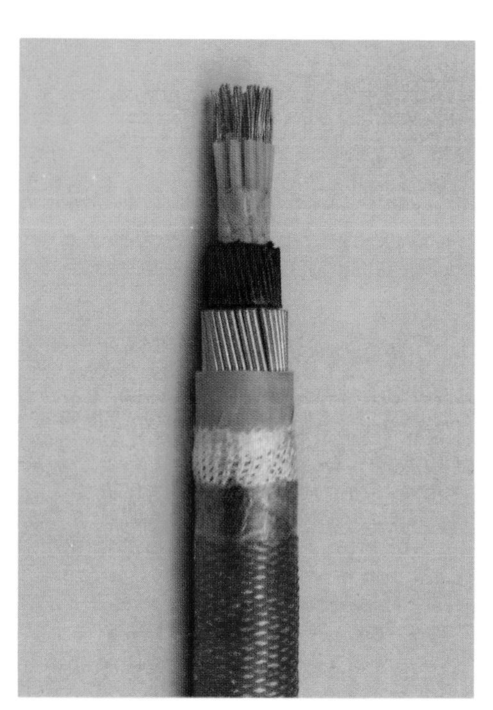

Copper is used in electrical wires and cables

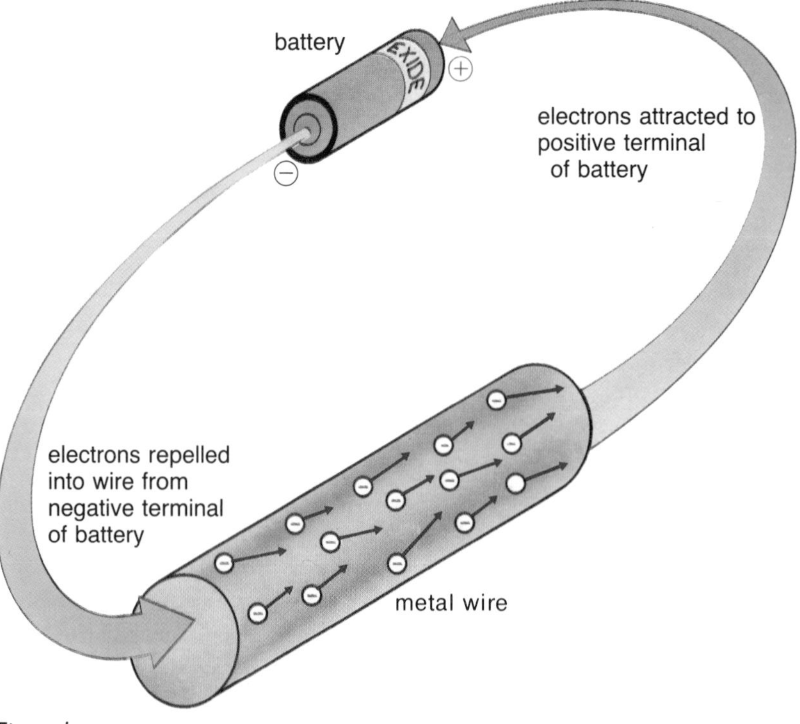

Figure 1

As we mentioned in section A, there are a few substances called *semiconductors*. Silicon and germanium containing a trace of impurity are two of the best known semiconductors. These substances allow electrons to flow easily through them in one direction, but not in the other. Semiconductors of this kind form an important part of any transistor. In some cases, hundreds of these tiny transistors are built up on a small flat plate forming a silicon chip.

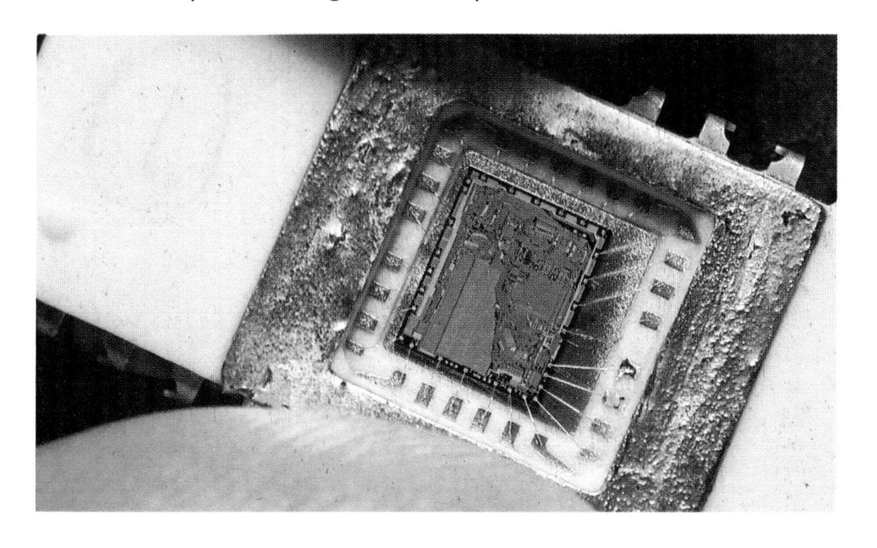

A silicon chip

Measuring electricity

We could measure the amount of electricity (electric charge) that has flowed along a wire by counting the number of electrons which pass a certain point. The charge on one electron is, however, much too small to be used as a practical unit in measuring the quantity of electricity. The practical unit normally used is the **coulomb** (C), which is about six million, million, million (6×10^{18}) electrons.

If one coulomb of charge passes along a wire in one second, then the rate of charge flow (i.e. the electric current) is 1 coulomb per second or 1 **ampere** (A). If 3 coulombs pass along the wire in 2 seconds, then the current is $3/2$ coulombs per sec or $3/2$ A. If Q coulombs flow along a wire in t seconds, the electric current (I) is given by,

$$I = \frac{Q}{t}$$

This equation can be rearranged to give

$$Q = I \times t$$

i.e. $\text{charge in coulombs} = \text{current in amps} \times \text{time in seconds}$

$\therefore$ 1 A for 1 sec = $1 \times 1 = 1$ C
2 A for 1 sec = $2 \times 1 = 2$ C
2 A for 2 sec = $2 \times 2 = 4$ C

The equation charge = current × time can be compared to the flow of water along a pipe since:

amount of water passed = rate of flow of water (i.e. current) × time

Questions

1 What is (i) an electric current; (ii) a conductor; (iii) an insulator?

2 (a) Name 4 elements which conduct electricity when solid.
(b) Name 4 elements which conduct electricity when liquid.
(c) Name 4 elements which do not conduct electricity when solid.

3 A current of 0.5 A flows for 1 hour. How many coulombs have passed?

4 The current in a small torch bulb is about 0.25 A. How much electricity flows if the torch is used for 15 minutes?

5 The current in a car headlamp is 4 A.
(a) What is the current in coulombs per sec?
(b) What is the current in electrons per sec? (Assume 1 coulomb = 6×10^{18} electrons.)

3 Which Substances Conduct Electricity

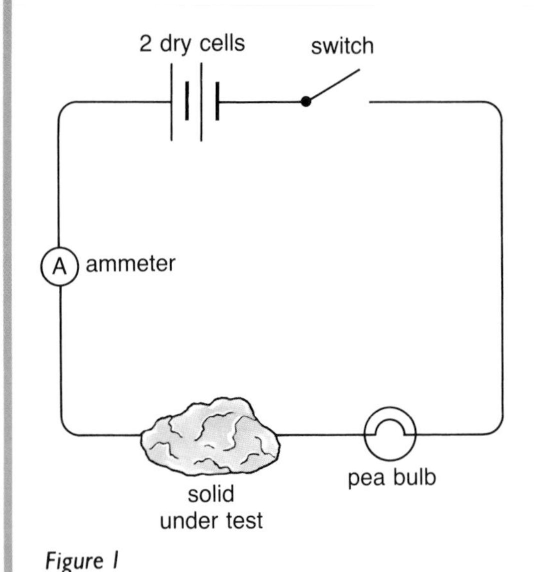

Figure 1

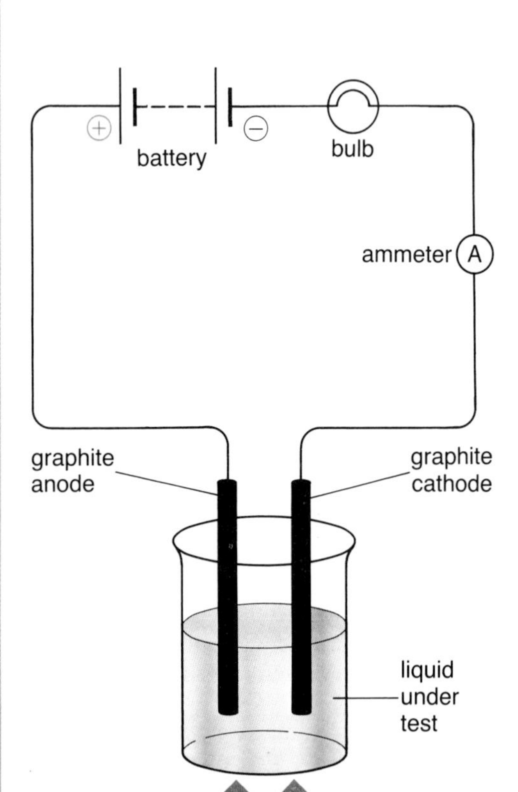

Figure 2

Which solids conduct?

The apparatus in figure 1 can be used to test whether a solid conducts electricity. If the solid conducts, what happens when the switch is closed?

Experiments show that *the only common solids which conduct electricity well are metals and graphite*. When metals and graphite conduct electricity, electrons flow through the material, but there is no chemical reaction. No new substances form and there is no change in weight. No solid *compounds* conduct electricity.

Which liquids conduct?

Water containing a little sulphuric acid will conduct electricity. Unlike metals, the water changes when it conducts electricity—it is decomposed into hydrogen and oxygen. Electricity is a form of energy, like heat. We can use it to boil water, to cook food and to cause chemical reactions. The decomposition of a substance, such as water, by electricity is called **electrolysis**. The compound which is decomposed is called an **electrolyte** and we say that it has been electrolysed.

Figure 2 shows how we can test the conductivity of liquids. The terminals through which the current enters and leaves the electrolyte are called **electrodes**. The electrode connected to the positive terminal of the battery is positive itself and is called the **anode**. The electrode connected to the negative terminal of the battery is negative itself and is called the **cathode**. Table 1 shows the results of tests on various liquids and aqueous solutions (solutions of substances in water).

Pure liquids	Does the liquid conduct?	Aqueous solutions	Does the solution conduct?
Bromine	No	Ethanol (C_2H_6O)	No
Mercury	Yes	Sugar ($C_{12}H_{22}O_{11}$)	No
Molten sulphur	No	Sulphuric acid (H_2SO_4)	Yes
Molten zinc	Yes	Acetic acid ($C_2H_4O_2$)	Yes
Water (H_2O)	No	Copper sulphate ($CuSO_4$)	Yes
Ethanol (C_2H_6O)	No	Sodium chloride (NaCl)	Yes
Tetrachloromethane (CCl_4)	No	Potassium iodide (KI)	Yes
Molten sodium chloride (NaCl)	Yes		
Molten lead bromide ($PbBr_2$)	Yes		

Table 1: testing to see which liquids and aqueous solutions conduct

Aluminium is manufactured by electrolysis. The workman is tapping molten aluminium from the cell

Look at the results in table 1.

1 Do the liquid metals conduct electricity?
2 Do the liquid non-metals conduct electricity?
3 Do the compounds containing only non-metals (non-metal compounds) conduct electricity (i) when liquid; (ii) in aqueous solution?
4 Do the compounds containing both metals and non-metals (metal/non-metal compounds) conduct electricity (i) when liquid; (ii) in aqueous solution?

The answers to these questions and the results of the experiment are summarised in table 2.

	Elements		**Compounds**	
Substance	**Metals and graphite**	**Non-metals except graphite**	**Metal/ non-metal**	**Non-metal**
Examples	Fe, Zn	Br_2, S	NaCl, $CuSO_4$	C_2H_6O, CCl_4
Solid	Yes	No	No	No
Liquid	Yes	No	Yes	No
Aqueous solution	—	—	Yes (and acids)	No (except acids)

Table 2: the conduction of electricity by elements and compounds

Notice the following points from these results:

- *Pure water does not conduct electricity*
- *Metal/non-metal compounds conduct electricity when they are molten (liquid) and when they are dissolved in water (aqueous)*

These compounds are decomposed during electrolysis. We shall investigate this later in the section.

- *Non-metal compounds do not conduct in the liquid state or in aqueous solution (except aqueous solutions of acids)*

Questions

1 Explain the following:
electrolysis; electrolyte; electrode; anode; cathode.
2 *Calcium; carbon disulphide; copper sulphate solution; lead; carbon; water; methane (natural gas); phosphorus; dilute sulphuric acid.*
From this list name: (i) two metals; (ii) two non-metals; (iii) two electrolytes; (iv) two pure liquids at 20°C; (v) two elements which conduct; (vi) two compounds that are gases at 110°C; (vii) three compounds that are non-electrolytes.
3 Which solids conduct electricity?
4 Which liquids conduct electricity?

4 Investigating Electrolysis

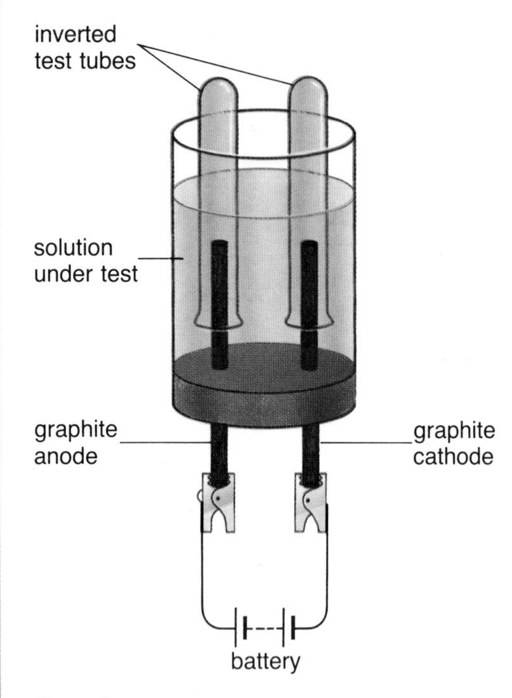

Figure 1

Products of electrolysis

When compounds are electrolysed, new substances are produced at the electrodes. For example, when electricity is passed through molten sodium chloride, pale green chlorine gas comes off at the anode and sodium forms at the cathode. When copper sulphate solution is electrolysed using the apparatus in figure 1, a pink deposit of copper appears on the cathode and bubbles of a colourless gas stream off the anode and collect in the inverted test tube. This gas relights a glowing splint. This shows that it is oxygen.

The table lists the products formed at each of the electrodes when various liquids and aqueous solutions are electrolysed. Remember that water in the aqueous solutions may be electrolysed.

Substance electrolysed	Product at anode	Product at cathode
Molten lead bromide	Brown fumes of bromine	Lead
Molten potassium iodide	Purple fumes of iodine	Potassium
Molten sodium chloride	Pale green chlorine gas	Sodium
Aqueous potassium iodide	Iodine which colours the solution brown	Hydrogen
Aqueous copper sulphate	Oxygen	Copper (deposited on the cathode)
Hydrochloric acid	Chlorine	Hydrogen
Aqueous zinc bromide	Bromine which colours the solution brown	Zinc (deposited on the cathode)
Aqueous copper chloride	Chlorine	Copper (deposited on the cathode)

The products formed at the electrodes when some liquids and aqueous solutions are electrolysed

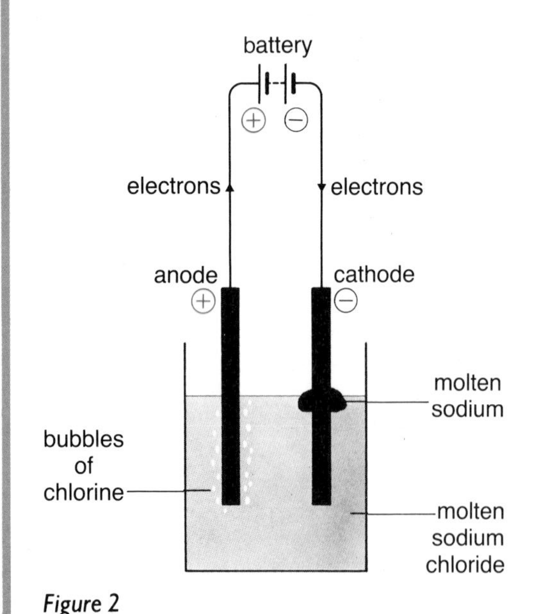

Figure 2

1 Which elements are produced at the anode?
2 Which elements are produced at the cathode?

When acids and metal/non-metal compounds conduct electricity
- *metals or hydrogen are formed at the cathode*
- *non-metals (except hydrogen) are formed at the anode*

The compound is decomposed by electrical energy and an element is produced at each electrode. This is very different from the conduction of electricity by metals which are not decomposed during conduction. The first three electrolyses in the table can be summarised in word equations as:

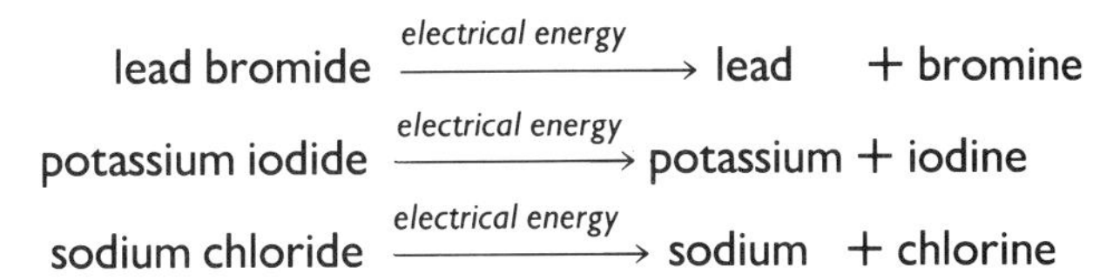

lead bromide $\xrightarrow{\text{electrical energy}}$ lead + bromine

potassium iodide $\xrightarrow{\text{electrical energy}}$ potassium + iodine

sodium chloride $\xrightarrow{\text{electrical energy}}$ sodium + chlorine

Figure 3(a)

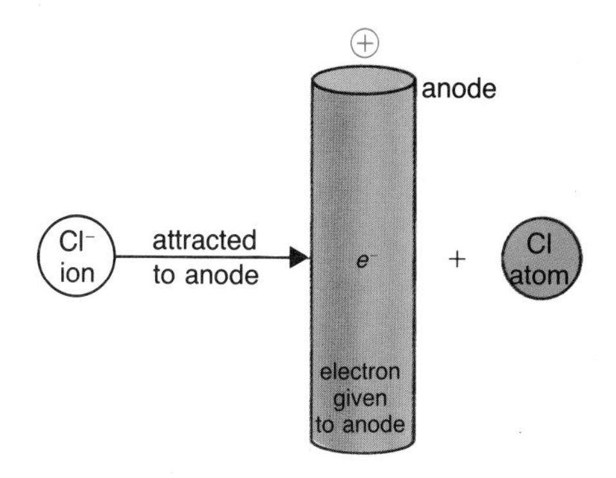

Figure 3(b)

Explaining electrolysis

When an electric current passes through molten sodium chloride, a shiny bead of sodium is produced at the cathode and chlorine gas forms at the anode (figure 2). This decomposition is caused by electrical energy in the current, but how does this happen? Sodium particles in the electrolyte must be positive since they are attracted to the negative cathode. At the same time, chlorine is produced at the anode, so chloride particles in the electrolyte are probably negative.

The formula of sodium chloride is NaCl so we can think of this as positive Na^+ particles and negative Cl^- particles. Since NaCl is neutral, the positive charge on one Na^+ must balance the negative charge on one Cl^-. These charged particles, like Na^+ and Cl^-, which move to the electrodes during electrolysis are called **ions**.

When Na^+ ions reach the cathode, they combine with negative electrons on the cathode forming neutral sodium atoms (figure 3a):

Na^+	+	e^-	$\longrightarrow$	Na
sodium ion in sodium chloride electrolyte		electron on cathode from battery		sodium atom in metal

At the anode, Cl^- ions lose an electron to the positive anode and form neutral chlorine atoms (figure 3b):

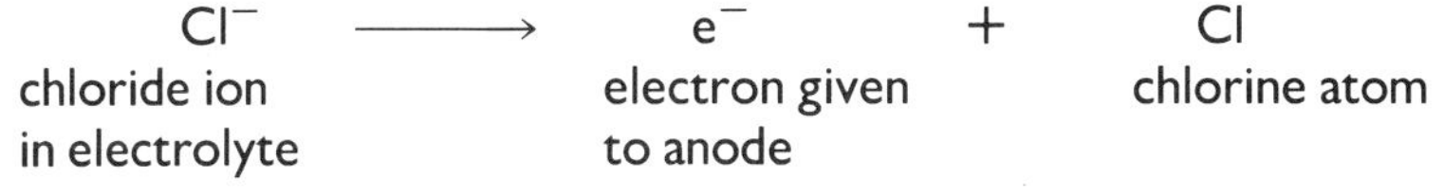

Cl^-	$\longrightarrow$	e^-	+	Cl
chloride ion in electrolyte		electron given to anode		chlorine atom

The Cl atoms immediately join up in pairs to form molecules of chlorine gas, Cl_2:

Cl	+	Cl	$\longrightarrow$	Cl_2
chlorine atom		chlorine atom		chlorine molecule

These equations show that Na^+ ions remove electrons from the cathode, and Cl^- ions give up electrons to the anode. The electric current is being carried through the molten sodium chloride by ions. The electrolysis of other molten and aqueous substances can also be explained in terms of ions.

Questions

Write *true* or *false* to each answer in questions 1 to 3.

1 When pure lead bromide is electrolysed
A it must be molten
B lead forms at the anode
C decomposition occurs
D a brown gas forms at the anode
E the process is exothermic.

2 The following substances are electrolytes:
A copper sulphate solution
B sugar solution
C dilute sulphuric acid
D copper
E molten wax
F liquid sulphur.

3 When potassium chloride solution is electrolysed, the products include:
A chlorine at the anode
B hydrogen at the cathode
C oxygen at the cathode
D potassium at the cathode
E chlorine at the cathode
F potassium at the anode.

5 The Faraday Constant

This electrolytic reduction cell produces aluminium

During electrolysis
- *a metal or hydrogen forms at the negative cathode*
- *a non-metal (except hydrogen) forms at the positive anode*

Since the cathode will attract only positive charges and the anode will attract only negative charges:

- *metals and hydrogen have positive ions. These ions are called* **cations** *because they are attracted to the cathode*
- *non-metals (except hydrogen) have negative ions. These ions are called* **anions** *because they are attracted to the anode*

How much charge is needed to deposit 1 mole of copper?

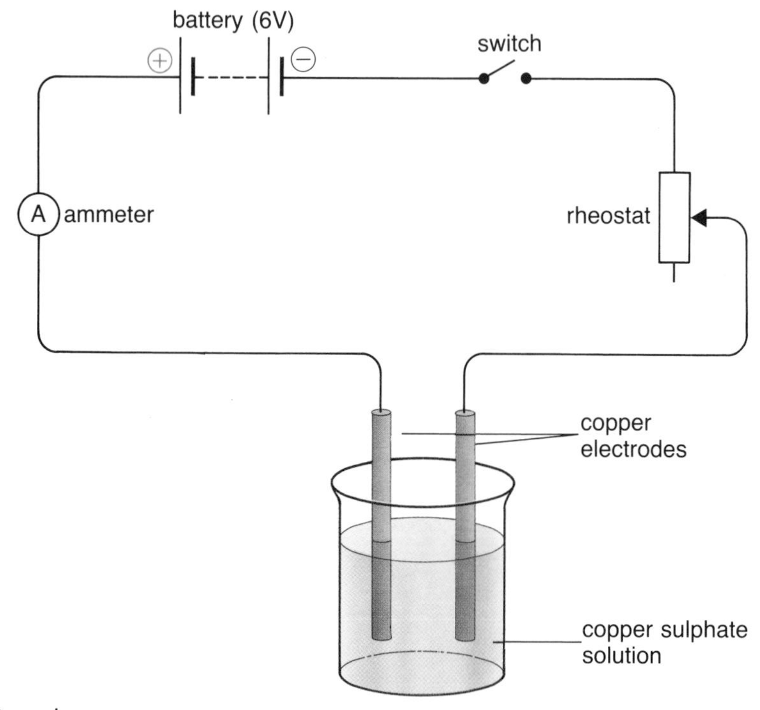

Figure 1

Using the apparatus in figure 1, we can find the amount of charge required to deposit 1 mole of copper (63.5 g) on the cathode during electrolysis. From this result, we can decide how much charge the copper ion has (i.e. whether the copper ion should be written as Cu^{+}, Cu^{2+}, Cu^{3+}, etc.). The rheostat (variable resistor) is used to keep the current constant and quite low. If the current is too large, the copper deposits too fast and drops off the cathode.

Make sure the copper cathode is clean and dry and then weigh it. Connect up the circuit and pass about 0.15 A for at least 45 minutes. Now, remove the cathode, wash it in distilled water and then acetone. When it is completely dry, reweigh it.

1 Why must the cathode be clean and dry when it is weighed before electrolysis?
2 Why is the cathode washed in distilled water and then acetone after electrolysis?

Here are the results of one experiment:

Mass of copper cathode before electrolysis $= 43.53$ g
Mass of copper cathode after electrolysis $= 43.66$ g

$\therefore$ Mass of copper deposited $= 0.13$ g

Time of electrolysis = 45 min $= 2700$ sec
Current $= 0.15$ A
Quantity of electric charge used $= I \times t$ (unit 2)
$= 0.15 \times 2700 = 405$ coulombs (C)

$\Rightarrow$ 0.13 g copper is produced by 405 C
$\therefore$ 1 g copper is produced by $\frac{405}{0.13}$ C
so, 1 mole of copper (63.5 g) is produced by $\frac{405}{0.13} \times 63.5$
$= 198\,000$ C

Accurate experiments show that 1 mole of copper is deposited by 193 000 coulombs. This amount of electricity would operate a 2-bar electric fire for about 6 hours.

A similar experiment to the last one was carried out using silver electrodes in silver nitrate solution. 0.45 g of silver was deposited in 45 minutes using a current of 0.15 A. Calculate the amount of charge needed to deposit 1 mole of silver (108 g) using a method similar to the one we used for copper.

The table shows the amount of charge required to produce 1 mole of atoms for 5 different elements. Notice that twice as much charge is required to produce one mole of copper as is required to produce one mole of sodium. Exactly three times as much charge is required to produce 1 mole of aluminium.

Element	**Number of coulombs required to produce 1 mole of atoms**
Copper	193000
Sodium	96 500
Silver	96 500
Aluminium	289 500
Lead	193000

The amount of charge needed to produce 1 mole of atoms, for 5 elements

When molten liquids and aqueous solutions are electrolysed, the quantity of electricity needed to produce one mole of an element is always a multiple of 96 500 coulombs (i.e. 96 500 or 193 000 ($2 \times 96\,500$) or 289 500 ($3 \times 96\,500$)). Because of this, 96 500 coulombs is called the **Faraday constant** (F), in honour of Michael Faraday. Faraday was the first scientist to measure the masses of elements produced during electrolysis.

Michael Faraday—one of the first scientists to investigate electrolysis

Questions

1 Explain the following:
cation; anion; the Faraday constant.
2 When nickel electrodes were used in a solution of nickel nitrate, 0.11 g of nickel was deposited in 60 minutes using a current of 0.10 A.
(a) What quantity of electric charge passes when a current of 0.10 A flows for 60 minutes?
(b) How many coulombs will deposit 1 g of nickel?
(c) How many coulombs will deposit 1 mole (59 g) of nickel?
3 Draw a clearly labelled diagram of the apparatus you would use to coat a graphite rod with copper.

6 Charges on Ions

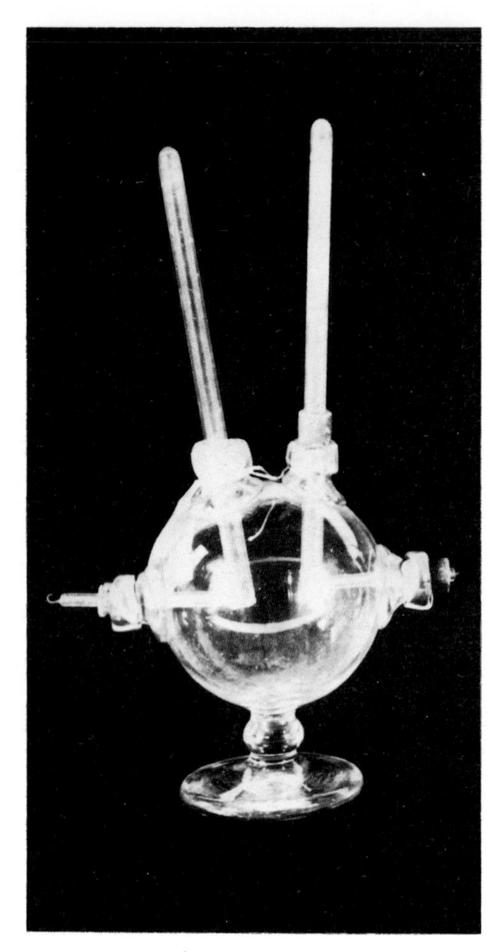

The electrolysis equipment in which Faraday studied the charges on ions. Where are the electrodes? What are the vertical tubes for?

During electrolysis, the positive charge on 1 mole of sodium ions is 'neutralized' by 96 500 C of negative charge from electrons on the cathode.

Since one electron has a charge of 1.6×10^{-19} C, the number of electrons carrying 96 500 C

$$= \frac{96\,500}{1.6 \times 10^{-19}} = 6 \times 10^{23} = 1 \text{ mole}$$

Thus, 1 mole of sodium ions requires 1 mole of electrons
⇒ 1 sodium ion requires 1 electron.
We can write this as:

$$\underset{\text{sodium ion}}{Na^+} + \underset{\text{electron}}{e^-} \rightarrow \underset{\text{sodium atom}}{Na}$$

Twice as much charge (193 000 C) is required to produce 1 mole of copper as is required to produce 1 mole of sodium (96 500 C). Thus, the charge on one copper ion, is twice as great as that on a sodium ion, and the copper ion is written as Cu^{2+}.
Look at the table in the last unit.

1 How much charge is required to produce 1 mole of silver?
2 What is the charge on one silver ion?
3 What is the formula for a silver ion?

In this way we can build up a list of ions with their charges, like those in the table on page 77. Notice that copper can form two ions, Cu^+ and Cu^{2+}. We show this in the names of the compounds by using the names copper(I) and copper(II). Thus copper forms two oxides, two chlorides, two sulphates, etc. The correct names for its two oxides are copper(I) oxide which is red, and copper(II) oxide which is black. Most of the common copper compounds are copper(II) compounds. These include copper(II) oxide and blue copper(II) sulphate. Iron can also form two different ions, Fe^{2+} and Fe^{3+}, and we use the names iron(II) and iron(III) for their respective compounds.

Most metal ions have a charge of 2+. All the *common* metal ions without a charge of 2+ are shown in the table. These are Ag^+, Na^+ and K^+ with a charge of 1+ (to remember this say 'AgNaK') and Cr^{3+}, Al^{3+} and Fe^{3+} with a charge of 3+ (to remember this say 'CrAlFe').

Copper has two oxides and two chlorides

Liming the soil. Lime is an ionic compound. Why is it added to the soil?

Notice also in the table that some negative ions are made from a group of atoms. For example, nitrate, NO_3^-, contains one nitrogen atom and three oxygen atoms. Groups of atoms like this are called **radicals**.

Positive ions (cations)		**Negative ions** (anions)	
Hydrogen	H^+	Chloride	Cl^-
Sodium	Na^+	Bromide	Br^-
Potassium	K^+	Iodide	I^-
Silver	Ag^+	Nitrate	NO_3^-
Copper(I)	Cu^+	Hydroxide	OH^-
Copper(II)	Cu^{2+}	Oxide	O^{2-}
Magnesium	Mg^{2+}	Carbonate	CO_3^{2-}
Calcium	Ca^{2+}	Sulphide	S^{2-}
Zinc	Zn^{2+}	Sulphate	SO_4^{2-}
Iron(II)	Fe^{2+}	Sulphite	SO_3^{2-}
Iron(III)	Fe^{3+}		
Aluminium	Al^{3+}		
Chromium	Cr^{3+}		

Common ions and their charges

Bonds between ions

Metal/non-metal compounds which are made of ions are called **ionic compounds**. They include common salt (sodium chloride), lime (calcium oxide) and iron ore (iron(III) oxide).

When ionic compounds form, the charges on the positive ions just balance the charges on the negative ions. So, because the sodium ion is Na^+ and the chloride ion is Cl^-, we might have predicted that the formula of sodium chloride would be Na^+Cl^- or simply NaCl. By balancing the charges in this way, we can work out the formulas of other ionic compounds. For example, the formula of lime is $Ca^{2+}O^{2-}$(CaO) and that of iron(III) oxide is $(Fe^{3+})_2(O^{2-})_3$ or simply Fe_2O_3. In iron(III) oxide, the six positive charges on two Fe^{3+} ions have been balanced by six negative charges on three O^{2-} ions.

Questions

1 Explain the terms:
ionic compound; radical.

2 (a) How much charge is required to produce 1 mole of (i) aluminium; (ii) lead? (See the table in the last unit.)
(b) What is the formula for (i) an aluminium ion; (ii) a lead ion?

3 Write the symbols for the ions in the following compounds and then work out their formulas:
calcium chloride; copper(II) sulphide; aluminium oxide; potassium hydroxide; iron(III) iodide; chromium bromide.

7 Ionic Compounds

Chalk cliffs are composed of an ionic compound containing calcium ions (Ca^{2+}) and carbonate ions (CO_3^{2-})

Forming ionic compounds—electron transfer

Ionic compounds form when metals react with non-metals. For example, when sodium burns in chlorine, sodium chloride is formed. This contains sodium ions (Na^+) and chloride ions (Cl^-).

Na	+	Cl	→	Na^+	Cl^-
sodium atom		chlorine atom		sodium ion	chloride ion

These ions form by transfer of electrons. During the reaction, each sodium atom gives up one electron and forms a sodium ion:

$$Na \rightarrow Na^+ + e^-$$

The electron is taken by a chlorine atom to form a chloride ion:

$$Cl + e^- \rightarrow Cl^-$$

When ionic compounds form, metal atoms *lose* electrons and form *positive* ions, whereas non-metal atoms *gain* electrons and form *negative* ions. This transfer of electrons from metals to non-metals explains the formation of ionic compounds. Figure 1 shows what happens when calcium reacts with oxygen to form calcium oxide. In this case, two electrons are transferred from each calcium atom to each oxygen atom.

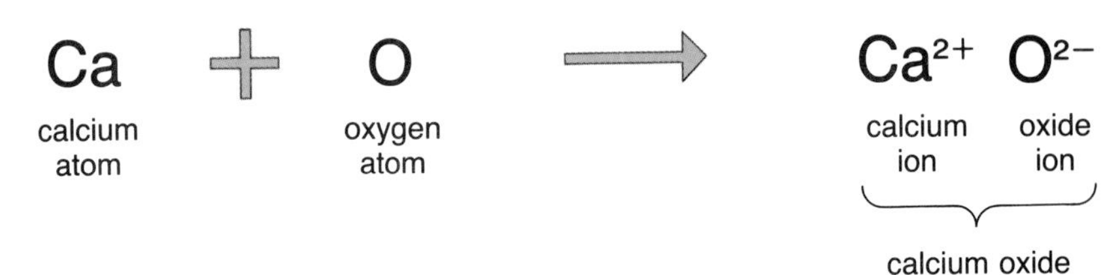

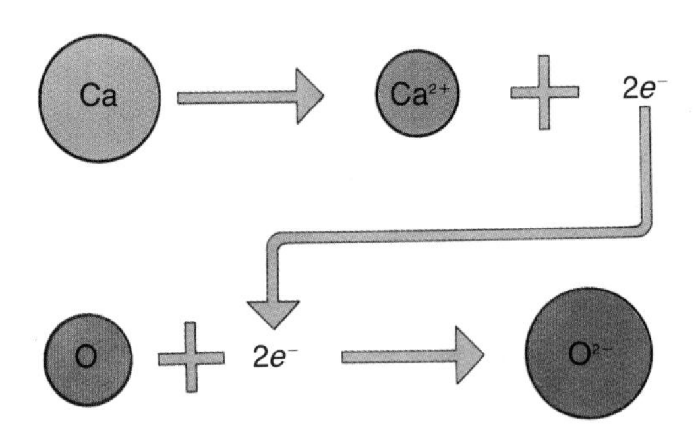

Figure 1

Bonding and properties of ionic compounds

In solid ionic compounds, the ions are held together by the attraction between positive ions and negative ions. Figure 2 shows how the ions are arranged in sodium chloride. Notice that Na^+ ions are surrounded by Cl^- ions and that the Cl^- ions are surrounded by Na^+ ions.

This kind of arrangement in which a large number of atoms or ions are packed together in a regular pattern is called a **giant structure**. The force of attraction between oppositely-charged ions in ionic compounds is called an **ionic or electrovalent bond**. The strong ionic bonds hold the ions together very firmly. This explains why ionic compounds:

1 are solids at room temperature with high melting points;
2 are hard substances;
3 conduct electricity when molten or aqueous;
4 cannot conduct when solid because the ions cannot move freely.

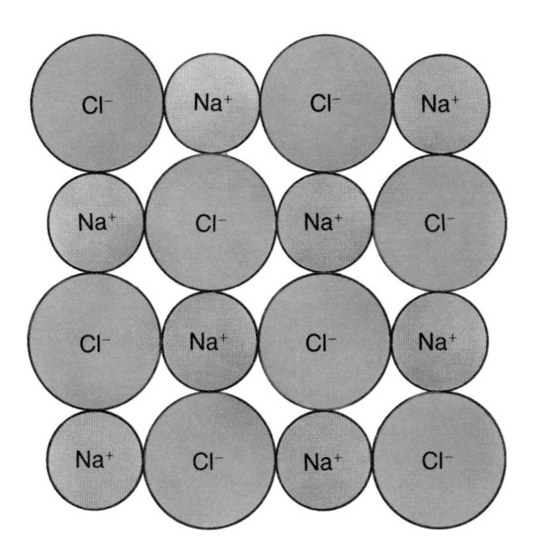

Figure 2

Formulas of ionic compounds—balancing charges

In the last unit, we found that the formulas of ionic compounds, like sodium chloride (NaCl) and calcium oxide (CaO), could be obtained by balancing the charges on the positive ions with those on the negative ions. For example, the formula of calcium chloride is $Ca^{2+}(Cl^-)_2$ or simply $CaCl_2$. Here, two Cl^- ions balance the charge on one Ca^{2+} ion. The formula $CaCl_2$ has a small 2 after the Cl to show that 2 Cl^- ions are needed for one Ca. These formulas show the ratio of the numbers of ions present in the ionic compound.

Can you see that the number of charges on an ion is a measure of its **combining power** or **valency**? Na^+ has a combining power of 1, whereas Ca^{2+} has a combining power of 2. Na^+ can combine with only one Cl^- to form Na^+Cl^-, whereas Ca^{2+} can combine with two Cl^- ions to form $Ca^{2+}(Cl^-)_2$.

Elements such as iron, which have two different ions (Fe^{2+} and Fe^{3+}), have two valencies. Thus iron can form two different compounds with chlorine—iron(II) chloride, $FeCl_2$, and iron(III) chloride, $FeCl_3$.

The table shows the names and formulas of some salts. Notice that the formula of calcium nitrate is $Ca(NO_3)_2$. The brackets around NO_3^- show that it is a single unit containing one nitrogen and three oxygen atoms with one negative charge. Thus, two NO_3^- ions balance one Ca^{2+} ion. Other ions like SO_4^{2-}, CO_3^{2-} and OH^- must also be regarded as single units and put in brackets when there are 2 or 3 of them in a formula.

Name of salt	Formula
calcium nitrate	$Ca^{2+}(NO_3^-)_2$ or $Ca(NO_3)_2$
zinc sulphate	$Zn^{2+}SO_4^{2-}$ or $ZnSO_4$
iron(III) chloride	$Fe^{3+}(Cl^-)_3$ or $FeCl_3$
copper(II) bromide	$Cu^{2+}(Br^-)_2$ or $CuBr_2$
sodium carbonate	$(Na^+)_2CO_3^{2-}$ or Na_2CO_3
potassium iodide	K^+I^- or KI

The names and formulas of some salts

Questions

1 Which of the following substances conduct electricity (i) when liquid; (ii) when solid:
diamond; potassium chloride; copper; carbon disulphide; sulphur?
2 Explain the following:
ionic compound; ionic bond; giant structure; valency.
3 Write the symbols for the ions and the formulas of the following compounds:
potassium hydroxide; iron(III) nitrate; barium chloride; sodium carbonate; silver sulphate; calcium hydrogencarbonate; aluminium oxide; zinc bromide; copper(II) nitrate; magnesium sulphide.

8 Molecular Compounds

If the nucleus of an atom was magnified a million, million times, it would be as big as a pea and the total volume of the atom in which the electrons move would be as large as Westminster Abbey

Metals can be mixed to form alloys, but they *never* react with each other to form compounds. For example, zinc and copper will form the alloy brass, but the two metals cannot react chemically because they both want to lose electrons to form positive ions. Unlike metals, two non-metals can react with each other and form a compound even though they both want to gain electrons. These *non-metal compounds* are made of molecules, and they do not contain ions. They are therefore called **molecular compounds**. Water (H_2O), carbon dioxide (CO_2), sugar ($C_{12}H_{22}O_{11}$) and ammonia (NH_3) are examples of simple molecular compounds.

Forming molecular compounds—electron sharing

All atoms have a small positive centre called a **nucleus**, surrounded by a larger region of negative charge. The negative charge consists of electrons; it is balanced by positive charge in the nucleus, so that the whole atom is neutral. Almost all the mass of the atom is concentrated in the nucleus (figure 1). Different atoms have different numbers of electrons. Hydrogen atoms are the smallest with only one electron, helium atoms have two electrons and oxygen atoms have eight electrons.

When two non-metals react to form a molecule, the regions of electrons in their atoms overlap so that each atom gains negative charge. The positive nuclei of both atoms attract the electrons in the region of overlap and this holds the atoms together (figure 2). This type of bond formed by *electron sharing* between non-metals is known as a **covalent bond**. Notice that covalent bonding, like ionic bonding, involves attraction between opposite charges. Covalent bonds hold the atoms together *within* a molecule but there must also be bonds holding the separate molecules together in molecular liquids like water and molecular solids like sugar. These bonds *between* separate molecules are known as **intermolecular bonds**. For example, in water there are strong covalent bonds between the two hydrogen atoms and the oxygen atom *within* each molecule of H_2O and also weak intermolecular bonds between the different molecules of H_2O.

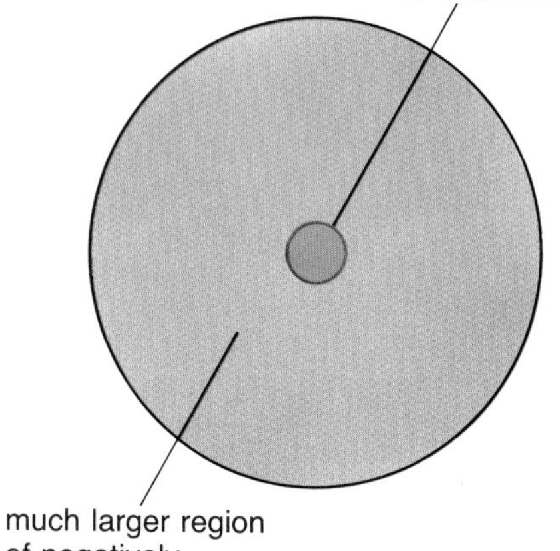

Figure 1
A simple picture of atomic structure

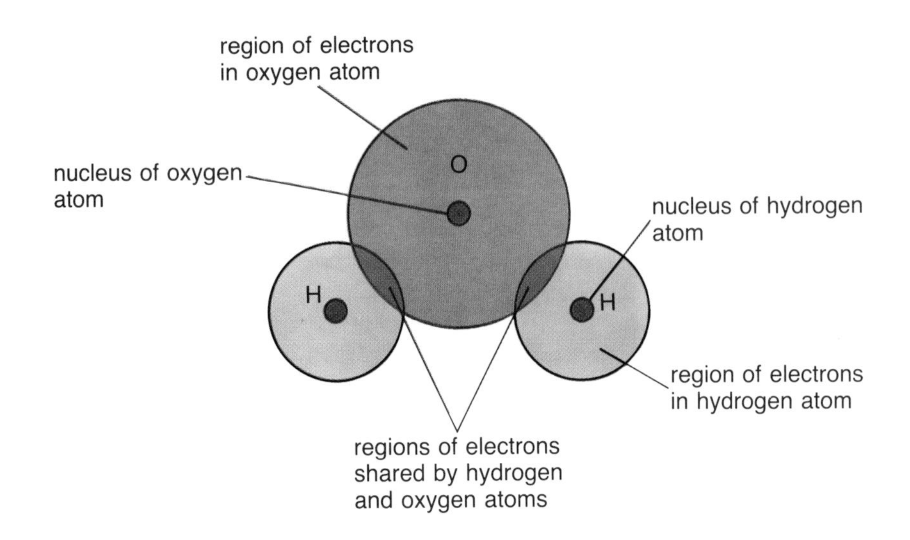

Figure 2
The simple structure of a molecule of water

Formulas of molecular compounds

The table shows the formulas and structures of some well-known molecular compounds. The structures are written so that the number of covalent bonds (drawn as a line —) to each atom can be seen. Notice that hydrogen can form 1 bond with other atoms (H—), so its combining power or valency is 1. The combining powers of chlorine and bromine are also 1. Oxygen and sulphur both form 2 bonds with other atoms (—O— and —S—). Their combining power is therefore 2. Nitrogen atoms form 3 bonds and carbon atoms form 4 bonds, so their valencies are 3 and 4.

Although we can predict the formulas of molecular compounds from the number of bonds which the atoms form, the only sure way of knowing a formula is by chemical analysis. For example, carbon forms 4 bonds and oxygen forms 2 bonds, so we would predict that carbon and oxygen will form a compound O=C=O. (Each bond is represented by a single line, so two lines show that there is a double bond between the atoms.) This compound, carbon dioxide, does exist, but so does carbon monoxide, CO, which we would not predict.

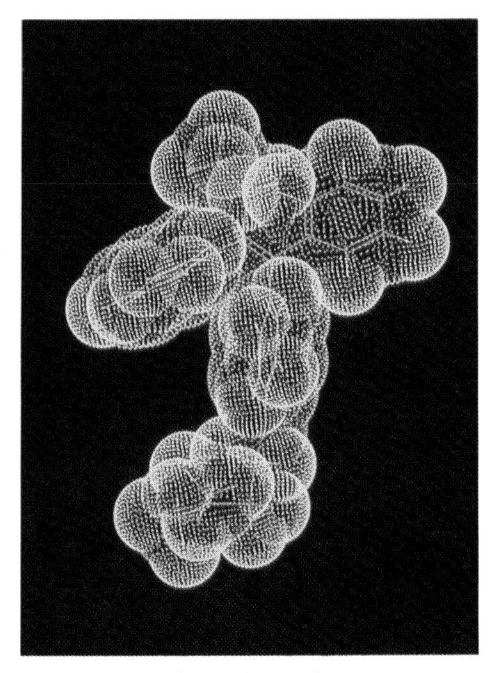

This model of a molecule was drawn using a computer to analyse the results obtained during the investigation of a complex organic substance

Compound	**Formula**	**Structure**
Hydrogen chloride	HCl	H—Cl
Hydrogen bromide	HBr	H—Br
Hydrogen iodide	HI	H—I
Water (hydrogen oxide)	H_2O	H—O—H
Hydrogen sulphide	H_2S	H—S—H
Ammonia (hydrogen nitride)	NH_3	H\N/H with N—H below (N bonded to three H)
Methane	CH_4	H above, H—C—H, H below (C bonded to four H)
Carbon dioxide	CO_2	O=C=O
Carbon disulphide	CS_2	S=C=S
Tetrachloromethane (carbon tetrachloride)	CCl_4	Cl above, Cl—C—Cl, Cl below (C bonded to four Cl)

The formulas and structures of some well-known molecular compounds

Questions

1 Explain the following:
non-metal compound; molecular compound; covalent bond; nucleus; intermolecular bond.

2 Assuming the usual combining powers of the elements, draw the structures of the following compounds: (Show each bond as a line —.)
sulphur dichloride; dichlorine oxide; tetrabromomethane (carbon tetrabromide); nitrogen triiodide; *hydrogen peroxide*, H_2O_2; *ethane*, C_2H_6; *ethene* (C_2H_4).

9 Electrolysis in Industry

Obtaining metals by electrolysis

Reactive metals like sodium, magnesium and aluminium cannot be obtained by reducing their oxides to the metal with carbon (coke). These metals can only be obtained by electrolysis of their molten (fused) compounds. We cannot use electrolysis of their aqueous compounds since hydrogen (from the water), and *not* the metal, is produced at the cathode. For example, during the electrolysis of aqueous sodium chloride, hydrogen (not sodium) is produced at the cathode.

Metals low in the reactivity series, such as copper and silver, can be obtained by reduction of their compounds or by electrolysis of their aqueous compounds. When their aqueous compounds are electrolysed, the metal is produced at the cathode rather than hydrogen (from the water).

Aluminium is obtained from bauxite, impure aluminium oxide. The picture shows bauxite being mined

Manufacturing aluminium by electrolysis

Aluminium is manufactured by the electrolysis of *molten* aluminium oxide obtained from bauxite. Although clay is the most abundant source of aluminium, the metal cannot be extracted from clay because it would cost too much.

Pure aluminium oxide does not melt until 2045°C and this makes its electrolysis uneconomic. The aluminium oxide is therefore dissolved in molten cryolite (Na_3AlF_6) which melts below 1000°C. Figure 1 shows a diagram of the electrolytic method used. Aluminium ions in the electrolyte are attracted to the carbon cathode lining the tank where they accept electrons to form aluminium:

Cathode (−) $Al^{3+} + 3e^- \rightarrow Al$

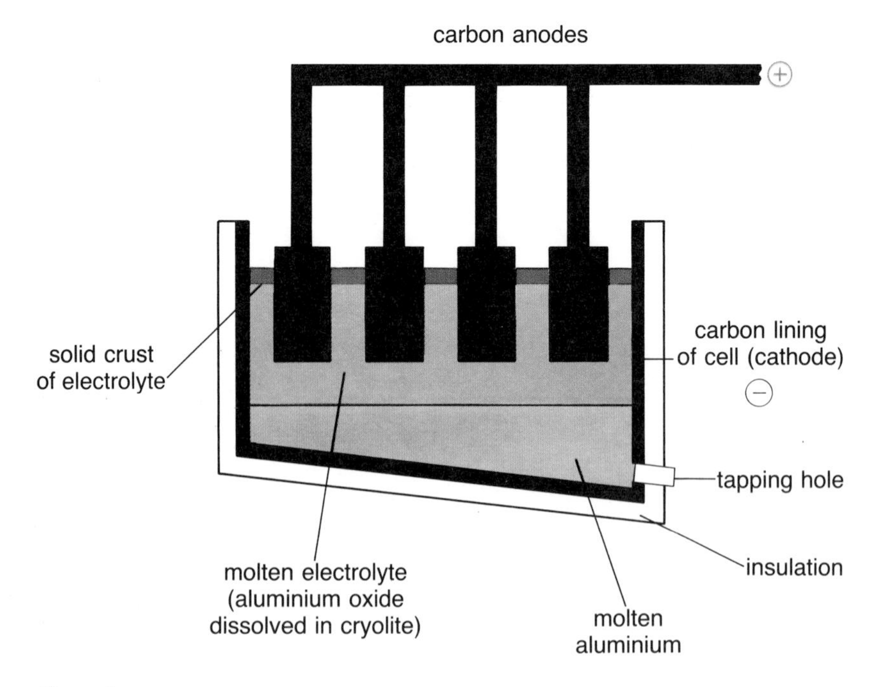

Figure 1
The electrolytic cell for aluminium manufacture

Molten aluminium collects at the bottom of the cell and is tapped off at intervals. It takes about 16 kilowatt-hours of electricity to produce 1 kg of aluminium. The extraction plants are therefore sited near sources of cheap electricity.

Oxide ions (O^{2-}) are attracted to the carbon anodes to which they give their electrons forming oxygen atoms:

Anode (+) $O^{2-} \rightarrow O + 2e^-$

The oxygen atoms then combine in pairs to form oxygen gas (O_2).

Purifying copper by electrolysis

When copper sulphate solution is electrolysed with copper electrodes, copper is deposited on the cathode and the copper anode loses weight (figure 2).

The aqueous copper sulphate contains copper ions (Cu^{2+}) and sulphate ions (SO_4^{2-}). During electrolysis, Cu^{2+} ions are attracted to the cathode where they gain electrons and deposit on the cathode:

Cathode (−) $Cu^{2+} + 2e^- \rightarrow Cu$

SO_4^{2-} ions are attracted to the anode, but they are not discharged. Instead, copper atoms, which make up the anode, give up two electrons each and go into solution as Cu^{2+} ions:

Anode (+) $Cu \rightarrow Cu^{2+} + 2e^-$

The overall result of this electrolysis is that the anode loses weight and the cathode gains weight—copper metal is transferred from the anode to the cathode.

This method is used industrially to purify crude copper. The impure copper is the anode of the cell, the cathode is a thin sheet of pure copper and the electrolyte is copper sulphate solution. The impure copper anode dissolves away and pure copper deposits on the cathode.

Impure copper anodes being transferred to an electrolysis tank for purification

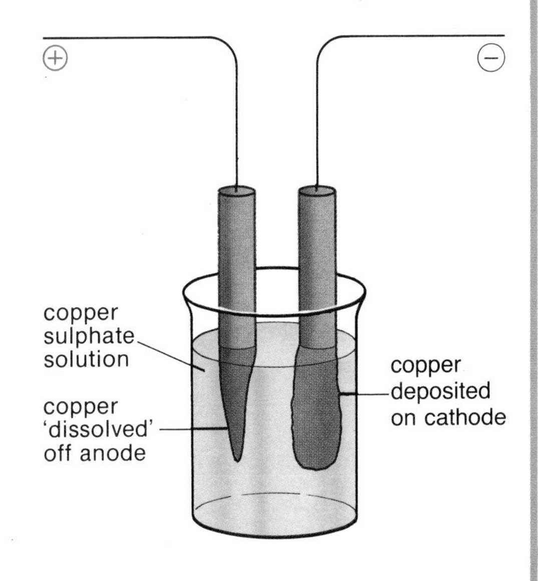

Figure 2
Purifying copper by electrolysis

Questions

Write *true* or *false* to each of the answers suggested in questions 1 and 2.

1 When electrolysis occurs, you *may* observe, at the negative electrode, the formation of

A a metal
B a non-metal
C oxygen
D hydrogen
E a gas
F a liquid.

2 Before an article can be electroplated it *must* be

A wiped over with a thin film of oil;
B dipped into molten metal;
C wrapped with wire of the metal to be coated;
D made the cathode in an electrolytic cell;
E dipped in concentrated acid.

3 Molten sodium chloride is electrolysed using carbon electrodes.

(a) Give the symbols and charges on the ions in the electrolyte.
(b) Draw a diagram to show the direction of migration of the ions and the direction of flow of electrons in the circuit.
(c) Write equations for the processes which occur at the anode and cathode.

10 Electroplating

This picture shows the bodywork of a car rising out of an electroplating bath. Notice the wire carrying the current from the bodywork

The method for purifying impure copper described in the last unit can be used industrially to coat (plate) articles with copper. The process is called *electroplating*. Several metals can be used for electroplating articles. The most commonly used metals, apart from copper, are chromium, silver and tin. The article to be plated is the cathode of the cell, the anode is a piece of copper and the electrolyte is copper sulphate solution. During the electroplating process, the copper anode dissolves away and copper deposits on the article at the cathode.

Although it is easy to deposit a metal during electrolysis, the conditions must be carefully controlled so that the metal sticks to the object to be plated. For example, chromium will not stick to iron (steel) or copper during electroplating. Thus, steel articles to be chromium plated (such as car bumpers and electric kettles) are first plated with nickel. This forms a firm deposit on the steel which can then be plated with chromium.

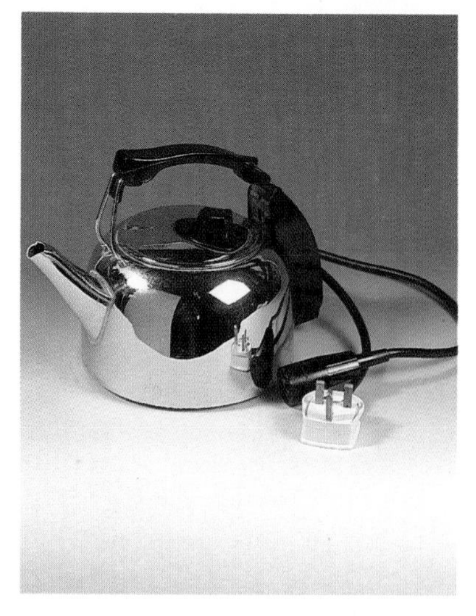

These three objects have been electroplated with either copper or chromium

In order to obtain a good coating of metal during electroplating:

1 the object to be plated must be clean and free of grease;

2 the object to be plated should be rotated to give an even coating;

3 the electric current must not be too large or the 'coating' will form too rapidly and will flake off;

4 the temperature and concentration of the electrolyte must be carefully controlled, otherwise the 'coating' will be deposited too rapidly or too slowly.

The metal coating is deposited on the cathode and so the object to be plated must be used as the cathode during electrolysis. The electrolyte must contain a compound of the metal which forms the coating.

Figure 1 shows a nickel alloy fork being electroplated with silver. Notice that the fork is attached to a rotatable arm and that the electrolyte contains silver nitrate ($AgNO_3$).

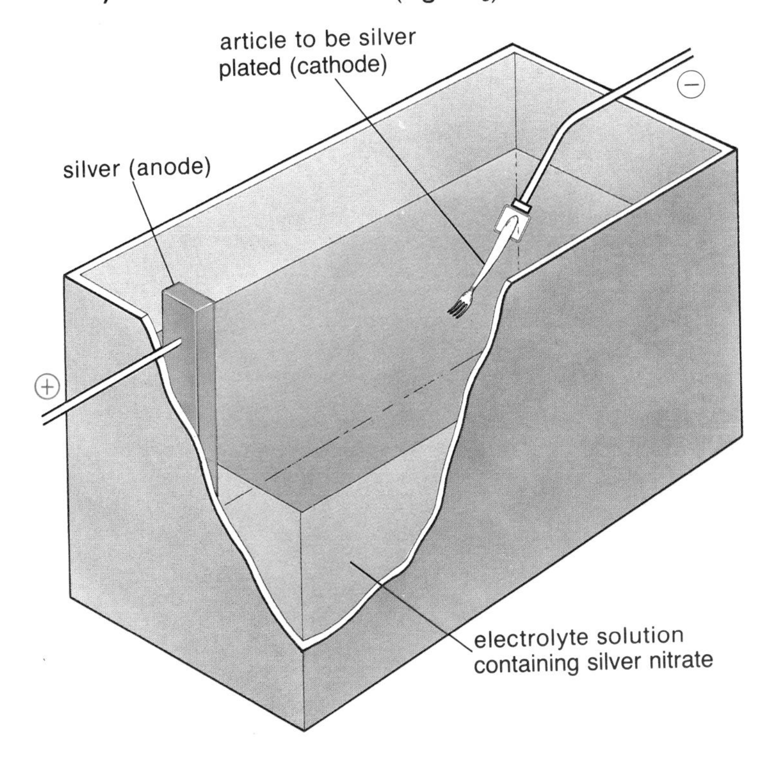

Figure 1
A fork being electroplated with silver

During electrolysis, silver ions (Ag^+) in the electrolyte are attracted to the cathode (the fork) where they gain electrons and form a deposit of silver:

Cathode (−) $Ag^+ + e^- \rightarrow Ag$

The anode is a piece of silver. Nitrate ions (NO_3^-) in the electrolyte are attracted to the anode, but they are not discharged. Instead, silver atoms in the anode give up electrons to the anode and go into solution as Ag^+ ions.

Anode (+) $Ag \rightarrow Ag^+ + e^-$

Questions

1 What precautions should be taken to obtain a good coating of metal during electroplating?

2 (a) The object to be electroplated is made the cathode in an electrolytic cell. Why is this?
(b) Suppose you want to nickel plate an article. What substance would you choose for (i) the anode; (ii) the electrolyte?

3 Steel bumpers on cars are electroplated with nickel and then with chromium.
(a) Why is the steel plated with a layer of nickel before chromium?
(b) Give two reasons for electroplating steel.

4 Some cutlery is stamped 'EPNS'. This stands for *e*lectro*p*lated *n*ickel *s*ilver. How is this plating done?

5 Articles of jewellery (e.g. bracelets) are often electroplated with gold, silver or copper. Why?

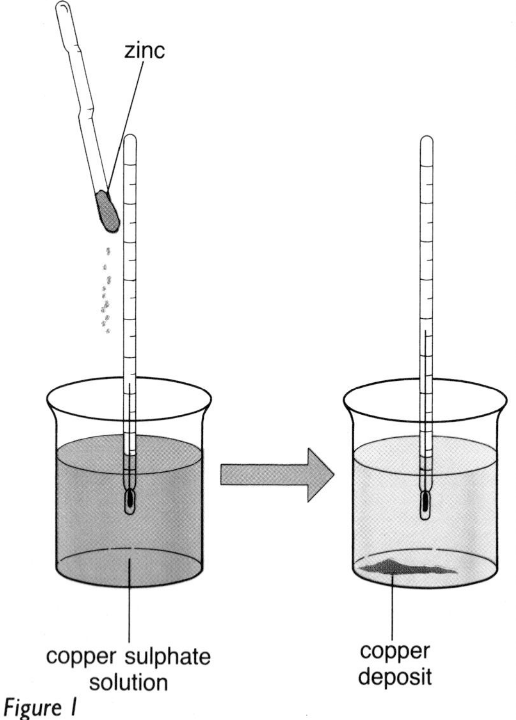

Figure 1

Simple cells

When zinc powder is added to copper(II) sulphate solution, an exothermic reaction occurs and the mixture gets warm. At the same time, the zinc is coated with copper and the blue colour of the solution fades (figure 1). The zinc has reacted with the blue copper sulphate solution to give zinc sulphate solution and copper metal.

zinc + copper sulphate → zinc sulphate + copper + heat

$$Zn(s) + CuSO_4(aq) \rightarrow ZnSO_4(aq) + Cu(s) + heat$$

Zinc is more reactive than copper, so it reacts to form zinc ions whilst copper ions are converted back to copper metal, i.e.

$$Zn(s) + Cu^{2+}(aq) \rightarrow Zn^{2+}(aq) + Cu(s) + heat$$

In this reaction, chemical energy in the reactants has been converted into heat energy. By carefully arranging the reacting materials, it is possible to convert the chemical energy of the reactants into electrical energy rather than into heat energy. This can be done using the circuit in figure 2. The bulb lights and the electric current can be measured using an ammeter. Arrangements like this, which generate electric currents, are called **cells**. Two or more cells joined together form a **battery**, although the term 'battery' is often used to describe both a cell and a battery.

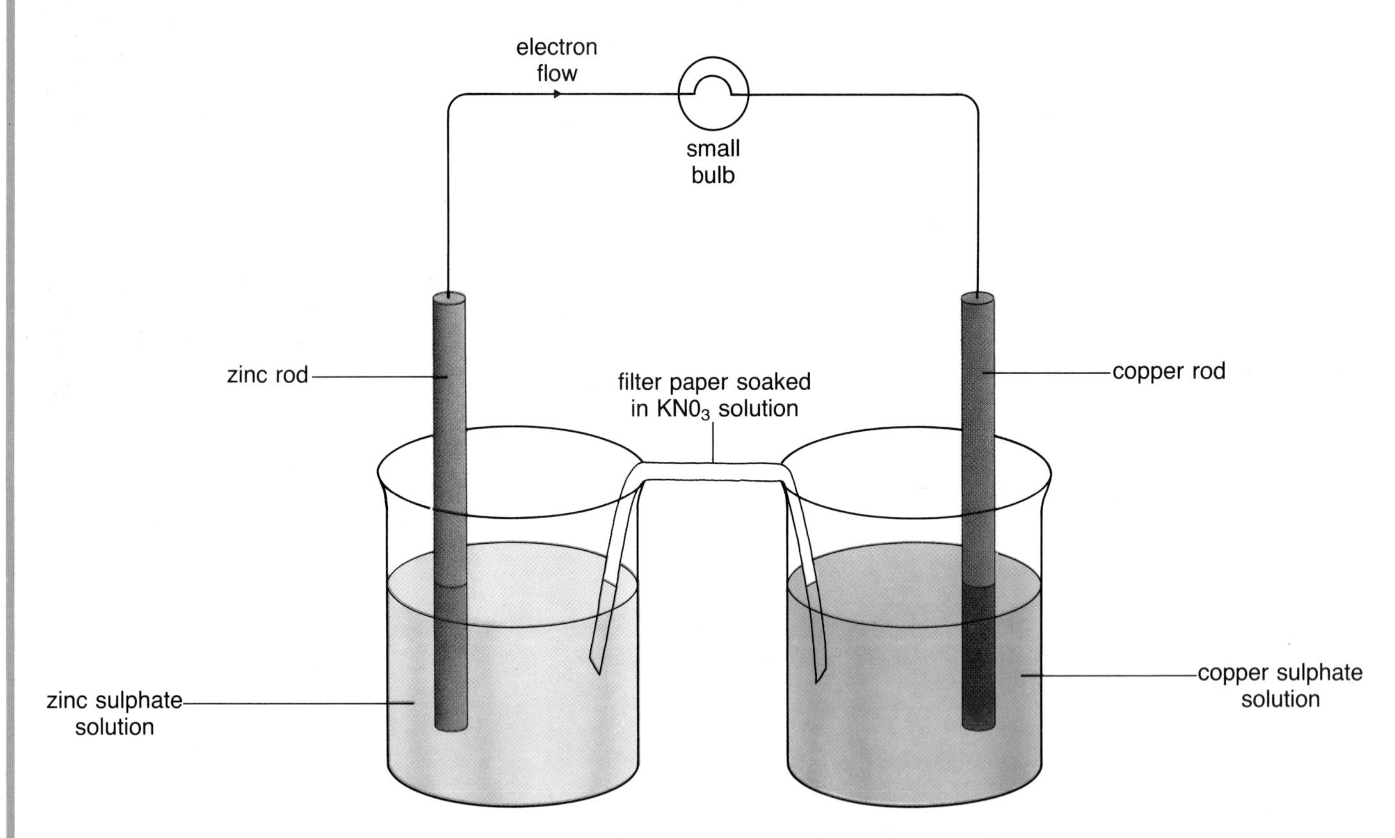

Figure 2

When the circuit in figure 2 is complete, zinc dissolves away from the zinc rod whilst copper is deposited on the copper rod. Thus, the overall reaction which occurs is

$$Zn(s) + Cu^{2+}(aq) \rightarrow Zn^{2+}(aq) + Cu(s) + electrical\ energy$$

This is the same as the reaction which happens when zinc is added to copper sulphate solution. In this experiment, the energy of the reaction is given out as electrical energy in an electric current. When the reactants are mixed together, the energy is given out as heat.

Using the apparatus in figure 2, the overall chemical process has been split into two separate reactions.

At the zinc rod, zinc atoms have given up electrons and gone into solution as zinc ions ($Zn^{2+}(aq)$).

$$Zn(s) \rightarrow Zn^{2+}(aq) + 2e^-$$

The electrons have moved through the circuit via the bulb to the copper rod. Here the electrons are taken by aqueous copper ions ($Cu^{2+}(aq)$) to form copper metal.

$$Cu^{2+}(aq) + 2e^- \rightarrow Cu(s)$$

The electrons which flow from the zinc rod to the copper rod make up the electric current.

Chemists say that the reaction now takes place **at a distance** because it has been separated into two halves. The two halves of the whole cell are called **half-cells**. These are written as Zn/Zn^{2+} and Cu/Cu^{2+}. When the cell is set up, Zn becomes Zn^{2+} and Cu^{2+} becomes Cu.

The Daniell cell

One of the first practical cells was the *Daniell cell* (figure 3). This was invented by J. F. Daniell in 1836. It can produce a voltage of about 1 volt and it was used to operate door bells. Notice how closely the Daniell cell resembles the apparatus in figure 2 by using the same half-cells, Zn/Zn^{2+} and Cu/Cu^{2+}. When the Daniell cell is working, the same reactions take place as in the apparatus in figure 2. The only difference is that the Daniell cell has a firm porous pot (to allow movement of ions between the two solutions) instead of a damp filter paper soaked in potassium nitrate solution.

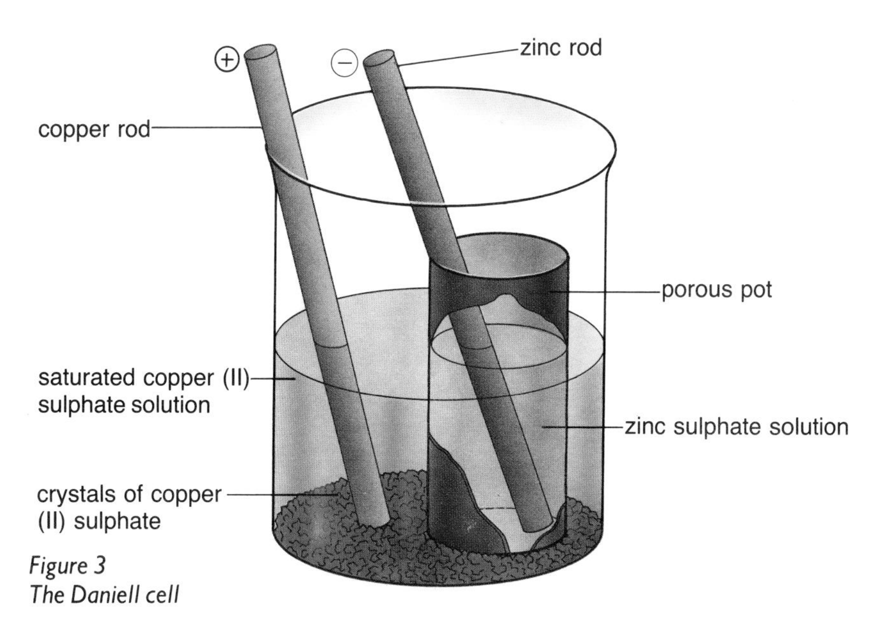

Figure 3
The Daniell cell

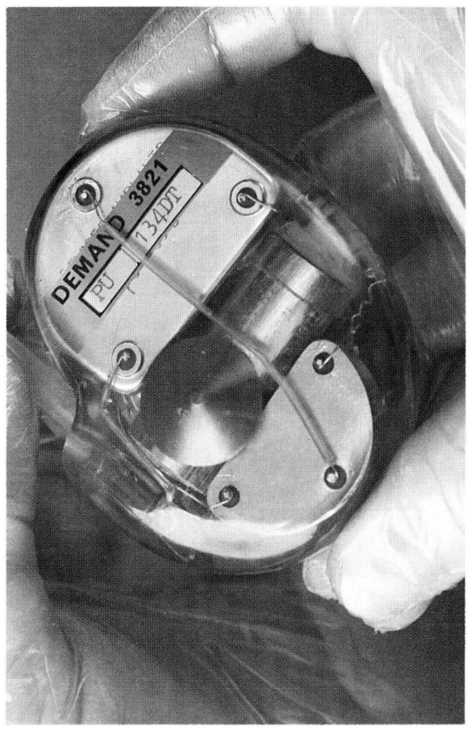

The electrical energy generated in zinc/silver cells and other cells, using body fluids as electrolytes, has been used to regulate the heart beat in patients with heart disorders

Questions

1 Explain the following:
cell; battery; reaction at a distance; half-cell.

2 Write equations for the reactions in a Daniell cell and explain how an electric current is produced.

3 Daniell cells have been replaced by cheaper and more convenient cells. Give *two* reasons why Daniell cells are no longer used.

12 Cells and Batteries

Most milk floats are powered by electricity. Notice the large battery between the wheels. Why are batteries like this not used in most cars?

Electrical cells and batteries are a very convenient source of electrical energy. They are used in torches, door bells, calculators, radios and milk floats. When electric cells are used, chemical energy stored in the materials of the cell is converted into electrical energy. This is the reverse of electrolysis in which electrical energy is converted into chemical energy in the products at the electrodes.

$$\text{Chemical energy} \underset{\text{electrolysis}}{\overset{\text{cells/batteries}}{\rightleftharpoons}} \text{Electrical energy}$$

The dry cell

One of the most widely used cells is the **dry cell** (figure 1). This is used in torches, radios and bicycle lamps. It is called a *dry* cell because the electrolyte is a thick paste of ammonium chloride. This means that there is no danger of any liquid spilling out. When the cell is working, the zinc case acts as the negative terminal giving up electrons which form the electric current from the cell:

$$Zn \rightarrow Zn^{2+} + 2e^-$$

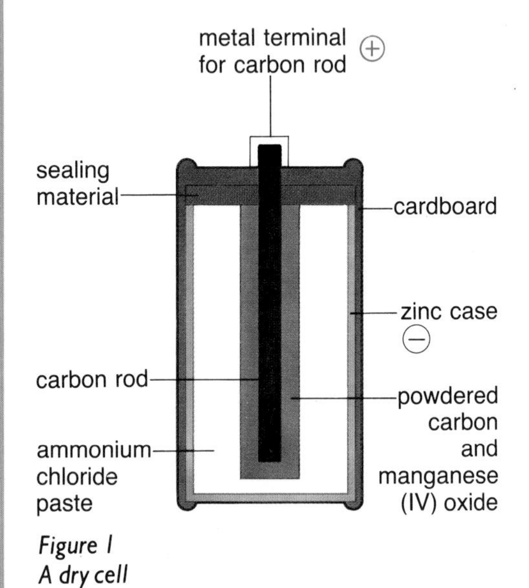

Figure 1
A dry cell

The positive terminal is a carbon rod. The surface area of the positive terminal is increased by surrounding it with a mixture of powdered carbon and manganese(IV) oxide. At this terminal, ammonium ions (NH_4^+) take electrons and form ammonia and hydrogen:

$$2NH_4^+ + 2e^- \rightarrow 2NH_3 + H_2$$

One dry cell can produce about 1.5 volts, but batteries of these cells giving 100 volts are in common use. When the cell is used, the zinc case becomes Zn^{2+} ions which dissolve in the paste. When the zinc case gets worn away, the paste leaks out. This is very corrosive to any metals which it touches, so old batteries should be replaced as soon as they have run down.

Rechargeable cells

When simple cells, like Daniell cells and dry cells, produce electricity, their chemicals are used up. These cells cannot be *recharged* and used again. Cells like this are called **primary cells** in contrast to **secondary cells** or **rechargeable cells** that can be recharged and used again.

The best known rechargeable cell is the **lead-acid cell** which is used in car batteries. The negative terminal in a lead-acid cell is a lead plate, whereas the positive terminal is a lead plate covered in lead(IV) oxide (PbO_2). The electrolyte is fairly strong sulphuric acid. One lead-acid cell can produce about 2 volts.

When the lead-acid cell produces electricity (discharges), lead atoms on the negative terminal give up electrons to form lead ions:

$$Pb(s) \xrightarrow{\text{discharges}} Pb^{2+}(aq) + 2e^-$$

Lead(IV) oxide, on the positive terminal, takes electrons and reacts with H^+ ions in the electrolyte to form lead ions and water.

These items all use dry cells and batteries

1 support grid
2 negative plate
3 positive plate
4 separator
5 injection moulded container
6 lid
7 vent to prevent build-up of gases
8 inter-cell connecting bar
9 battery terminal
10 label

The parts of a car battery—two of the six cells can be seen

A collection of cells and batteries

$$PbO_2(s) + 4H^+(aq) + 2e^- \xrightarrow{\text{discharges}} Pb^{2+}(aq) + 2H_2O(l)$$

The equations show that lead ions form at both plates during discharge. These Pb^{2+} ions react with sulphate ions (SO_4^{2-}) in the sulphuric acid to form white insoluble lead sulphate:

$$Pb^{2+}(aq) + SO_4^{2-}(aq) \rightarrow PbSO_4(s)$$

If the battery is discharged for too long, a thick layer of lead sulphate forms on the plates and this stops the reactions at the terminals. The battery is said to be '*sulphated*' and cannot be recharged.

As the battery discharges, H^+ ions are used up and this causes the acid concentration to drop and the density of the electrolyte to fall. Thus, the state of charge of the cell can be checked by measuring the density of the acid solution. Lead-acid cells must be recharged before they get 'flat' (fully discharged). When the cell or the battery is recharged during a long journey, electricity is passed through the cell in the opposite direction and the reactions at the terminals are reversed. Thus, recharging is an example of electrolysis in which electrical energy is converted into chemical energy.

Questions

1 Explain the following:
primary cell; rechargeable cell; discharge; recharge.

2 Write equations for the reactions in (i) a dry cell; (ii) a lead-acid cell. Explain how each produces an electric current.

3 Most car batteries consist of 6 lead-acid cells connected one after another.
(a) What total voltage will this give?
(b) How is the state of charge in a car battery checked?
(c) Milk floats are often powered by lead-acid batteries. Why are other vehicles not run on lead-acid cells?
(d) Why are battery-powered vehicles likely to increase in the future?

4 What are the advantages of (i) a dry cell over a Daniell cell; (ii) a lead-acid cell over a dry cell?

Section D: Study Questions

1 When aluminium is exposed to the air it becomes coated very quickly with a thin layer of oxide about 10^{-6} cm thick. This layer does not flake off, nor does it increase in thickness on standing. In order to protect the aluminium even more than its natural oxide layer does, it is possible to thicken the layer to 10^{-3} cm by a process called anodizing. The aluminium is first degreased and then anodized by making it the anode during the electrolysis of sulphuric acid. The oxygen released at the anode combines with the aluminium and increases the thickness of the oxide layer.

(a) Why is anodizing useful?
(b) Why does aluminium not corrode away like iron, even though aluminium is coated very quickly with a layer of oxide?
(c) Write a word equation for the reaction which takes place when aluminium is exposed to the air.
(d) Carbon tetrachloride (tetrachloromethane) can be used to degrease aluminium before anodizing. Name one other liquid which would be a suitable degreasing agent.
(e) Why is water not used to degrease aluminium?
(f) Why is it necessary to degrease the aluminium?
(g) How many times thicker is the oxide coating after anodizing?
(h) Why is anodized aluminium especially useful as a building material?

2 Question 2 concerns the properties of substances *A* to *E*.

	Electrical conductivity (of the pure substance) at room temperature	Solubility in water	Properties of the solution in water
A	Non-conductor	Soluble	A neutral solution which conducts electricity
B	Non-conductor	Soluble	A neutral solution which does NOT conduct electricity
C	Non-conductor	Insoluble	—
D	Conductor	Insoluble	—
E	Non-conductor	Soluble	An alkaline solution which conducts electricity

Select, from *A* to *E*, the appropriate set of properties for:
(a) graphite (b) sodium hydroxide
(c) copper (d) potassium nitrate
(e) glucose (f) ammonia.

LEAG

3 A solution of copper sulphate is electrolysed using copper electrodes.

(a) Draw a clearly labelled diagram of the circuit.
(b) What happens during electrolysis
(i) at the anode
(ii) at the cathode
(iii) to the solution?
(c) Mention two practical uses of this type of electrolysis.
(d) If the two copper electrodes are replaced by two platinum electrodes state what happens now
(i) at the anode
(ii) at the cathode
(iii) to the solution.

4

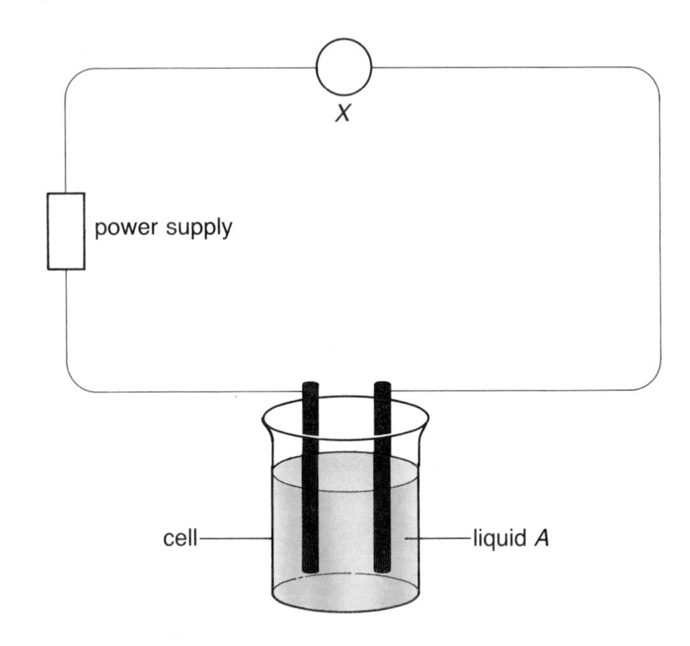

Liquid *A* is to be electrolysed.

(a) What type of power supply is suitable for this purpose? (1)
(b) What piece of apparatus could be connected at *X* to show that liquid *A* is a conductor? (1)
(c) Give the name of the charged particles which conduct the electric current in
(i) the connecting wires
(ii) the electrolyte. (2)
(d) A metal kettle is to be coated with copper using electrolysis. State what you would use in the cell as the
(i) anode
(ii) cathode
(iii) electrolyte. (3)

Total [7] **LEAG**

SECTION E

Patterns and Properties

The periodic table is like a jigsaw. In a jigsaw, the pieces must be arranged correctly to give a pattern or a picture. In the periodic table, elements are arranged in order of relative atomic mass to give a pattern in properties

1 Looking for Patterns

How are the books arranged in a library?

Metals, non-metals and metalloids

Books in a library are divided into two sections—fiction and non-fiction. These two sections are then divided into smaller sections. There are history books, poetry books, science books and so on. The books are classified in this way so that you will know what the books on a particular shelf will be like. Chemists have also worked in a similar way to librarians. They have tried to classify elements by grouping those with similar properties. One of the most useful ways of classifying elements is as metals and non-metals (see section A, unit 4). Unfortunately, there are limitations in classifying some elements in this way. Take, for example, silicon and germanium. These two elements have high melting points and high boiling points (like metals), but they have low densities and they are brittle (like non-metals). They conduct electricity better than non-metals but not as well as metals. Elements with some properties like metals and other properties like non-metals are called **metalloids**.

Because of the difficulty of classifying elements neatly as metals and non-metals, chemists looked for patterns in the properties and reactions of smaller groups of elements.

Why are graphite strips used around the nose cone of rockets such as the Space Shuttle Columbia in this photograph?

Families of elements

Early in the nineteenth century, a German chemist called Johann Wolfgang Döbereiner noticed that several elements could be arranged in groups of three. Each group has similar properties. These families of three elements became known as **Döbereiner's Triads** (figure 1).

Lithium, sodium and potassium are three metals with very similar properties (see the table). Döbereiner put them in one triad.

Nowadays we call them **alkali metals**. Another of Döbereiner's triads was chlorine, bromine and iodine. Can you think of two properties in which these three elements are similar?

Property	Character
Appearance	Shiny but tarnish rapidly in air forming a layer of oxide
Strength	Soft metals—easily cut with a knife
M.pt. and b.pt.	Low compared with other metals
Density	Less than 1.0 g/cm^3, float on water
Reaction with air	Burn vigorously forming white oxides, e.g. sodium + oxygen → sodium oxide $4Na + O_2 \rightarrow 2Na_2O$
Reaction with water	React very vigorously with water producing hydrogen and an alkaline solution of the hydroxide, e.g. sodium + water → sodium hydroxide + hydrogen $2Na + 2H_2O \rightarrow 2NaOH + H_2$
Valency	All have a valency of one, e.g. oxides are Li_2O, Na_2O, K_2O.

Similarities of lithium, sodium and potassium

Döbereiner also noticed that when the three elements in a triad were arranged in order of relative atomic mass, the RAM of the middle element was very close to the average of the other two elements. Thus, the relative atomic mass of sodium (23.0) is the average of the relative atomic masses of lithium and potassium (figure 2).

i.e. $\frac{6.9 + 39.1}{2} = \frac{46.0}{2} = 23.0$

The discovery of these triads and the link between the properties of elements and their relative atomic mass encouraged other chemists to search for patterns.

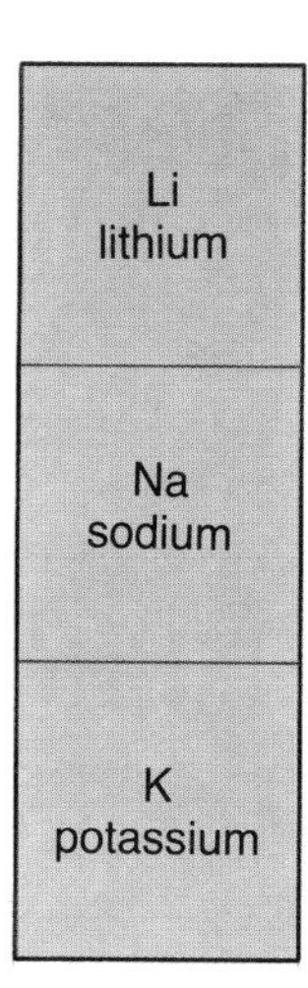

Figure 1
Two of Döbereiner's Triads

Johann Wolfgang Döbereiner

Li
6.9
Na
23.0
K
39.1

Figure 2

Questions

1 What is (i) a metal; (ii) a non-metal; (iii) a metalloid; (iv) a triad?

2 (a) Write down 2 properties in which chlorine, bromine and iodine resemble one another.
(b) Calculate the average of the relative atomic masses of chlorine (35.5) and iodine (126.9). How close is this average to the relative atomic mass of bromine (79.9)?

3 Write word equations and balanced chemical equations for the following reactions:
sodium tarnishing in air; lithium burning in oxygen; potassium reacting with chlorine; lithium reacting with water.

2 Patterns of Elements

H	Li	Be	B	C	N	O
F	Na	Mg	Al	Si	P	S
Cl	K	Ca	Cr	Ti	Mn	Fe

Figure 1
Newlands' Octaves

Newlands' Octaves

In 1864, John Newlands, an English chemist, arranged all the known elements in order of their relative atomic masses. He found that one element had properties like those of the element 8 places in front of it in the list. Newlands called this a **'Law of Octaves'**. He said that 'the eighth element is a kind of repetition of the first, like the eighth note of an octave in music'.

Figure 1 shows the first three of Newlands' Octaves. Notice that similar elements often occur in the same column. For example, the second column contains lithium, sodium and potassium.

> *The regular or periodic repetition of elements with similar properties led to the name* **periodic table**.

Unfortunately, Newlands' classification grouped together some elements which were very different. For example, iron (Fe) was placed in the same family as oxygen (O) and sulphur (S). Because of this, many of Newlands' ideas were criticised and rejected.

Dmitri Mendeléev

Mendeléev's periodic table

In spite of the criticism that Newlands received, chemists kept searching for a pattern linking the properties and relative atomic masses of the elements.

In 1869, the Russian chemist, Dmitri Mendeléev, produced new ideas to support the theories which Newlands had suggested five years earlier. Figure 2 shows part of Mendeléev's periodic table. Notice that elements with similar properties, such as lithium, sodium and potassium, fall in the same vertical column.

	GROUP							
	I	**II**	**III**	**IV**	**V**	**VI**	**VII**	**VIII**
Period 1	H							
Period 2	Li	Be	B	C	N	O	F	
Period 3	Na	Mg	Al	Si	P	S	Cl	
Period 4	K Cu	Ca Zn	* *	Ti *	V As	Cr Se	Mn Br	Fe Co Ni
Period 5	Rb Ag	Sr Cd	Y In	Zr Sn	Nb Sb	Mo Te	* I	Ru Rh Pd

Figure 2

The vertical columns of similar elements (i.e. the chemical families) are called **groups**.
The horizontal rows are called **periods**.

Mendeléev was more successful than Newlands because of two important steps which he took in writing his table.

1 He suggested that some elements were still undiscovered and left gaps for them in his table. In some cases, he had to leave gaps so that similar elements were in the same vertical group. There are 4 gaps (shown by asterisks) in figure 2. These 4 gaps are now filled by the elements scandium, gallium, germanium and technetium.

2 He predicted the properties of the missing elements from the properties of the elements above and below them in the table. Within 15 years of Mendeléev's predictions, three of the missing elements had been discovered. Their properties were very similar to his predictions. Look at figure 3. This compares the properties that Mendeléev predicted for the element he called eka-silicon (meaning 'below silicon') with its actual properties when it was found in 1886 and named germanium (Ge).

Mendeleev's predictions for eka-silicon in 1871	The observed properties of germanium after its discovery in 1886
1 Grey metal	1 Greyish white metal
2 Density about 5.5g cm^{-3}	2 Density = 5.47g cm^{-3}
3 Relative atomic mass of 'Ek' = Mean of relative atomic mass of Si and Sn = $\frac{28.1+118.7}{2} = 73.4$	3 Relative atomic mass = 72.6
4 Melting point of 'Ek' higher than that of tin perhaps about 800°C	4 Melting point = 958°C
5 The metal will react with oxygen forming an oxide EkO_2, density 4.7g cm^{-3} The oxide EkO may also exist	5 Ge reacts with oxygen forming GeO_2, density 4.7g cm^{-3} GeO is also known

Figure 3

The success of Mendeléev's predictions provided good evidence that his ideas were correct. His periodic table was quickly accepted as an important summary of the properties of elements.

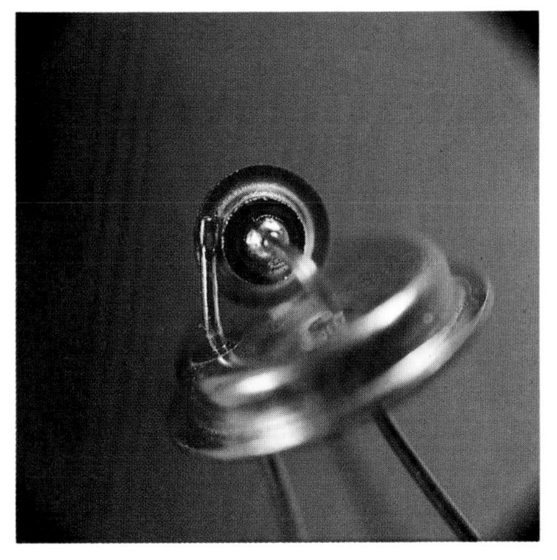

Germanium is used as the semiconductor in these transistors

Questions

1 Explain the following words in relation to the periodic table: *periodic properties; group; period.*

2 Look at figure 1.
(a) Why did Newlands use the word 'octaves'?
(b) Name one triad from one column of Newlands' Octaves.
(c) The relative atomic mass of hydrogen is 1.0 and that of chlorine is 35.5. Predict the relative atomic mass of fluorine. Explain your prediction.

3 Why was Mendeléev more successful with his periodic table than Newlands had been?

3 Modern Periodic Tables

Group ► / Period ▼	I Alkali metals	II Alkaline —earth metals											III	IV	V	VI	VII Halogens	0 Noble Gases
1									H Hydrogen 1									He Helium 2
2	Li Lithium 3	Be Beryllium 4											B Boron 5	C Carbon 6	N Nitrogen 7	O Oxygen 8	F Fluorine 9	Ne Neon 10
3	Na Sodium 11	Mg Magnesium 12	Transition elements										Al Aluminium 13	Si Silicon 14	P Phosphorus 15	S Sulphur 16	Cl Chlorine 17	Ar Argon 18
4	K Potassium 19	Ca Calcium 20	Sc 21	Ti 22	V 23	Cr Chromium 24	Mn Manganese 25	Fe Iron 26	Co 27	Ni 28	Cu Copper 29	Zn Zinc 30	Ga 31	Ge 32	As 33	Se 34	Br Bromine 35	Kr Krypton 36
5	Rb 37	Sr 38	Y 39	Zr 40	Nb 41	Mo 42	Tc 43	Ru 44	Rh 45	Pd 46	Ag Silver 47	Cd 48	In 49	Sn Tin 50	Sb 51	Te 52	I Iodine 53	Xe 54
6	Cs 55	Ba 56	57–71 See below	Hf 72	Ta 73	W 74	Re 75	Os 76	Ir 77	Pt Platinum 78	Au Gold 79	Hg Mercury 80	Tl 81	Pb Lead 82	Bi 83	Po 84	At 85	Rn 86
7	Fr 87	Ra 88	89–103 See below	Ku 104	Ha 105													

Lanthanides	La Lanthanum 57	Ce 58	Pr 59	Nd 60	Pm 61	Sm 62	Eu 63	Gd 64	Tb 65	Dy 66	Ho 67	Er 68	Tm 69	Yb 70	Lu 71
Actinides	Ac Actinium 89	Th 90	Pa 91	U Uranium 92	Np 93	Pu 94	Am 95	Cm 96	Bk 97	Cf 98	Es 99	Fm 100	Md 101	No 102	Lr 103

Figure 1
The modern periodic table

All modern periodic tables are based on the one produced by Mendeléev in 1869. A modern periodic table is shown in figure 1. If the elements are numbered along each period, starting with period 1, then period 2, etc., then the number given to each element is called its **atomic number**. Thus, hydrogen has an atomic number of 1, helium 2, lithium 3, etc. You will learn more about atomic numbers in section L.

There are several points to note about the modern periodic table.

1 The most obvious difference between modern periodic tables and Mendeléev's is the position of the **transition elements**. These are 10 or more elements in each period which have been taken out of the simple groups. Period 4 is the first to contain a series of transition elements. These include chromium, iron, nickel, copper and zinc.

2 Some groups have names as well as numbers. These are summarised in the table.

Group number	Group name
I	alkali metals
II	alkaline-earth metals
VII	halogens
O	noble (inert) gases

The names of groups in the periodic table

3 Metals are clearly separated from non-metals. The 20 or so non-metals are packed into the top right-hand corner above the red stepped line in figure 1. Some elements close to the steps are metalloids. These elements have some properties like metals and some properties like non-metals.

4 Apart from the noble gases, the most reactive elements are near the left- and right-hand sides of the periodic table. The least reactive elements are in the centre. Sodium and potassium, two very reactive metals, are at the left-hand side. The next most reactive metals, like calcium and magnesium, are in group II, whereas less reactive metals (like iron and copper) are in the centre of the table. Carbon and silicon, unreactive non-metals, are in the centre of the periodic table. Sulphur and oxygen, which are nearer the right-hand edge, are more reactive. Fluorine and chlorine, the most reactive non-metals, are very close to the right-hand edge.

5 Besides the vertical groups of similar elements, we can pick out five blocks of elements with similar properties. These blocks are coloured differently in figure 2.

	Group I	Group II											Group III	IV	V	VI	VII	O
Period 1					H													He
Period 2	Li	Be											B	C	N	O	F	Ne
Period 3	Na	Mg											Al	Si	P	S	Cl	Ar
Period 4	K	Ca	Sc	Ti	V	Cr	Mn	Fe	Co	Ni	Cu	Zn	Ga	Ge	As	Se	Br	Kr
Period 5	Rb	Sr	Y	Zr	Nb	Mo	Tc	Ru	Rh	Pd	Ag	Cd	In	Sn	Sb	Te	I	Xe
Period 6	Cs	Ba	La	Hf	Ta	W	Re	Os	Ir	Pt	Au	Hg	Tl	Pb	Bi	Po	At	Rn
Period 7	Fr	Ra	Ac	Ku	Ha													

Figure 2
Blocks of similar elements in the periodic table

(a) *Reactive metals* in Group I and Group II, including sodium, potassium, magnesium and calcium.
(b) *Transition metals* in a shallow rectangle between Group II and Group III. The transition metals include chromium, iron, copper, zinc, silver and gold. In this block, the elements resemble each other across the series as well as the usual vertical similarities.
(c) *Poor metals* (including tin and lead) in a triangle below the steps separating metals from non-metals. Most of these metals are fairly unreactive and some of them have some properties like non-metals.
(d) *Non-metals* in a triangle above the steps separating metals from non-metals.
(e) *Noble gases.* These elements are very unreactive. The first noble gas compound was not made until 1962. Because they are so unreactive, the noble gases were once called 'inert gases'. Nowadays several compounds of them are known and the name 'inert' has been replaced by 'noble'. The word 'noble' was chosen because very unreactive metals like silver and gold are sometimes called noble metals.

Questions

1 What do you understand by the following:
atomic number; transition element; noble gas?
2 Where in the periodic table would you find (i) metals; (ii) non-metals; (iii) metalloids; (iv) elements with atomic numbers 11 to 18 inclusive; (v) the alkaline-earth metals; (vi) the halogens; (vii) the most reactive metals; (viii) the most reactive non-metals, (ix) the transition elements; (x) the noble gases?
3 Why are there no noble gases in Mendeléev's periodic table?

4 The Alkali Metals

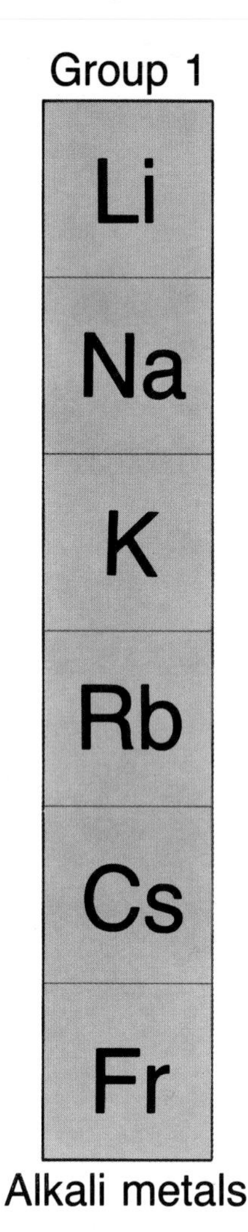

Figure 1

The elements in Group 1 are called the alkali metals because they react with water to form alkaline solutions. Lithium, sodium and potassium are the best known alkali metals. The properties of these three elements were considered in unit 1 of this section. The other elements in Group 1 are rubidium, caesium and francium (figure 1). The alkali metals are so reactive that they must be stored under oil to protect them from oxygen and water vapour in the air. Some of the physical properties of lithium, sodium and potassium are given in table 1.

Element	Relative atomic mass	Melting point/°C	Boiling point/°C	Density /g cm^{-3}
Lithium	6.9	180	1330	0.53
Sodium	23.0	98	892	0.97
Potassium	39.1	64	760	0.86

Table 1: the physical properties of lithium, sodium and potassium

Alkali metals have some unusual properties for metals:

- their melting points and boiling points are unusually low;
- their densities are so low that they can float on water;
- they are soft enough to be cut with a knife.

The properties in table 1 also illustrate another important feature of the periodic table—*although the elements within a group are similar, there is usually a steady change in property from one element to the next.* The graph in figure 2 shows the steady change in the melting points of the alkali metals very neatly.

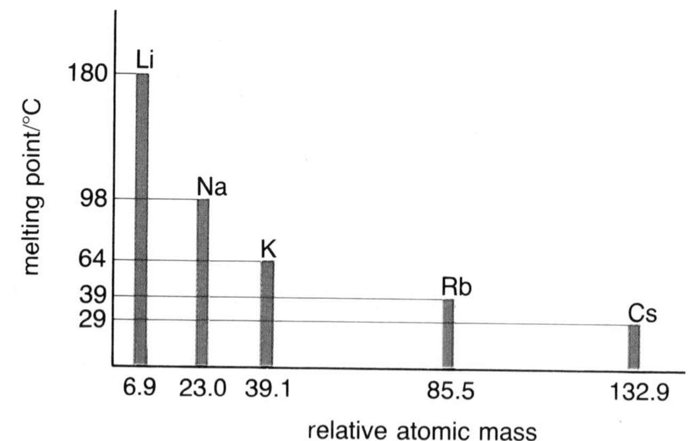

Figure 2

Table 2 summarises the chemical properties of lithium, sodium and potassium. Notice how the elements react more vigorously with water and chlorine from lithium to potassium. This relates to the increase in metallic character down the group.

The ions of all alkali metals have a charge of +1 and so all their compounds will have similar formulas.

Once we know the general properties of the elements in a group, and how these properties vary from one element to the next, we can predict the properties of other elements in the group.

Element	Reaction with cold water	Reaction with chlorine	Formula of oxide	Symbol of ion	Salts (chlorides, nitrates, carbonates)
Lithium	Reacts steadily to give $H_2(g)$ and $LiOH(aq)$	Reacts on heating to give white $LiCl(s)$	Li_2O	Li^+	
Sodium	Reacts vigorously to give $H_2(g)$ and $NaOH(aq)$	Reacts vigorously on heating to give white $NaCl(s)$	Na_2O	Na^+	All salts are white, ionic, solids, soluble in water
Potassium	Reacts violently to give $H_2(g)$ and $KOH(aq)$	Reacts violently on heating to give white $KCl(s)$	K_2O	K^+	

Table 2: the chemical properties of lithium, sodium and potassium

A workman harvesting salt from sea-water in Southern Spain

The uses of sodium compounds

Most sodium compounds are made from sodium chloride. In hot countries, impure sodium chloride is left if sea-water is allowed to evaporate. Sodium chloride also occurs in salt beds beneath the Earth's surface. The sodium chloride can be obtained from these salt beds by *solution mining*. This involves piping hot water down to the salt bed to dissolve the salt. Concentrated salt solution is then pumped to the surface.

Impure salt is used for de-icing roads. Pure salt is used in cooking. Sodium and chlorine can be produced by electrolysing molten salt. Liquid sodium is used as a coolant in fast nuclear reactors and sodium vapour provides the yellow glow in street lamps.

The large cylinder carries liquid sodium which is used as a coolant for nuclear reactors

Questions

1 Write word equations and then balanced equations with symbols for the reaction of sodium with (i) oxygen; (ii) chlorine; (iii) water.
2 Rubidium (Rb) is in Group I below potassium. Use the information in tables 1 and 2 to predict: (i) its melting point; (ii) its boiling point; (iii) its density relative to that of water; (iv) the symbol of its ion; (v) the formula of its oxide and chloride; (vi) its reaction with water; (vii) its reaction with chlorine; (viii) the properties of its salts.

5 The Transition Metals

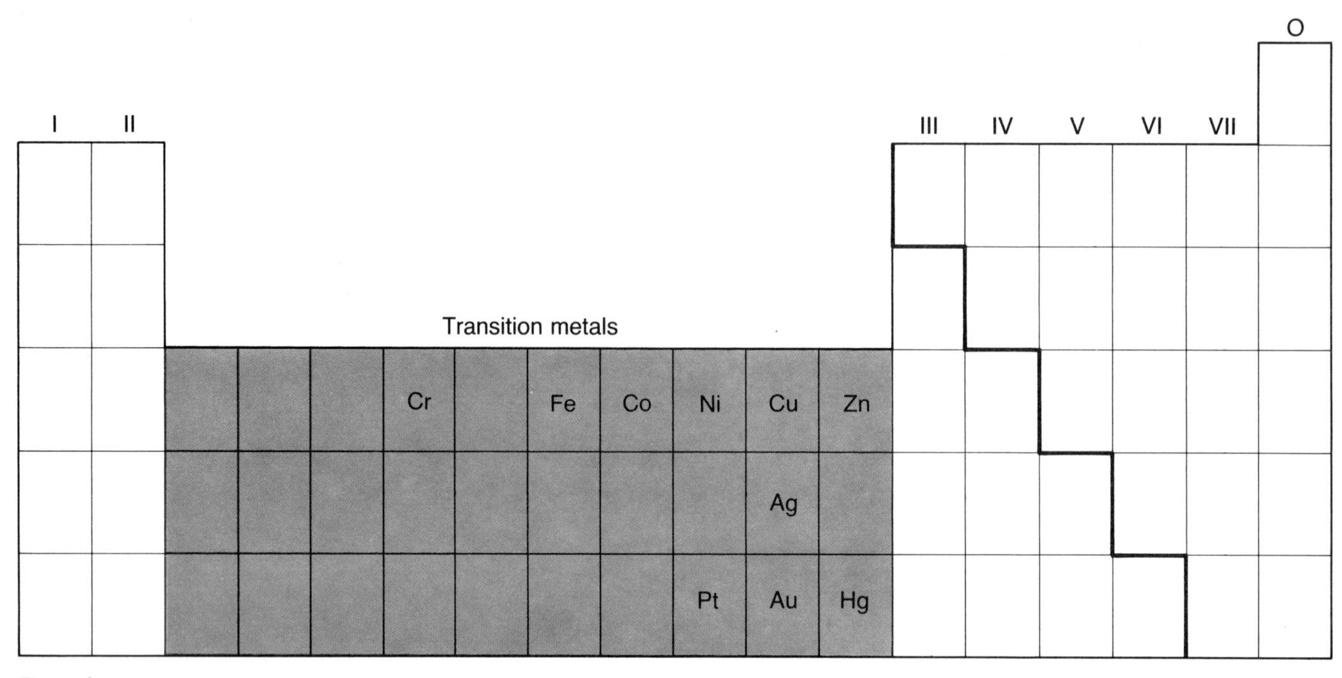

Figure 1

Figure 1 shows some of the common transition metals in the periodic table. The transition metals separate the reactive metals in Groups I and II from the poor metals in the triangle below the steps separating metals and non-metals.

> *The transition metals have very similar properties. Unlike the alkali metals, there are similarities in the transition metals across the periods and down the groups.*

The most important transition metals are iron and copper. Iron is the most widely used metal. More than 700 million tonnes of it are manufactured every year throughout the world. It is used as steel in girders and supports for bridges and buildings, in vehicles, in engines and in tools.

After iron and aluminium, copper is the third most widely used metal. About 9 million tonnes are manufactured each year. Copper is a good conductor of heat and electricity. It is also malleable (it can be made into different shapes) and ductile (it can be drawn into wires). Because of these properties, copper is used in electrical wires and cables and in hot water pipes and radiators. The uses of copper are increased by *alloying* it with other metals. Alloying copper with zinc produces brass which is harder and stronger than pure copper. Alloying copper with tin produces bronze. This is stronger and easier to cast into moulds than pure copper. There is more about alloys in section F.

General properties of transition metals

Some properties of iron and copper are shown in the table (above right). Compare these with the properties of alkali metals in tables 1 and 2 in the last unit.

Element	Melting point /°C	Boiling point /°C	Density /g cm^{-3}	Reaction with water	Formulas of oxides	Symbols of ions	Salts
Iron	1540	3000	7.9	Does not react with pure water. Reacts slowly with steam.	FeO Fe_2O_3	Fe^{2+} Fe^{3+}	Fe^{2+} salts are green. Fe^{3+} salts are yellow or brown.
Copper	1080	2600	8.9	No reaction with water or steam.	Cu_2O CuO	Cu^{+} Cu^{2+}	Cu^{2+} salts are blue or green.

Properties of iron and copper

The information in the table illustrates some typical properties of transition metals.

1 High melting points and boiling points—much higher than alkali metals.

2 High densities—much higher than alkali metals.

3 Hard strong metals (high tensile strength), unlike the soft alkali metals.

Iron is used as steel in girders and supports such as those in this oil rig under construction

4 Fairly unreactive with water, unlike alkali metals. None of the transition metals react with cold water, but a few of them react slowly with steam, like iron.

5 More than one valency. Most of the transition metals can have more than one valency: Iron forms Fe^{2+} ions (valency 2) and Fe^{3+} ions (valency 3); Copper forms Cu^{+} ions (valency 1) and Cu^{2+} ions (valency 2). Alkali metals only form ions of 1+ charge (valency 1).

6 Coloured compounds. Transition metals usually have coloured compounds with coloured solutions in contrast to alkali metals which have white salts with colourless solutions.

7 Catalytic properties. Transition metals and their compounds can act as **catalysts** (see section K). Iron or iron(III) oxide (Fe_2O_3) is used as a catalyst in the *Haber process* to manufacture ammonia (NH_3).

Questions

1 What are the transition metals?
2 Make a list of the characteristic properties of transition metals.
3 Make a table contrasting the properties of transition metals with those of alkali metals.
4 Write equations with symbols for the following word equations:
(a) iron + oxygen → iron(III) oxide
(b) iron + chlorine → iron(III) chloride
(c) iron + hydrogen chloride → iron(II) chloride + hydrogen.
5 What is (i) an alloy; (ii) brass; (iii) bronze?

6 The Noble Gases

	III	IV	V	VI	VII	O
						He
	B	C	N	O	F	Ne
		Si	P	S	Cl	Ar
					Br	Kr
					I	Xe
						Rn

Figure 1

Figure 1 shows the position of the noble gases in the periodic table. It also includes some other well-known non-metals. *The discovery of the noble gases showed the real value of the periodic table.* In 1868, an orange line was noticed in the spectrum of light from the Sun. This showed that the Sun's atmosphere contained an element that had not been found on the Earth. The element was named helium which comes from the Greek word *helios* meaning Sun.

In 1890, Raleigh and Ramsay prepared what they thought was pure nitrogen by removing oxygen, water vapour and carbon dioxide from fresh air. They found that the density of this gas differed by 0.5% from that of pure nitrogen obtained by decomposing nitrogen compounds. Further experiments on the impure nitrogen from the air showed that it contained a new element, argon, with a relative atomic mass of 40.

Later, Ramsay found helium on Earth and showed that it was very similar to argon. However, these two elements were very different from any of the other elements in the periodic table. This suggested that there was another group in the periodic table. The group was called Group O and the search began for the other four elements in it.

The fractional distillation of liquid air (section B, unit 1) led to the discovery of three elements (neon, krypton and xenon) by Ramsay in 1898. The remaining element, radon, was discovered in 1900 as a product from the breakdown of the radioactive element, radium.

Sir William Ramsay helped in the discovery and identification of five of the noble gases

Element	Relative atomic mass	Melting point /°C	Boiling point /°C	Density at 20°C and atm. pressure /g dm^{-3}
Helium	4.0	−270	−269	0.17
Neon	20.2	−249	−246	0.83
Argon	40.0	−189	−186	1.7
Krypton	83.8	−157	−152	3.5
Xenon	131.3	−112	−108	5.5

Properties of the noble gases

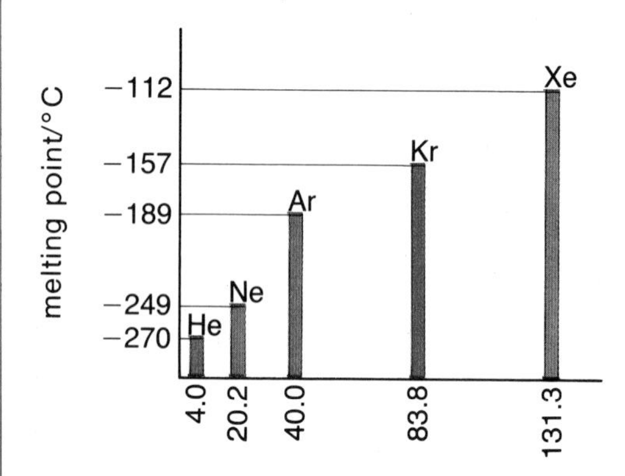

Figure 2
Variation of melting points of the noble gases as their relative atomic masses increase

Properties of the noble gases

The table lists some properties of the noble gases. They are all colourless gases at room temperature with low melting points and boiling points. As expected, their melting points, boiling points and densities show a steady change as the relative atomic mass increases. The graph in figure 2 shows the steady increase in melting point with relative atomic mass.

The noble gases all exist as separate single atoms (i.e. monatomic molecules). Until 1962, no compounds of the noble gases were known. Chemists thought they were completely unreactive. Hence they were called the *inert* gases. Nowadays, several compounds of them are known and the name *inert* has been replaced by *noble*.

Obtaining the noble gases

Neon, argon, krypton and xenon are obtained industrially during the fractional distillation of liquid air. There are only minute traces of helium in air. Therefore, it is more economical to extract helium from the natural gas in oil wells.

Uses of the noble gases

Helium is used in meteorological balloons because of its low density and because it is non-flammable. Helium, mixed with oxygen, is also used as the gas breathed by divers.

Neon is used in neon lights, and argon and krypton are used in electric light bulbs. If there was a vacuum inside the bulbs, metal atoms would evaporate from the very, very hot tungsten filament. To reduce this evaporation and to prolong the life of the filament, the bulb is filled with an unreactive gas which cannot react with the hot tungsten filament.

Argon is also used during certain welding processes. The argon provides an inert atmosphere and prevents any reaction between the metals being welded and the oxygen in the air.

The cylinders of this diving apparatus contain oxygen and helium

Helium being used to fill a weather balloon

Questions

1 (a) Which elements make up the noble gases?
(b) Why are they called *noble* gases?
(c) Why were they once called inert gases?

2 Use the values in the table to plot a graph of the boiling points of the noble gases (vertically) against their relative atomic masses (horizontally).
(a) How do the boiling points vary with relative atomic mass?
(b) Explain the pattern shown by the graph.
(c) Use the graph to predict the boiling point of radon (RAM = 222).

3 Suppose that liquid air contains oxygen (b.pt. −183°C), nitrogen (b.pt. −196°C), water (b.pt. 100°C), neon, argon, krypton and xenon. If the liquid air is fractionally distilled, what is the order in which the constituents will boil away from the liquid air?

4 Make a list summarising the properties of the noble gases.

7 The Halogens

VI	VII	0
		He
O	F	Ne
S	Cl	Ar
	Br	Kr
	I	Xe
	At	Rn

Figure 1

The elements in Group VII of the periodic table are called **halogens** (figure 1). The common elements in the group are fluorine (F), chlorine (Cl), bromine (Br) and iodine (I). The final element, astatine (At), does not occur naturally. It is an unstable radioactive element which was first synthesized in 1940.

The elements in Group VII are so reactive that they never occur free in nature. They occur combined with metals in salts such as sodium chloride (NaCl), calcium fluoride (CaF_2) and magnesium bromide ($MgBr_2$). This gave rise to the name for the group as halogens which means 'salt-formers'.

Sources of the halogens

The halogens occur in compounds with metals as negative ions: fluoride (F^-), chloride (Cl^-), bromide (Br^-) and iodide (I^-). The most widespread compound containing fluorine is fluorspar or fluorite (CaF_2). The commonest chlorine compound is sodium chloride (NaCl) which occurs in sea-water and in rock salt. Each kilogram of sea-water contains about 30 g of sodium chloride. Sea-water also contains small amounts of bromides and traces of iodides. Certain seaweeds also contain iodine but the main source of iodine is sodium iodate ($NaIO_3$). This is found in Chile mixed with larger quantities of sodium nitrate ($NaNO_3$).

Laminaria is a type of seaweed which contains iodine

Pale yellow cubic crystals of fluorite (CaF_2). Fluorite is a common source of fluorine. The small black crystals are galena (PbS)

Patterns in physical properties

The table lists some physical properties of chlorine, bromine and iodine.

How do the following properties of the halogens change as their relative atomic mass increases?

1 state at room temperature;
2 colour of vapour;
3 melting points (see figure 2);
4 boiling points (see figure 2).

Element	Relative Atomic Mass	Colour and state at room tempera-ture	Colour of vapour	Structure	M.pt. /°C	B.pt. /°C
Chlorine	35.5	Pale green gas	Pale green	Cl_2 molecules	−101	−35
Bromine	79.9	Red brown liquid	Orange	Br_2 molecules	−7	58
Iodine	126.9	Dark grey solid	Purple	I_2 molecules	114	183

Physical properties of chlorine, bromine and iodine

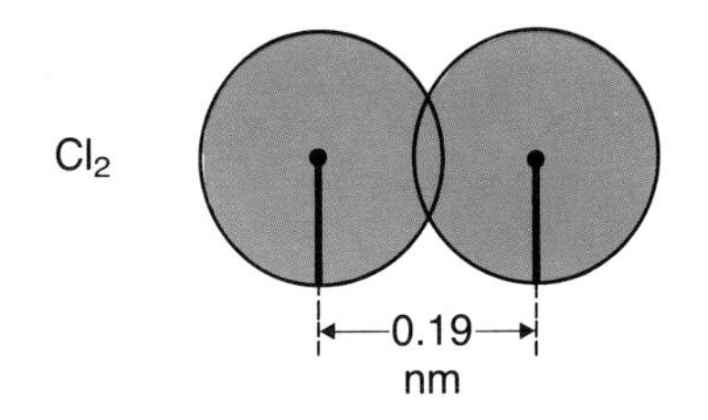

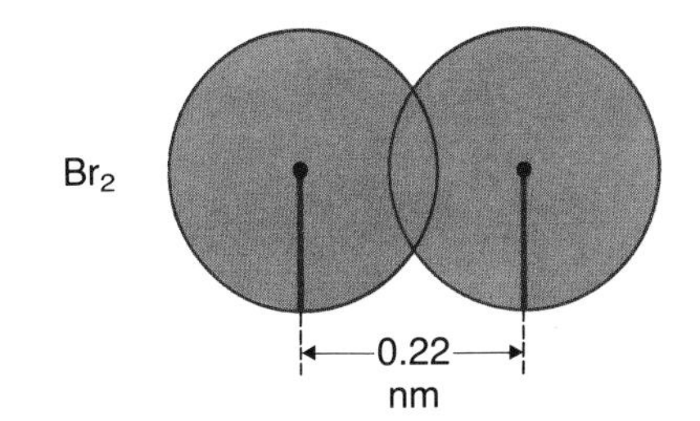

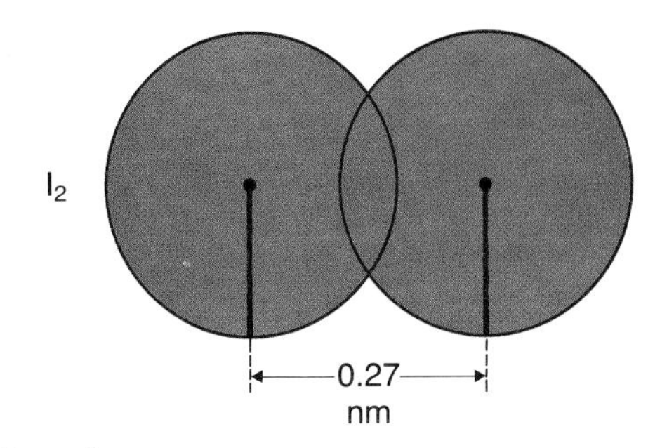

Figure 3

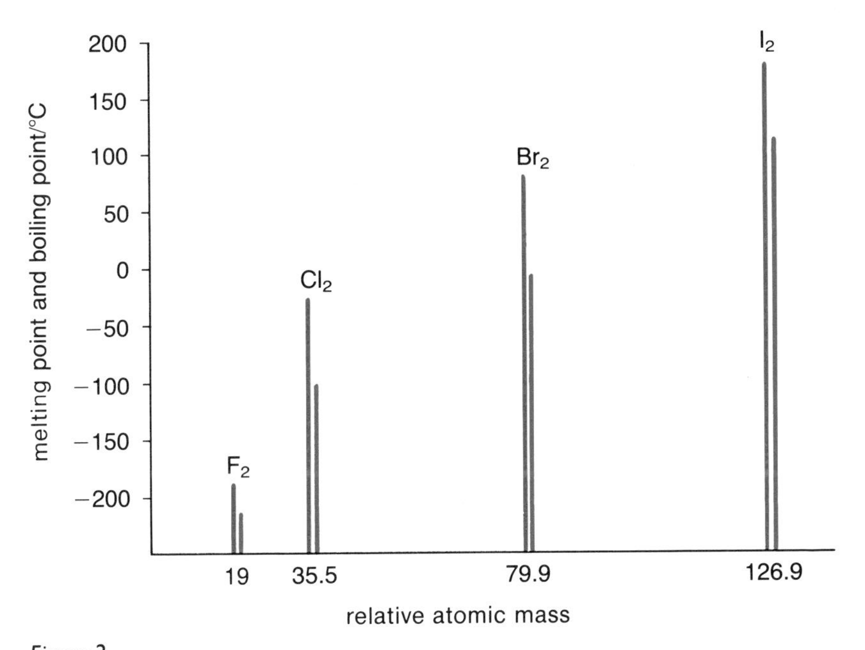

Figure 2
Melting points and boiling points of the halogens

The halogens form diatomic molecules, F_2, Cl_2, Br_2 and I_2. The relative sizes of these molecules (figure 3) can explain the changes in volatility (i.e. melting points and boiling points) with increasing relative atomic mass.

The larger the molecule, the harder it is to break up the orderly arrangement within the solid to form the liquid. It is also more difficult to separate larger molecules and form the gas. This is because there are stronger intermolecular forces between the larger molecules than the smaller ones. Thus, more energy is needed to melt and to boil the larger molecules. Hence the melting points and boiling points increase with relative molecular mass. F_2 and Cl_2, the smallest halogen molecules, are gases at room temperature, Br_2 is a liquid and I_2 is a solid.

Questions

1 What are the halogens?
2 What are the most important sources of (i) chlorine; (ii) bromine; (iii) iodine?
3 List the characteristic physical properties of the halogens.
4 Use the information in the table and figure 2 to predict the following properties of fluorine:
colour; state at room temperature; structure; melting point; boiling point.

8 Reactions of the Halogens

Reactions with water

When chlorine is bubbled into water, it reacts to form a mixture of hydrochloric acid and hypochlorous acids:

$$\text{chlorine} + \text{water} \rightarrow \text{hydrochloric acid} + \text{hypochlorous acid}$$
$$Cl_2 + H_2O \rightarrow HCl + HClO$$

The solution produced is very acidic. It is also a strong bleach which quickly turns litmus paper or universal indicator paper white. Bromine reacts less easily, but in a similar way. The solution produced is orange red and acidic. It bleaches less rapidly than chlorine water.

Iodine dissolves in water only very slightly. The solution is a pale yellow colour. It is only slightly acidic and it bleaches very slowly.

Household bleaches, like Domestos, usually contain chlorine itself and chlorine compounds. These compounds are so reactive that they remove coloured stains by reacting with them to make colourless substances.

A piece of cloth which has been bleached by chlorine compared with an unbleached piece

Reactions with iron

The reaction between chlorine and iron can be studied using the apparatus in figure 1. The iron wool is heated to start the reaction and then the Bunsen is removed. The iron glows as it reacts with the chlorine. The product is brown iron(III) chloride which sublimes along the tube:

$$\text{iron} + \text{chlorine} \rightarrow \text{iron(III) chloride}$$
$$2Fe + 3Cl_2 \rightarrow 2FeCl_3$$

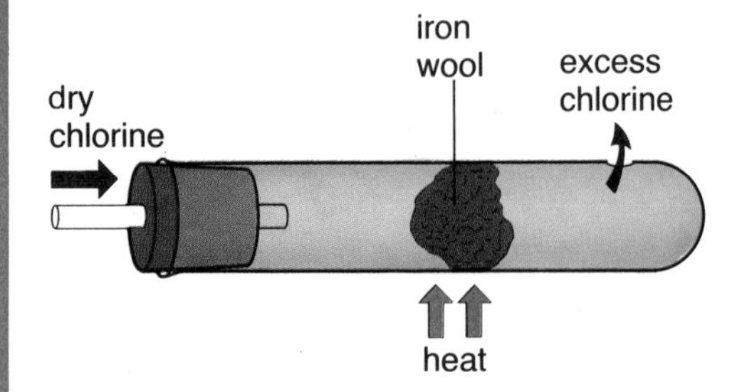

Figure 1

Figure 2 shows how we can investigate the reactions of iron with bromine or iodine. When iron is heated in bromine vapour, the reaction is less vigorous than that with chlorine. The iron glows but it must be heated all the time. The product is iron(III) bromide. When the experiment is repeated with iodine, the iron reacts very slowly to form iron(II) iodide.

Comparing reactivities

The reactions of halogens with water, with iron and as bleaching agents shows their relative reactivities very clearly. Chlorine is more reactive than bromine, and bromine is more reactive than iodine. Notice that *the halogens get less reactive as their relative atomic mass increases.* This is opposite to the trend in Group I. The alkali metals get *more* reactive with increasing relative atomic mass. The relative reactivities of halogens can be checked by studying the displacement of one halogen by another.

Household bleaches, like Domestos, usually contain chlorine itself and chlorine compounds

Displacement reactions

When chlorine water (water saturated with chlorine) is added to aqueous potassium iodide, the solution goes brown as iodine is produced. The chlorine has reacted with the potassium iodide to displace iodine and to form potassium chloride:

chlorine	+	potassium iodide	→	potassium chloride	+	iodine
Cl_2	+	$2KI$	→	$2KCl$	+	I_2

Chlorine is more reactive than iodine, so chlorine forms chloride ions in potassium chloride at the expense of iodide ions which are converted back to iodine. We can test for iodine by adding starch solution which gives a dark blue colour with the iodine:

$$Cl_2 + 2I^- \rightarrow 2Cl^- + I_2$$

If iodine is added to potassium chloride solution, nothing happens. Iodine is not reactive enough to convert chloride ions to chlorine:

iodine + potassium chloride → no reaction

But, what happens when bromine water is added to potassium iodide solution? In this case, bromine is more reactive than iodine. So, bromine forms bromide ions at the expense of iodide ions which are converted to iodine:

bromine	+	potassium iodide	→	potassium bromide	+	iodine
Br_2	+	$2KI$	→	$2KBr$	+	I_2

The orange solution turns dark brown as iodine is produced.

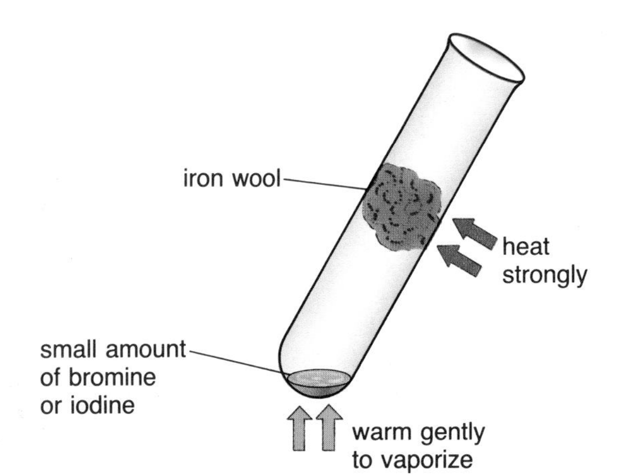

Figure 2

Questions

1 Describe experiments which show the relative reactivity of bromine, chlorine and iodine.
2 Write the five halogen elements in order of reactivity (the most reactive first).
3 Write word equations and balanced equations with symbols for the reaction of (i) bromine with water; (ii) bromine with iron; (iii) iodine with iron; (iv) chlorine with potassium bromide solution.
4 Arrange the following pairs of elements in order of decreasing vigour of reaction under the same conditions: *lithium and iodine; potassium and chlorine; potassium and fluorine; sodium and chlorine; sodium and iodine.*

9 Chlorine

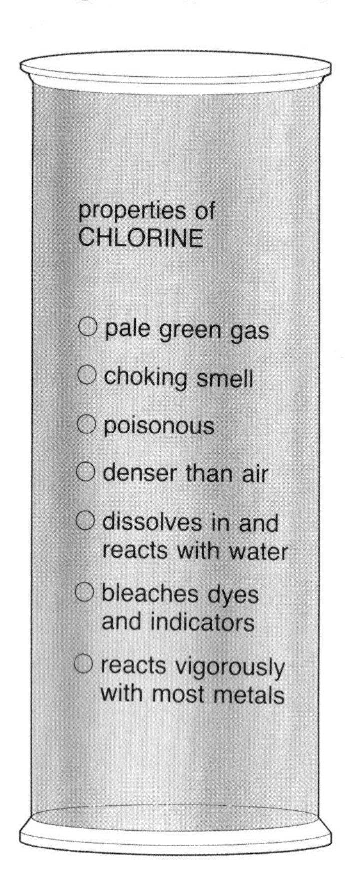

Figure 1

Chlorine is the most important element in Group VII. Chlorine and its compounds have many more uses than the other halogens and we shall look at these in the next unit. The important properties of chlorine are summarised in figure 1.

Remember that chlorine has a choking smell and that it is poisonous (toxic). Always treat it with care.

Tests for chlorine and chloride ions

The best test for chlorine is to show that it will bleach moist litmus paper. If blue litmus paper is used, the paper first turns red (because chlorine reacts with water to form an acidic solution) and then goes white.

When silver nitrate solution, containing silver ions, $Ag^+(aq)$, is added to a solution containing chloride ions, white, insoluble silver chloride is produced.

$$Ag^+(aq) + Cl^-(aq) \rightarrow AgCl(s)$$

The white silver chloride slowly turns purple/grey in sunlight. This reaction is used as a **test for chloride ions**.

Manufacturing chlorine

The main source of chlorine is sodium chloride in rock salt and in sea-water. About one million tonnes of chlorine are manufactured from sodium chloride each year in the UK. The chlorine is obtained by the electrolysis of saturated sodium chloride solution (brine). The photograph below shows a large number of the electrolysis cells in an industrial plant.

In these cells, the anodes are made of graphite. During electrolysis, chloride ions in the saturated brine are attracted to the graphite anodes and converted to chlorine gas:

Anode (+) $2Cl^-(aq) \rightarrow 2e^- + Cl_2(g)$

The cell room in a plant which produces chlorine by the electrolysis of brine

Preparing chlorine in the laboratory

The easiest way to make chlorine in the laboratory is from concentrated hydrochloric acid. The concentrated hydrochloric acid is oxidized to chlorine using either manganese(IV) oxide or potassium manganate(VII) (figure 2). *The preparation must be done in a fume cupboard.*

If potassium manganate(VII) is used, the mixture reacts quickly at room temperature. If manganese(IV) oxide is used, then the mixture needs gentle heating. As the chlorine is produced, it fills the gas jar from the bottom upwards and pushes the air out. This method of collecting a gas is called *downward delivery*.

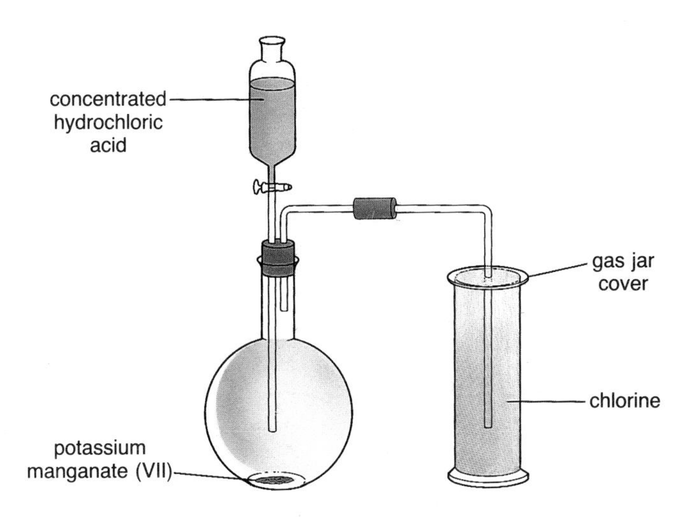

Figure 2

Chlorine as an oxidizing agent

When chlorine reacts with metals, it gains electrons from them, forming negative chloride ions. For example,

$$2Na + Cl_2 \rightarrow 2Na^+Cl^-$$

Compare this with the reaction of oxygen and metals to form oxide ions. For example,

$$2Mg + O_2 \rightarrow 2Mg^{2+}O^{2-}$$

The oxygen acts as an oxidizing agent by accepting electrons from the metal. In the same way, chlorine can be regarded as an oxidizing agent since it accepts electrons from the metal. Chlorine also acts as an oxidizing agent when it:

- bleaches dyes and indicators;
- reacts with iodides and bromides (see unit 8);
- reacts with green solutions of Fe^{2+} ions forming pale yellow Fe^{3+} ions:

$$2Fe^{2+} \rightarrow 2Fe^{3+} + 2e^- \qquad Cl_2 + 2e^- \rightarrow 2Cl^-$$

Questions

1 What is the test for (i) chlorine; (ii) chloride ions; (iii) hydrogen chloride?

2 Make a list of the important properties of chlorine.

3 (a) How is chlorine manufactured?
(b) Write an equation for the reaction used.

4 This question is about the laboratory preparation of chlorine.
(a) Why is the preparation done in a fume cupboard?
(b) Why does the chlorine sink to the bottom of the gas jar?
(c) Why is it easy to tell when the gas jar is full of chlorine?
(d) Why is the method of collecting the chlorine called downward delivery?
(e) When manganese(IV) oxide (MnO_2) is used, the products are manganese(II) chloride, water and chlorine. Write a word equation for the reaction and then a balanced equation using symbols.

10 The Uses of Chlorine

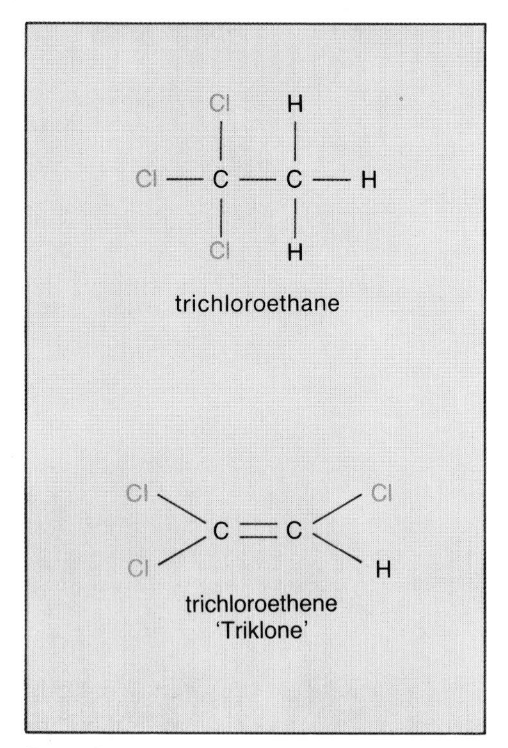

Figure 1
Solvents containing chlorine

Chlorine has far more uses than the other halogens. The element itself has very few uses, but its compounds are very important in industry, agriculture, medicine and the home. Two important uses of the element are in water sterilization and in bleaches. Chlorine is added in small quantities to our water supplies. The chlorine kills bacteria in the water, but it does not affect humans or animals. Larger quantities of chlorine are added to swimming baths. This kills one person's germs and stops them from affecting anyone else. The use of chlorine in domestic bleaches was discussed in unit 8 of this section.

Uses of chlorine compounds

Large quantities of chlorine are used to make hydrochloric acid. Chlorine is first burnt in hydrogen to form hydrogen chloride:

$$\text{hydrogen} + \text{chlorine} \rightarrow \text{hydrogen chloride}$$
$$H_2 + Cl_2 \rightarrow 2HCl$$

The hydrogen chloride is then dissolved in water to produce hydrochloric acid. This is the cheapest industrial acid. It is used to clean the rust from steel sheets before galvanizing and to produce ammonium chloride. Chlorine compounds are also used as solvents, plastics, antiseptics, disinfectants, artificial rubbers and pesticides.

- **Solvents.** Many of the solvents used to remove oil, fat and grease from clothing and machinery are compounds containing chlorine (figure 1). Tetrachloroethene is used in dry-cleaning clothes and trichloroethene ('Triklone') is used to degrease machinery. The 'thinner' (solvent) for Tippex is trichloroethane.

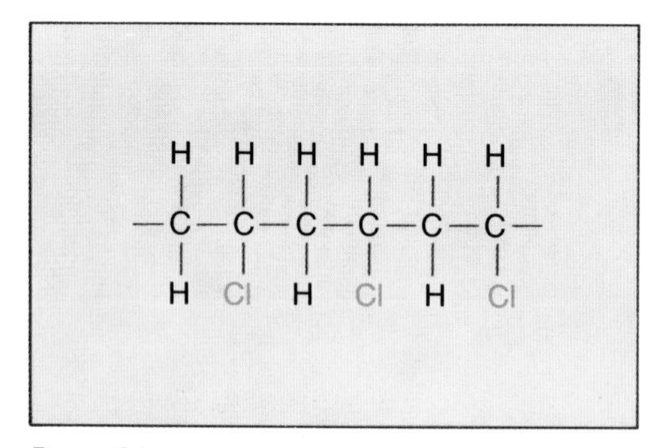

Figure 2
A section of PVC polymer

Circuit boards being degreased and cleaned by dipping in Arklone, a chlorine-containing solvent

trichlorophenol
(TCP)

benzene hexachloride
(BHC)

Dichlorodiphenyl
trichloroethane
(DDT)

Figure 3
Disinfectants and insecticides containing chlorine

- **Plastics.** One of the most important plastics is PVC (polyvinyl chloride). The correct chemical name for this is polychloroethene (figure 2). PVC plastics are used for raincoats, coverings for tables and shelves ('Fablon'), floor tiles ('Vinyl' tiles), upholstery, records and electrical insulation.
- **Disinfectants.** Two of the most widely used general disinfectants are TCP (trichlorophenol—figure 3) and Dettol, which is a similar chlorine-containing compound. These are used in many homes for cleaning cuts, gargling and cleaning sinks and toilets. During the Second World War, DDT (dichlorodiphenyltrichloroethane—figure 3) was produced and used as a disinfectant. It killed lice, flies and mosquitoes which were causing typhoid, dysentery and malaria. When the allied armies liberated France in 1944, every British soldier wore a DDT-impregnated shirt. The success of DDT was remarkable. For every one louse-infected British soldier, there were 8000 louse-infected enemy soldiers. DDT has also been used to delouse soldiers in more recent wars. The use of DDT has now been restricted. It was used a great deal as a pesticide for crops during the 1950s, but its poisonous nature led to the deaths of birds and mammals.

All these products are made from PVC

DDT being used to delouse captured soldiers during the Korean War

Questions

1 What are the main uses of chlorine itself?

2 Make a list of three uses of chlorine-containing compounds and draw the structure of the chemicals being used.

3 (a) How is hydrochloric acid manufactured?
(b) What are its uses?

4 Why is the use of DDT now banned in some countries?

11 Patterns in the Periodic Table

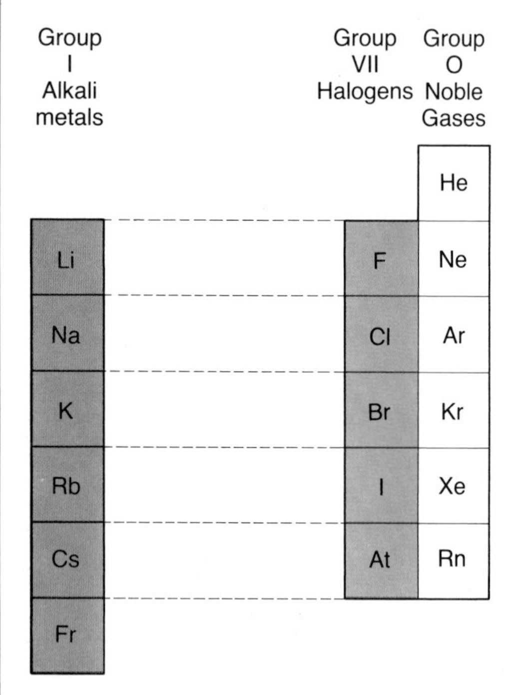

Figure 1

Down the groups

So far we have concentrated on the patterns within the groups of the periodic table. We have studied three groups in this way (figure 1):

> *In each group, the elements have similar properties with a gradual trend in properties down the group.*

The groups are like families of similar elements. In Group I for example, the metals are all reactive, but they get more reactive with increase in relative atomic mass. Lithium reacts steadily with water, but potassium reacts violently. Imagine how francium will react with water! This trend shows that *metallic character is increasing down the group.* Potassium is more metallic than lithium, and francium is more metallic than potassium.

The increasing metallic character down a group is most noticeable in Group IV. Carbon, at the top of the group, is normally classed as a non-metal. Silicon and germanium, in the middle of the group, are metalloids. Tin and lead, at the bottom of the group, are metals.

Across the periods

There are important patterns across the periods of the periodic table as well as down the groups. Some of these patterns are summarised in table 1. This shows some of the properties of the elements from sodium to argon across period 3.

The most obvious pattern across each period is the change from metals to non-metals. In period 3, sodium, magnesium and aluminium are metals with high boiling points and good electrical conductivity. Silicon, in the middle of the period, is a metalloid with a high boiling point and moderate conductivity. Phosphorus, sulphur, chlorine and argon, on the right-hand side of the period, are non-metals with low boiling points and poor conductivity.

Element	Na	Mg	Al	Si	P	S	Cl	Ar
State at room temperature	s	s	s	s	s	s	g	g
Boiling point /°C	890	1120	2450	2680	280	445	−34	−186
Electrical conductivity	Good	Good	Good	Moderate	Poor	Poor	Poor	Poor
Type of element	← METALS		→	METALLOID	← NON-METALS			→
Reactivity	Reactive metals		Moderately reactive metal	Very unreactive	Moderately reactive non-metals		Very reactive non-metal	Extremely unreactive

Table 1

Notice how the reactivity of the elements changes across a period. Apart from the noble gases, the most reactive elements are near the edges of the periodic table and the least reactive elements are in the centre.

Patterns in the properties of oxides

Table 2 shows some properties of the oxides in period 3.

Element	Na	Mg	Al	Si	P	S	Cl	Ar
Formula of oxide	Na_2O	MgO	Al_2O_3	SiO_2	P_4O_{10} (P_4O_6)	SO_3 (SO_2)	Cl_2O_7 (Cl_2O)	No oxide
Valency of element with oxygen	1	2	3	4	5 (3)	6 (4)	7 (1)	0
State of oxide at room temp.	s	s	s	s	s (s)	l (g)	l (g)	
Nature of oxide	BASIC		Basic and Acidic	ACIDIC				
Type of structure	GIANT IONIC			Giant Molec.	SIMPLE MOLECULAR			

Table 2

Notice the following trends from table 2.

1 The valencies of the elements with oxygen rise from 1 to 7. (The lower valencies with oxygen rise from 1 to 4 and then fall to 1.)

2 The oxides become more volatile across the period. Those on the left are solids at room temperature; those on the right are gases or liquids.

3 The structure of the oxides changes from ionic to simple molecular. Na_2O, MgO and Al_2O_3 are giant ionic structures which are solids at room temperature. P_4O_{10}, P_4O_6, SO_3, SO_2, Cl_2O_7 and Cl_2O are simple molecules—mainly gases or liquids.

4 The nature of the oxides changes from basic to acidic. The oxides of the metals on the left are **basic**. These basic oxides react with acids to form salts. For example,

sodium oxide + hydrochloric acid → sodium chloride + water

$$Na_2O + 2HCl \rightarrow 2NaCl + H_2O$$

magnesium oxide + hydrochloric acid → magnesium chloride + water

$$MgO + 2HCl \rightarrow MgCl_2 + H_2O$$

The oxides of the non-metals on the right are **acidic**. These acidic oxides react with alkalis (soluble bases) to form salts:

sulphur trioxide + sodium hydroxide → sodium sulphate + water

$$SO_3 + 2NaOH \rightarrow Na_2SO_4 + H_2O$$

Notice that Al_2O_3 is both basic and acidic. It reacts with acids to form salts and also with alkalis to form salts. Oxides which are both basic and acidic are called **amphoteric oxides**.

Questions

1 Summarise the important trends in the periodic table, (i) down the groups; (ii) across the periods.

2 Consider the elements Li, Be, B, C, N, O, F and Ne in period 2.
Which of these elements: (i) are metals; (ii) exist as simple molecules; (iii) has the highest boiling point; (iv) is the most reactive non-metal; (v) are gases at room temperature; (vi) exist as diatomic molecules at room temperatures?

3 Make a list of the trends in the properties of the oxides of the elements across period 3.

Section E: Study Questions

1 (a) How was Mendeléev's periodic table similar to that of Newlands?
(b) How did Mendeléev's periodic table differ from that of Newlands?
(c) Why was Mendeléev's periodic table so successful and so readily accepted by other scientists?
(d) How do modern periodic tables differ from the one suggested by Mendeléev?

2 This question concerns the following families of elements in the periodic table.
A Group I—the alkali metals
B Group IV—containing carbon and lead
C Group VII—the halogens
D Group O—the noble gases
E The transition elements
Select from *A* to *E*, the *one* family containing an element which:
(i) melts at 25 K and boils at 27 K.
(ii) reacts with water forming a solution of pH 12.
(iii) forms a chloride, nitrate and sulphate all of which are pink.
(iv) is a liquid at room temperature and forms a white soluble sodium salt.
(v) forms no compounds.
(vi) is a solid, a poor conductor of electricity and boils above 3000°C.
(vii) is a good conductor of electricity and floats on water.
(viii) is a metalloid and melts on heating with a Bunsen.
(ix) has two crystalline forms at room temperature.
(x) forms an oxide which catalyses the decomposition of hydrogen peroxide very effectively.

3 This question concerns the outline periodic table shown below. The letters are *not* the chemical symbols of the elements.

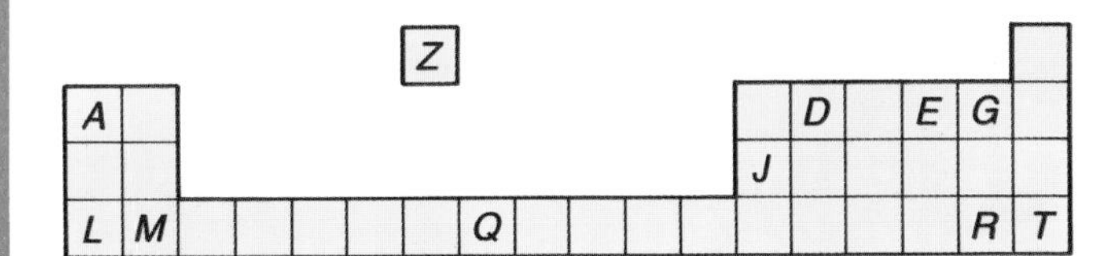

In answering the following questions, the same letter may be used once, more than once or not at all.
(a) Give the letters of any *two* elements in the same group. (1)
(b) Which letter represents a transition metal? (1)
(c) Give *one* way in which an alkali metal is different from a transition metal. (1)
(d) What is the name of element *Z*? (1)
(e) What is the term used to describe a row of elements such as *L* to *T*? (1)
(f) Give the letter of an element which is a gas consisting of single atoms. (1)
(g) Give the letter of the most reactive metal listed. (1)
Total [7] **LEAG**

4 (a) One important requirement for an anaesthetic is that its boiling point should be between about 40°C and 60°C.
(i) What is the disadvantage of a boiling point below 40°C for an anaesthetic?
(ii) What is the disadvantage of a boiling point above 60°C?
(b) The table below shows the anaesthetic effect, toxicity (i.e., harmful effect on the body) and flammability of methane and its four chlorine-containing derivatives. How do
(i) the anaesthetic effect,
(ii) the toxicity,
(iii) the flammability
depend on the number of chlorine atoms introduced into methane?
(c) Suggest how the chlorine atoms influence the toxicity.

Compound	Anaesthetic effect	Toxicity	Flammability
Methane, CH_4	None	None	Flammable
Chloro-methane, CH_3Cl	Weak	Little	Flammable
Dichloro-methane, CH_2Cl_2	Moderate	Little	Non-flammable
Trichloro-methane, $CHCl_3$	Strong	Toxic in large quantities	Non-flammable
Tetrachloro-methane, CCl_4	Strong	Very toxic	Non-flammable

5 The element rubidium (Rb) has similar properties to sodium and potassium and is in the same group in the periodic table.
(a) Write down the formula of:
(i) rubidium oxide
(ii) rubidium hydroxide
(iii) rubidium nitrate.
(b) Describe what happens and write equations for the reactions which occur when:
(i) rubidium is added to water,
(ii) rubidium hydroxide solution is added to hydrochloric acid.
(c) Is rubidium hydroxide an alkali? Explain your answer.

6 A metal *M* is in Group 3 of the periodic table. Its chloride has a formula mass of 176.5.
(a) Write the formula of the chloride of the metal. (1)
(b) Calculate the atomic mass of the metal. (Chlorine has relative atomic mass = 35.5) (2)
(c) Given that the atomic mass of the Group IV elements Si, Ge and Sn are 28, 73 and 119 respectively, identify the metal. (1)
SEG

SECTION F

Metals

What properties of metals and alloys made them useful in constructing Concorde?

1 Metals and Acids

Which of these foodstuffs contain acids?

Various foods, including vinegar, citrus fruits and rhubarb, contain acids. These acids will attack kitchen utensils and cutlery made of certain metals.

The table shows what happens when some metals are added to dilute hydrochloric acid (HCl) at room temperature (21 °C). This acid is more reactive than the acids in food, but it shows how easily different metals are attacked.

Look at the table.

1 Which metal reacts (i) most vigorously; (ii) least vigorously?

2 Write the metals in order of decreasing reactivity with dilute hydrochloric acid.

Hydrogen is produced from dilute hydrochloric acid by all the metals except copper. The other product of the reaction is the chloride of the metal.

$$\text{metal} + \text{hydrochloric acid} \rightarrow \text{metal chloride} + \text{hydrogen}$$

Using M as the symbol for the metal, and assuming that the metal forms M^{2+} ions, we can write a general equation as:

$$M(s) + 2HCl(aq) \rightarrow MCl_2(aq) + H_2(g)$$

A similar reaction occurs between metals and dilute sulphuric acid. This time, the products are hydrogen and a metal sulphate. In general, acids react with metals above copper in the activity series to give the metal compound and hydrogen.

$$\textit{metal} + \textit{acid} \rightarrow \textit{metal compound} + H_2$$

The reaction of zinc with dilute hydrochloric acid is used to produce hydrogen in the laboratory (figure 1). Zinc is chosen because it gives a steady stream of hydrogen. It does not react too fast or too slow. Why does the zinc react with the acid when the tap is opened? What happens when the tap is closed?

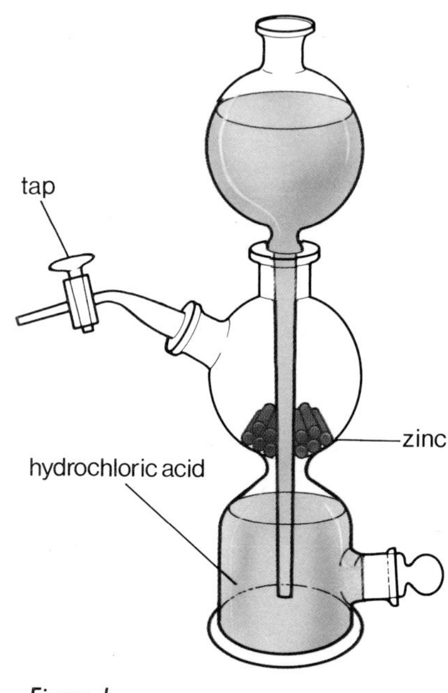

Figure 1
Preparing hydrogen in a Kipp's apparatus

Copper is so unreactive that a copper cooking pot will last for many years

Notice in the table that aluminium does not react at first with dilute hydrochloric acid. This is because its surface is protected by a layer of aluminium oxide. The oxide reacts slowly with the acid to expose aluminium, which reacts more vigorously.

The metals used most commonly for pans and cutlery are aluminium, copper, nickel and iron (steel). Copper is the only one which does not react with the acids in food. But, copper is so expensive that aluminium and steel are used in most saucepans nowadays. The thin oxide coating on aluminium helps to protect it from the weak acids in many foods. However, oxalic acid in rhubarb and acetic acid in vinegar do react with aluminium and 'clean' the saucepan during cooking. Iron and nickel, which are used in cutlery, are also attacked by acids in food. Ordinary kitchen cutlery becomes badly marked if it is left in lemon juice (which contains citric acid) or in vinegar.

Foods which contain acids are best stored in unreactive containers made of glass or plastic. Tin cans are also used to store acidic foods like pineapples and grapefruit. Tin cans are made of steel coated on both sides with tin and then lacquered on the inside. The lacquer gives an unreactive layer between the tin and its contents.

Metal used	Reaction with dilute hydrochloric acid	Highest temp. recorded
Aluminium	No reaction at first, but vigorous after a time. Hydrogen is produced rapidly	85°C
Copper	No reaction. No bubbles of hydrogen	21°C
Iron	Slow reaction, bubbles produced slowly from iron	35°C
Lead	Very slow reaction, a few bubbles appear on the lead	23°C
Magnesium	Very vigorous reaction, hydrogen is produced rapidly	95°C
Zinc	Moderate reaction, bubbles of hydrogen are produced steadily	55°C

Reactions of metals with dilute hydrochloric acid

Questions

1 Which acids are present in (i) lemon juice, (ii) vinegar; (iii) rhubarb; (iv) sour milk?

2 Look at the Kipp's apparatus in figure 1.
(a) What happens when the tap is opened?
(b) What happens when the Kipp's is supplying hydrogen and the tap is closed?
(c) Why is zinc used in the Kipp's and not magnesium?
(d) Write an equation for the reaction of zinc with hydrochloric acid.

3 (a) Why is copper a better metal for saucepans than aluminium?
(b) Why is copper not used for most pans nowadays?
(c) Why is aluminium less reactive to acidic foodstuffs than expected?
(d) How are tin cans protected from acids in their contents?

4 Write equations for the reactions of
(a) aluminium with dilute hydrochloric acid;
(b) magnesium with dilute sulphuric acid;
(c) zinc with dilute nitric acid.

2 The Reactivity Series

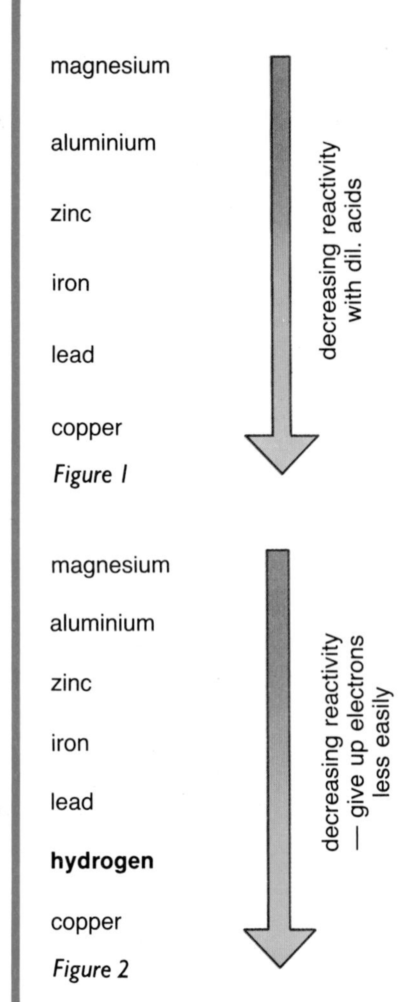

Figure 1

Figure 2

Figure 1 shows a reactivity series for the metals tested with dilute hydrochloric acid in the last unit. When a reaction occurs, metal atoms, which we can write as M, become metal ions (M^{2+}) in the metal chloride, and hydrogen ions (H^+) in the acid form hydrogen molecules (H_2):

$$M + 2H^+ \rightarrow M^{2+} + H_2$$

Magnesium reacts the most vigorously. It is keenest to form ions. Lead reacts very slowly because it is only a little keener than hydrogen to form ions. Copper does not react with dilute acids at all. This means that hydrogen will form its ions more easily than copper.

Can you see that the reactivity series shows how easily metals and hydrogen give up electrons to form ions? Metals at the top of the series form ions more easily than those lower down. Since hydrogen forms its ions more easily than copper and less easily than lead, it goes between these two metals in the reactivity series (figure 2).

Metals such as magnesium, at the top of the series, give up electrons easily to form their ions:

$$Mg \rightarrow Mg^{2+} + 2e^-$$

With dilute acids, these electrons are given to H^+ ions which form hydrogen gas:

$$2H^+ + 2e^- \rightarrow H_2$$

Metals at the bottom of the reactivity series, such as copper, do not give up electrons easily. They have a stronger hold on their electrons than hydrogen atoms. Thus, copper does not react with dilute acids.

Reactivity series	Reaction with air or oxygen	Reaction with cold water	Reaction with steam	Reaction with dilute acids
K Na	Burn brightly and very vigorously	Produce H_2 with decreasing vigour	Produce H_2 with decreasing vigour	Produce H_2 from dilute HCl, H_2SO_4 and HNO_3 with decreasing vigour
Ca Mg	Burn to form oxides with decreasing vigour			
Al Zn Fe				
Pb	React slowly to form a layer of oxide	Do not react with cold water		
Cu Hg			Do not react with steam	Do not react with dilute acids
Ag Au Pt	Do not react with air or oxygen			

The reactions of metals with oxygen, water, steam and acids

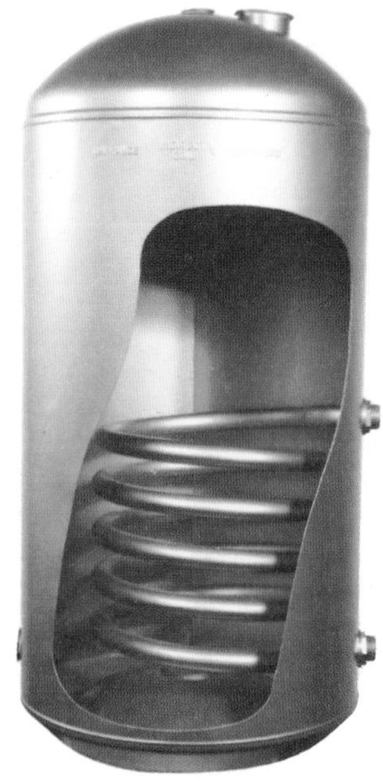

Why are hot water tanks made of copper and cold water tanks made of steel?

Summarising reactions of metals

The table summarises the reactions of metals with oxygen, water, steam and acids. Notice that the order of reactivity is the same in all four columns. This is not surprising because metal atoms react to form metal ions in each case:

$$M \rightarrow M^{2+} + 2e^-$$

The higher a metal is in the reactivity series, the more easily it forms its ions, so the more reactive it is.

- **Reaction with oxygen.** Metal atoms lose electrons to form metal ions. The electrons are taken by oxygen molecules (O_2) to form oxide ions (O^{2-}):

$$2M + O_2 \rightarrow 2M^{2+}O^{2-}$$

Metals below mercury do not form stable oxides so they do not react with oxygen.

- **Reaction with water and steam.** Here again the metals lose electrons to form ions. The electrons are taken by water molecules (H_2O) which form oxide ions (O^{2-}) and hydrogen (H_2):

$$M + H_2O \rightarrow M^{2+}O^{2-} + H_2$$

The oxides of metals high in the reactivity series (e.g. Na_2O and CaO) react with more water to form solutions of their hydroxides:

$$(Na^+)_2O^{2-}(s) + H_2O(l) \rightarrow 2Na^+(aq) + 2OH^-(aq)$$

The reactivity series is a very useful summary of the reactions of metals. It shows the relative reactivities of metals. Metals at the top of the series want to lose electrons and form ions. Atoms of these metals are reactive but their ions are stable (figure 3). Metals at the bottom of the reactivity series are just the opposite. Ions of these metals want to gain electrons and form atoms.

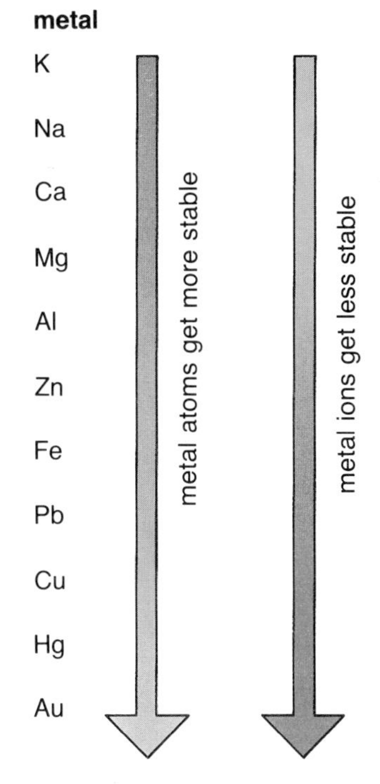

Figure 3

What is the disadvantage in using iron (steel) for the boilers in steam trains?

Questions

1 What is meant by the reactivity series?
2 Why is the order of reactivity of metals the same with oxygen, with water and with acids?
3 X is a metal which reacts with steam to form hydrogen and a white insoluble oxide. X does not react with water. Name two metals that X could be.
4 Write equations for the reaction of (i) copper with oxygen; (ii) aluminium with oxygen; (iii) calcium with water; (iv) zinc with steam; (v) lead with dilute nitric acid; (vi) magnesium with dilute sulphuric acid.

3 Studying Crystals

Crystals of impure rock salt

White quartz crystals on impure dolomite

Look at the crystals of rock salt and quartz in the photographs above. What do you notice about all the salt crystals? What do you notice about all the quartz crystals? All the salt crystals are roughly the same cubic shape. All the quartz crystals are roughly cylindrical with pointed tops. Further studies show that *all the crystals of one substance have similar shapes*. This suggests that the particles in the crystal are always packed in a regular fashion to give the same overall shape to the crystal. Sometimes, crystals grow unevenly and their shapes become a little distorted. Even so, it is easy to see their general shape. Solid substances which have a regular packing of particles are described as **crystalline**. The particles may be atoms, ions or molecules. Figure 1 shows how cubic crystals and hexagonal crystals can form. If the particles are always placed in parallel lines or at 90° to each other, the crystal will be cubic. If the particles are always placed at 120° in the shape of a hexagon, the final crystal will be hexagonal.

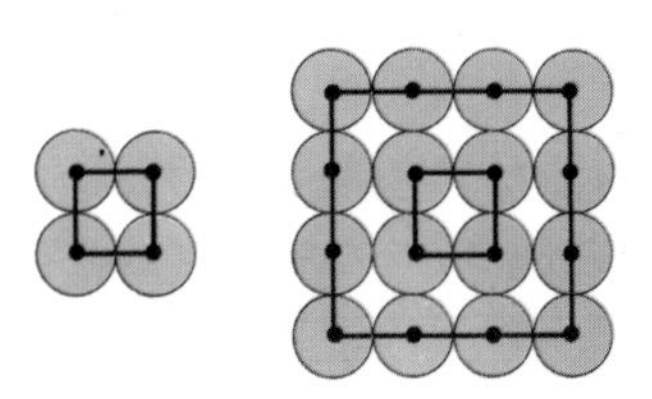

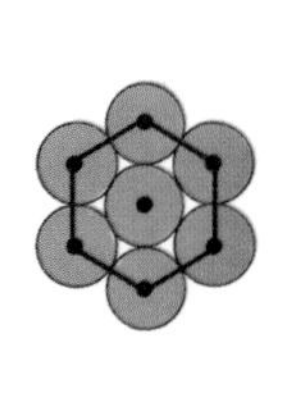

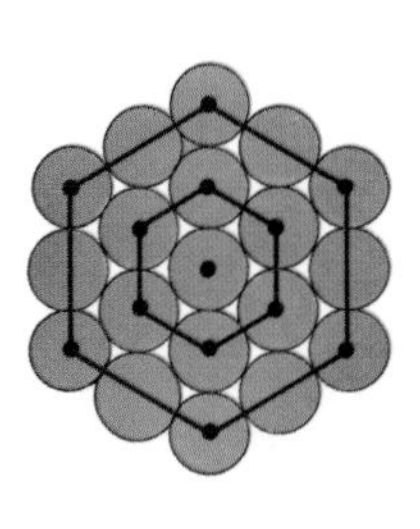

When the particles are arranged in a cubic fashion the final crystal will be cubic

When the particles are arranged in a hexagonal fashion the final crystal will be hexagonal

Figure 1

We can compare the way in which a crystal grows to the way in which a bricklayer lays bricks. If the bricklayer always places the bricks in parallel lines or at 90° to each other, then the final buildings will be like cubes or boxes. However, if some bricks are laid at 120° to make hexagons, then the final buildings will be hexagonal.

The overall shape of a crystal can only give a *clue* to the way in which the particles are arranged. X-rays give much *better evidence* of the way in which particles are packed.

Using X-rays to study crystals

Look through a piece of thin stretched cloth at a small bright light. The pattern you see is due to the deflection of the light as it passes through the regularly spaced threads of the fabric. This deflection of light is called **diffraction** and the patterns produced are **diffraction patterns**. If the cloth is stretched so that the threads in the fabric get closer, then the pattern spreads further out. From the diffraction pattern which we *can* see, we can work out the pattern of the threads in the fabric which we *cannot* see. The same idea is used to work out how the particles are arranged in a crystal.

A narrow beam of X-rays is directed at a well-formed crystal (figure 2). Some of the X-rays are diffracted by particles in the crystal onto X-ray film. When the film is developed, a regular pattern of spots appears. This is the diffraction pattern for the crystal. From the diffraction pattern which we *can* see, it is possible to work out the pattern of particles in the crystal which we *cannot* see. A regular arrangement of spots on the film indicates a regular arrangement of particles in the crystal. This regular arrangement of particles in the crystal is called a **lattice**.

X-rays have been used to study thousands of different crystals. An X-ray diffraction photograph is shown in figure 3. When crystals of various elements are studied using X-rays, the atoms appear to be arranged in regular patterns. Nowadays, we can also use rays of electrons, like X-rays, to study the way in which particles are arranged in crystals.

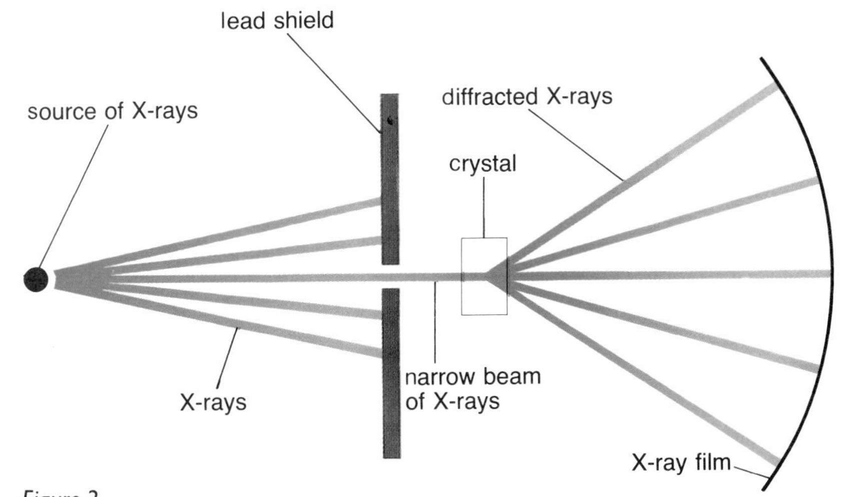

Figure 2
Diffraction of X-rays by a crystal

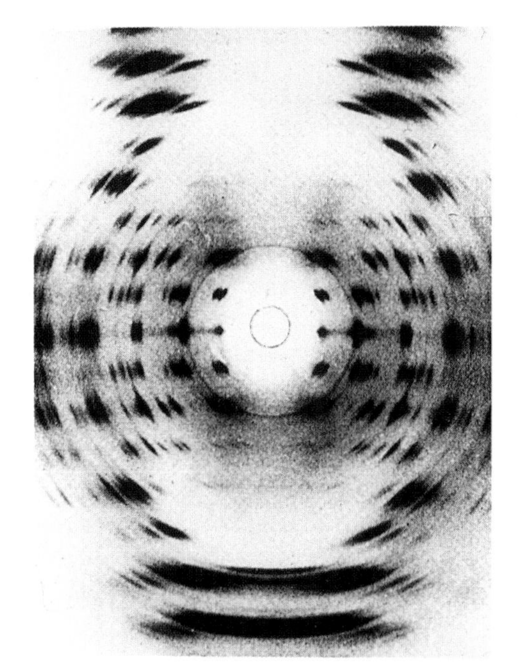

Figure 3
An X-ray diffraction photograph of DNA crystals. What general pattern do the dots form?

Snowflake crystals

Questions

1 Explain the words:
crystal; lattice; diffraction.

2 Look at the snowflake crystals in the photograph above.
(a) What substance makes up snowflakes?
(b) What particles do snowflakes contain?
(c) How do you think the particles are arranged in snowflakes?

3 How are X-rays used to give evidence for the arrangement of particles in a crystal?

4 Why do all crystals of one substance have roughly the same shape?

4 Metal Crystals

Zinc crystals on the surface of galvanised iron

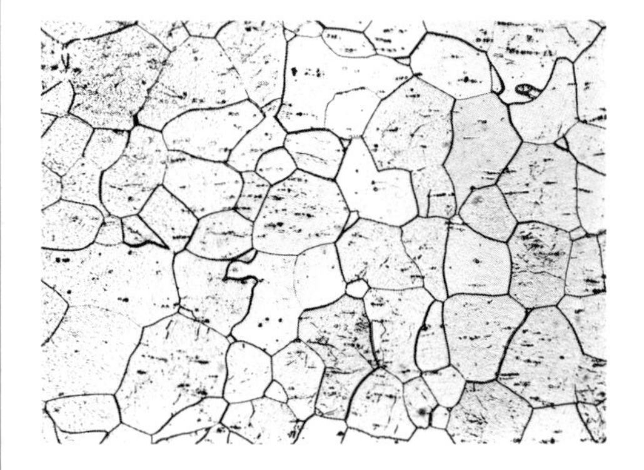

Grain boundaries in antimony

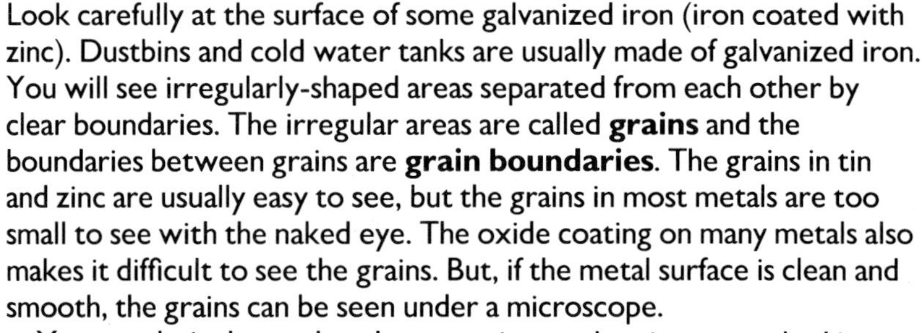

Look carefully at the surface of some galvanized iron (iron coated with zinc). Dustbins and cold water tanks are usually made of galvanized iron. You will see irregularly-shaped areas separated from each other by clear boundaries. The irregular areas are called **grains** and the boundaries between grains are **grain boundaries**. The grains in tin and zinc are usually easy to see, but the grains in most metals are too small to see with the naked eye. The oxide coating on many metals also makes it difficult to see the grains. But, if the metal surface is clean and smooth, the grains can be seen under a microscope.

X-ray analysis shows that the atoms in metal grains are packed in a regular order, but the grains themselves are irregularly-shaped crystals of the metal pushed tightly together.

Metals usually have a high density which suggests that the particles are packed close together. The high melting points and boiling points of most metals indicate that the atoms are held closely together by strong forces of attraction. In fact, X-ray studies show that the atoms of most metals are packed as close together as possible. This arrangement is called **close packing**. Figure 1 shows a few atoms in one layer of a metal crystal. Notice that each atom in the middle of the crystal touches 6 other atoms in the same layer. When a second layer is placed on top of the first layer, atoms in the second layer sink into the dips between atoms in the first layer (figure 2). This means that any one atom in the first layer can touch 3 atoms in the layer above it and 3 atoms in the layer below it. The total number of 'nearest neighbours' to one atom in a close-packed structure is therefore 12. This is made up from 6 in the same layer, 3 in the layer above and 3 in the layer below. The number of nearest neighbours to an atom or ion in a crystal is called its **coordination number**. Thus, we an say that the coordination number for atoms in a close-packed metal structure is 12.

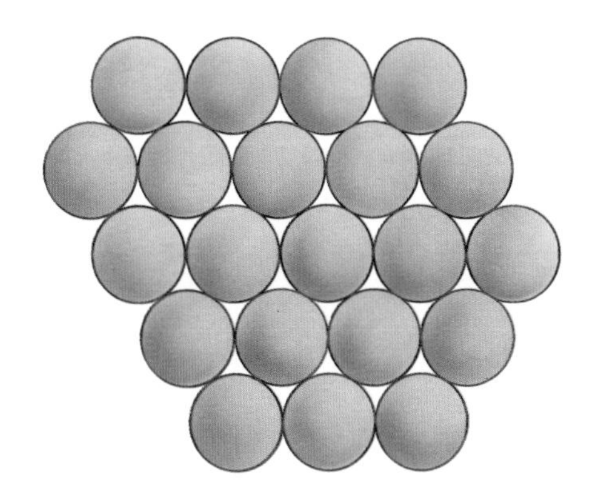

Figure 1

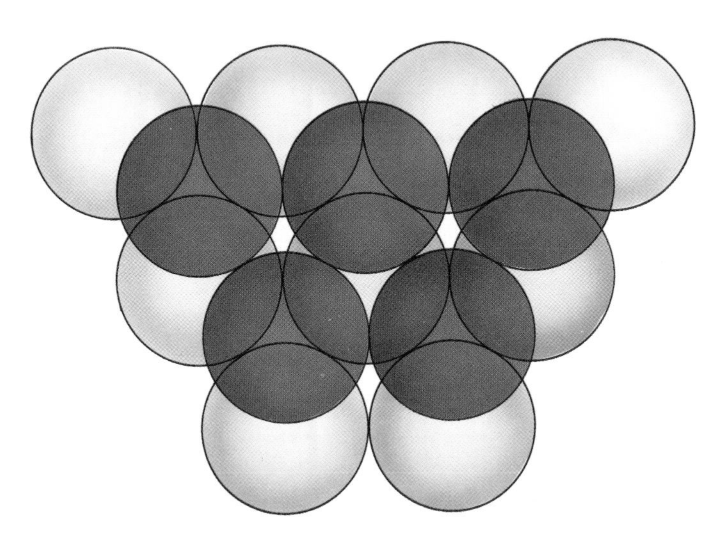

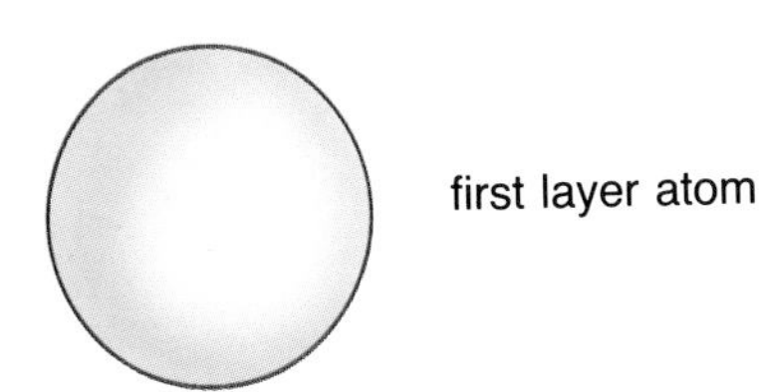

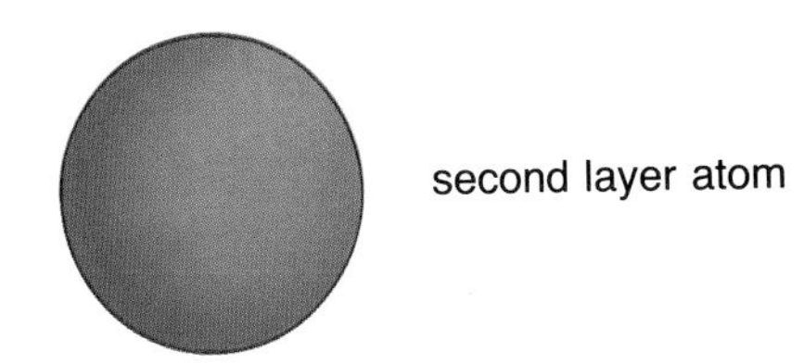

Figure 2

A good model of the close packing in a single layer of metal atoms can be made by blowing a slow stream of gas into a dish containing dilute soap solution (figure 3). The small bubbles act like atoms in a layer of metal.

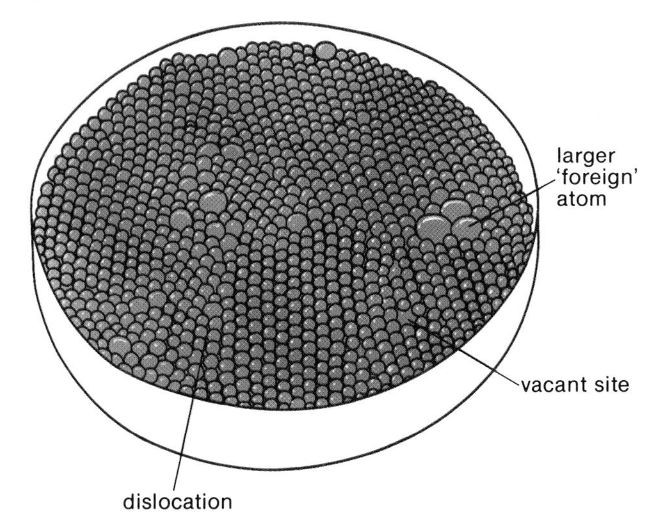

Figure 3
A bubble raft showing crystal grains, and also grain boundaries, dislocations, vacant sites and larger 'foreign' atoms

The large areas of regular packing in the 'bubble raft' represent a single crystal grain. Figure 3 also shows the four main 'flaws' (imperfections) in metal crystals.

1 Grain boundaries. These are the narrow areas of disorder between one crystal grain and another.

2 Dislocations. Occasionally, a row of atoms is displaced or it 'peters out' and the regular packing ceases. When this happens in a metal crystal, the flaw is called a **dislocation**.

3 Vacant sites. These occur where an atom is missing from the crystal structure.

4 'Foreign' atoms. When atoms of another element form part of the crystal structure, they break up the orderly arrangement. This is noticeable in the 'bubble raft' where there are 'foreign' bubbles which are larger or smaller than the rest.

Questions

1 Explain the following:
grain; grain boundary; close packing; coordination number.

2 Name and explain the four main kinds of flaw (imperfection) in metal crystals.

3 Explain why atoms in a close-packed structure have a coordination number of 12.

5 Properties of Metals

Metals are malleable

The properties of a metal depend on the way its atoms are arranged within the crystal and also on the size of the grains (crystals) in the metal.

The table below lists some physical properties of six metals. The average density for metals is about 8 g/cm^3. This is high compared to the densities of non-metals and compounds. But, notice the low density of magnesium. All the metals in Groups I and II, such as magnesium and sodium, have lower densities than most other metals.

Metal	**Density /g cm^3**	**Melting point /°C**	**Boiling point /°C**
Aluminium	2.7	659	2447
Copper	9.0	1083	2600
Iron	7.9	1540	3000
Magnesium	1.7	650	1110
Silver	10.5	961	2177
Zinc	7.1	420	908

Physical properties of six metals

Metals usually
- *have high densities* } *see the table*
- *have high melting points* }
- *have high boiling points* }
- *are good conductors of heat*
- *are good conductors of electricity*
- *are malleable (can be hammered into different shapes)*
- *are ductile (can be pulled into wires)*

- **Density.** The close packing of atoms in most metal crystals helps to explain their high densities.

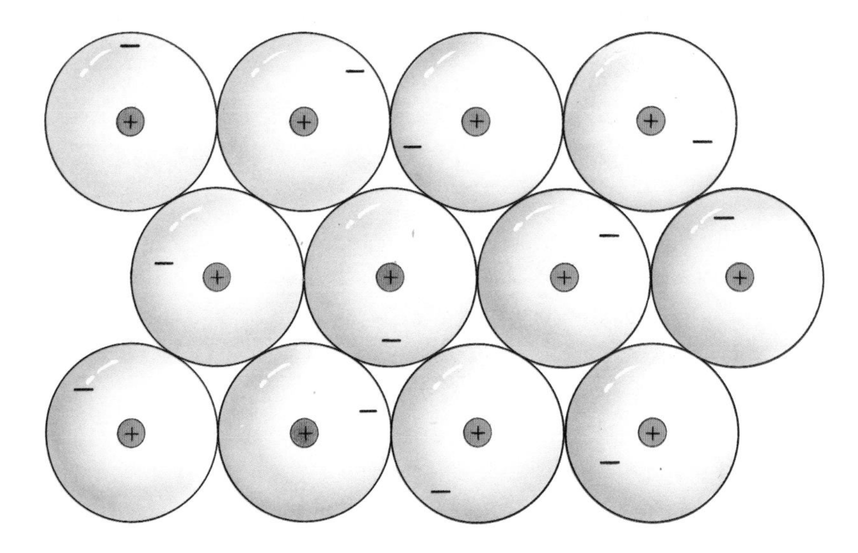

Figure 1
The outermost electrons of each atom move around freely in the metal structure

• **Melting and boiling points.** High melting points and high boiling points suggest that there are strong forces holding the atoms together in metal crystals. Chemists think that the outermost electrons of each atom can move about freely in the metal. So, the metal consists of positive ions surrounded by a 'sea' of moving electrons (figure 1). The negative 'sea' of electrons attracts *all* the positive ions and cements everything together. The strong forces of attraction between the moving electrons and the positive ions result in high melting points and high boiling points.

• **Conductivity.** When a metal is connected in a circuit, freely moving electrons in the metal move towards the positive terminal. At the same time, electrons can be fed into the other end of the metal from the negative terminal (figure 2). This flow of electrons through the metal forms the electric current.

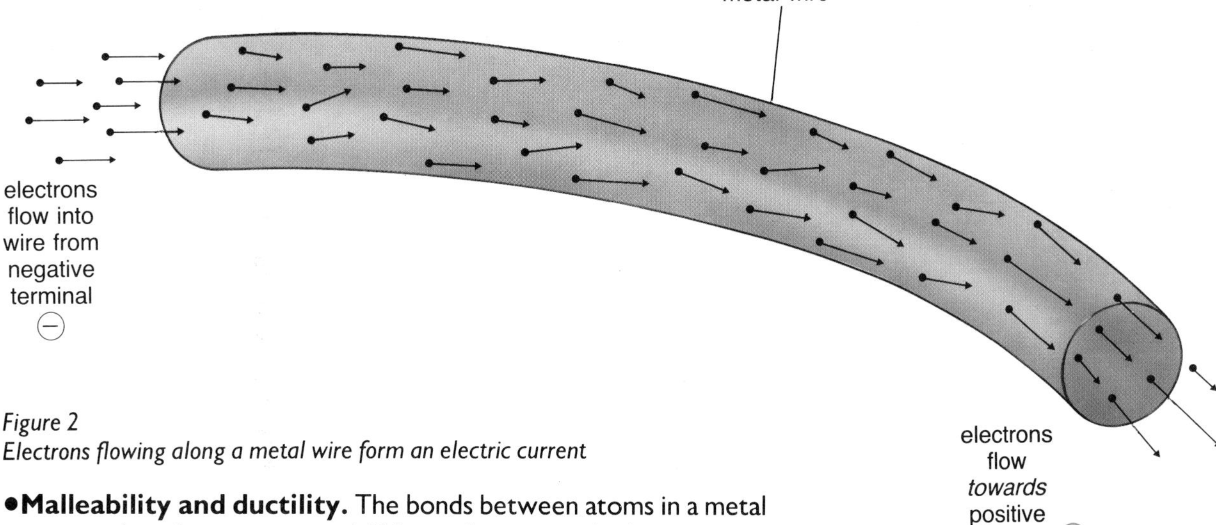

Figure 2
Electrons flowing along a metal wire form an electric current

• **Malleability and ductility.** The bonds between atoms in a metal are strong but they are *not* rigid. When a force is applied to a metal crystal, the layers of atoms can 'slide' over each other. This is known as **slip**. After slipping, the atoms settle into position again and the close-packed structure is restored. Figure 3 shows the positions of atoms before and after slip. It also shows what happens when a metal is hammered into different shapes or drawn into a wire.

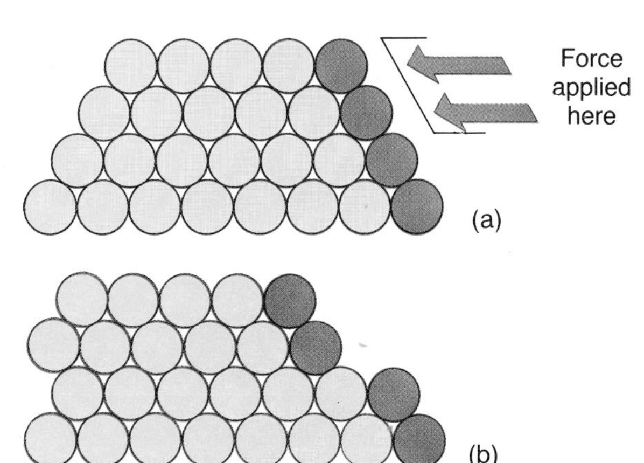

Figure 3
The positions of atoms in a metal crystal, (a) before and (b) after 'slip' has taken place

Questions

1 Explain the words:
slip; malleable; ductile.

2 What are the general properties of metals?

3 Explain why metals (i) have a high density; (ii) have a high melting point; (iii) are good conductors of heat; (iv) are malleable.

4 Answer true or false to parts A to F.
Some reasons for classifying magnesium as a metal are
A it burns to form an oxide.
B it reacts with non-metals.
C it reacts only with non-metals.
D it is magnetic.
E it conducts electricity.
F it has a high density.

6 Alloys

Metallurgists have found ways of making metals harder and stronger. They do this by preventing slip in the metal crystals. If slip cannot occur, then the metal is less malleable and less ductile. Two important methods of strengthening metals are reducing the size of crystal grains and alloying.

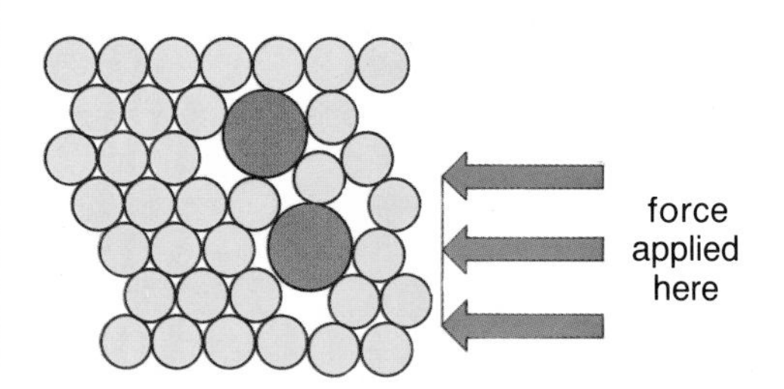

Figure 1
Slip cannot occur in the alloy because the atoms of different size cannot slide over each other

Reducing grain size

Slip does not occur across the grain boundaries in a metal. So, metals with small grains are stronger, harder and less malleable than metals with larger grains. Grains can be made smaller by treating the metal in a particular way. For example, if the molten metal solidifies very rapidly, it will have smaller grains than a sample which solidifies slowly.

Alloying

Metals can be made stronger by adding small amounts of another element. Brass is made by mixing copper and zinc. This alloy is much stronger than pure copper or pure zinc. The different-sized atoms break up the regular packing of metal atoms. This prevents slip and makes the metal tougher and less malleable (figure 1). In steel, the lattice of iron atoms is distorted by adding smaller, carbon atoms. These form crystals of iron carbide which are very hard. The regions of iron carbide in the softer iron make steel very strong (figure 2).

As soon as our ancestors had built furnaces that would melt metals, they began to make alloys. The first alloy to be used was probably bronze, a mixture of copper and tin. Bronze swords, ornaments and coins were being made as early as 1500 BC.

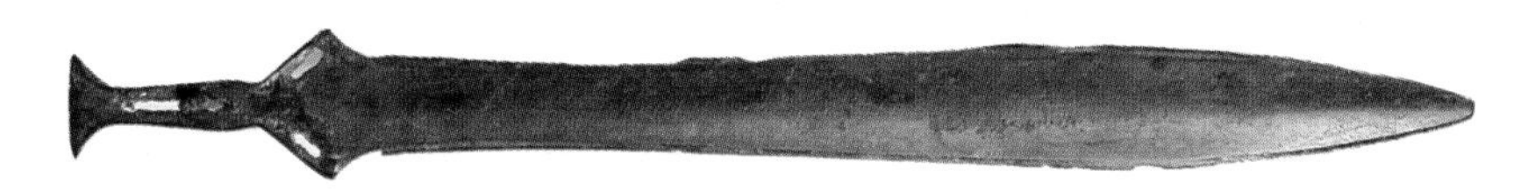

Bronze articles were made as early as 1500 BC

Figure 2
Only 0.2% carbon in iron produces steel strong enough for supports and girders in bridges like the Forth Road Bridge

Nowadays, we depend on alloys. Most of the metallic materials we use are alloys and not pure metals. Almost all the metal parts in a car are alloys. If the bodywork and components were pure iron, they would rust very rapidly compared to present-day steels. They would also be very malleable and would buckle under strain.

Using alloys

It is possible to make alloys with very specific properties. Some are made for hardness, some are resistant to wear and corrosion. Other alloys have special magnetic or electrical properties. Usually alloys are made by mixing the correct amounts of the constituents in the molten state. Solder is made by melting lead, mixing in tin and then casting the alloy into sticks.

The most important alloys are those based on steel. 'Manganese steel', containing 1% carbon and 13% manganese, can be made very tough by heating it to 1000°C and then 'quenching' (cooling quickly) in water. It is used for parts of rock-breaking machinery and railway lines. Stainless steel is another important alloy. Cutlery contains about 20% chromium and 10% nickel. Chromium stops the steel rusting; nickel makes it harder and less brittle.

Steel being quenched

2p coins are 97% copper, 2.5% zinc and 0.5% tin. 50p coins are 75% copper and 25% nickel

During the last 30 years, aluminium alloys have been used more and more. These include *duralumin* which contains 4% copper. Aluminium alloys are light, strong and corrosion resistant, but they may cost six times as much as steel. They are used for aircraft bodywork, window frames and lightweight tubing. The best-known alloys of copper are brass and bronze. Coins are also copper alloys.

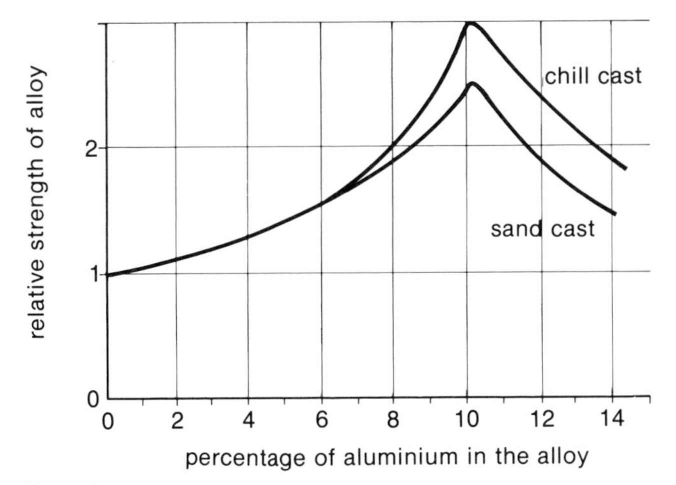

Figure 3
The effect of aluminium on the strength of copper alloys

Questions

1 Name and explain two ways of strengthening metals.

2 Make a table showing the elements present in the following alloys and the main uses of each alloy:
brass; bronze; stainless steel; solder; duralumin.

3 Look at figure 3.
(a) Which process produces the strongest alloy—chill casting (i.e. rapid cooling) or sand casting (i.e. slow cooling) of the liquid alloy?
(b) What percentage of aluminium produces the strongest copper/aluminium alloy?
(c) How many times stronger is this alloy than pure copper?
(d) What percentage of aluminium produces a sand-cast alloy twice as strong as pure copper?
(e) How could you compare the strengths of two wires made of copper/aluminium alloy?

7 Metal Oxides

In section B, unit 10, we studied the reduction of metal oxides with hydrogen. Metals low in the reactivity series, like copper and lead, have a weak attraction for oxygen. So, hydrogen will remove the oxygen from their oxides to leave the metal:

$$\text{lead oxide} + \text{hydrogen} \rightarrow \text{lead} + \text{hydrogen oxide (water)}$$
$$PbO + H_2 \rightarrow Pb + H_2O$$

Metals above lead in the reactivity series, like iron and aluminium, have a strong attraction for oxygen. Hydrogen cannot reduce the oxides of these metals.

In this unit, we will study the reduction of metal oxides using carbon. Put some metal oxide in a hollow on the carbon block (figure 1). Then use a blowpipe to direct a hot flame onto the metal oxide. Watch the metal oxide for any signs of reaction and reduction. Note the colour of the solid before and after heating. Table 1 shows some results.

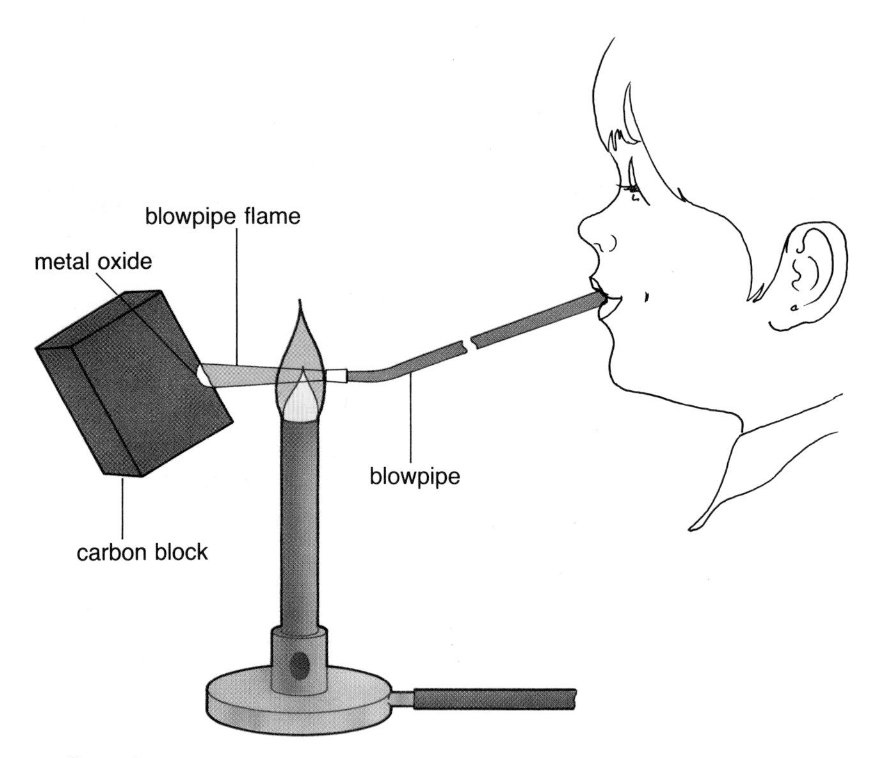

Figure 1

Iron(III) oxide is reduced to iron very slowly by heating with carbon. In fact, the reducing agent is probably carbon monoxide. Hot air from the blowpipe first reacts with carbon to form carbon monoxide:

$$\text{carbon} + \text{oxygen} \rightarrow \text{carbon monoxide}$$
$$2C + O_2 \rightarrow 2CO$$

The carbon monoxide then reacts with iron(III) oxide to form iron:

$$\text{iron(III) oxide} + \text{carbon monoxide} \rightarrow \text{iron} + \text{carbon dioxide}$$
$$Fe_2O_3 + 3CO \rightarrow 2Fe + 3CO_2$$

This reaction is used industrially to manufacture iron from iron ore.

Oxide tested	Colour of oxide before heating	Colour of solid after heating	Evidence for reduction
calcium oxide, CaO	white	white	none
aluminium oxide, Al_2O_3	white	white	none
iron(III) oxide, Fe_2O_3	red-brown	black	Black solid may be iron
lead(II) oxide, PbO	yellow	shiny grey globules	Shiny globules may be lead

Table 1: the reduction of metal oxides using carbon

Tinstone was once mined in Cornwall and reduced to tin in furnaces near the mines. The tin ore is now virtually all used up

Iron ore is impure iron(III) oxide. It is heated in a blast furnace with coke (impure carbon). More details of this process are given in the next unit.

Some metal ores, like tinstone (tin(IV) oxide), can be reduced by coke itself. There is no need to form carbon monoxide first. So, tin(IV) oxide can be reduced more easily than iron(III) oxide, and tin is below iron in the reactivity series.

$$\text{tin(IV) oxide} + \text{carbon} \rightarrow \text{tin} + \text{carbon monoxide}$$
$$SnO_2 + 2C \rightarrow Sn + 2CO$$

Some metals, such as copper and lead, are manufactured from their sulphides. Their sulphide ores, copper glance (Cu_2S) and galena (PbS), are first converted to oxides by heating in air. These oxides are then reduced to the metal using coke or carbon monoxide.

Table 2 summarises the ease of reducing metal oxides with hydrogen, carbon and carbon monoxide.

<table>
<tr><th>Metal oxide</th><th>Ease of reduction by heat alone</th><th>Ease of reduction with hydrogen</th><th>Ease of reduction with C and CO</th></tr>
<tr><td>Na_2O
CaO
MgO
Al_2O_3</td><td rowspan="3">Stable on heating to 1000°C (i.e. not decomposed on heating with a Bunsen)</td><td rowspan="2">Not reduced by H_2</td><td>Not reduced by C or CO</td></tr>
<tr><td>ZnO
Fe_2O_3</td><td rowspan="3">Reduced by C or CO
$MO + CO \rightarrow M + CO_2$</td></tr>
<tr><td>PbO
CuO</td><td rowspan="2">Reduced by hydrogen
$MO + H_2 \rightarrow M + H_2O$</td></tr>
<tr><td>Ag_2O</td><td>Decomposes to silver and oxygen</td></tr>
</table>

Table 2: The ease of reducing metal oxides

Questions

1 Which is the better reducing agent, hydrogen or carbon monoxide? Explain your answer.

2 Complete the following statements. All metal oxides, below and including .... oxide, are reduced by carbon. Thus, carbon has a greater affinity for oxygen than all metals below .... in the reactivity series.

3 Write out a reactivity series including carbon.

4 Write equations for the reactions which occur when (i) silver oxide is heated; (ii) copper oxide reacts with hydrogen; (iii) lead oxide reacts with carbon.

5 Indicate the oxidation process, the reduction process, the oxidizing agent and the reducing agent in the following equation:

$$Fe_2O_3 + 3CO \rightarrow 2Fe + 3CO_2$$

8 Iron for Industry

A blast furnace

Iron is the most important metal. Most of it is made into steel and used for machinery, tools, vehicles and large girders in buildings and bridges. The world production of iron is about 750 million tonnes per year. The main raw material for making iron is iron ore (haematite). This ore contains iron(III) oxide. The largest deposits of iron ore in the UK are in Northamptonshire and Lincolnshire, but most of this is of poor quality. The best quality ores are found in Scandinavia, America, Australia, North Africa and Russia.

Iron ore is converted to iron in a special furnace called a **blast furnace** (figure 1). This furnace is a tapered cylindrical tower about 30 metres tall. A mixture of iron ore, coke and limestone is added at the top of the furnace. At the same time, *blasts* of hot air (which give the furnace its name) are blown in through small holes near the bottom. Oxygen in the blast of air reacts with the coke to form carbon monoxide:

$$\underset{\text{(coke)}}{2C} + O_2 \rightarrow 2CO$$

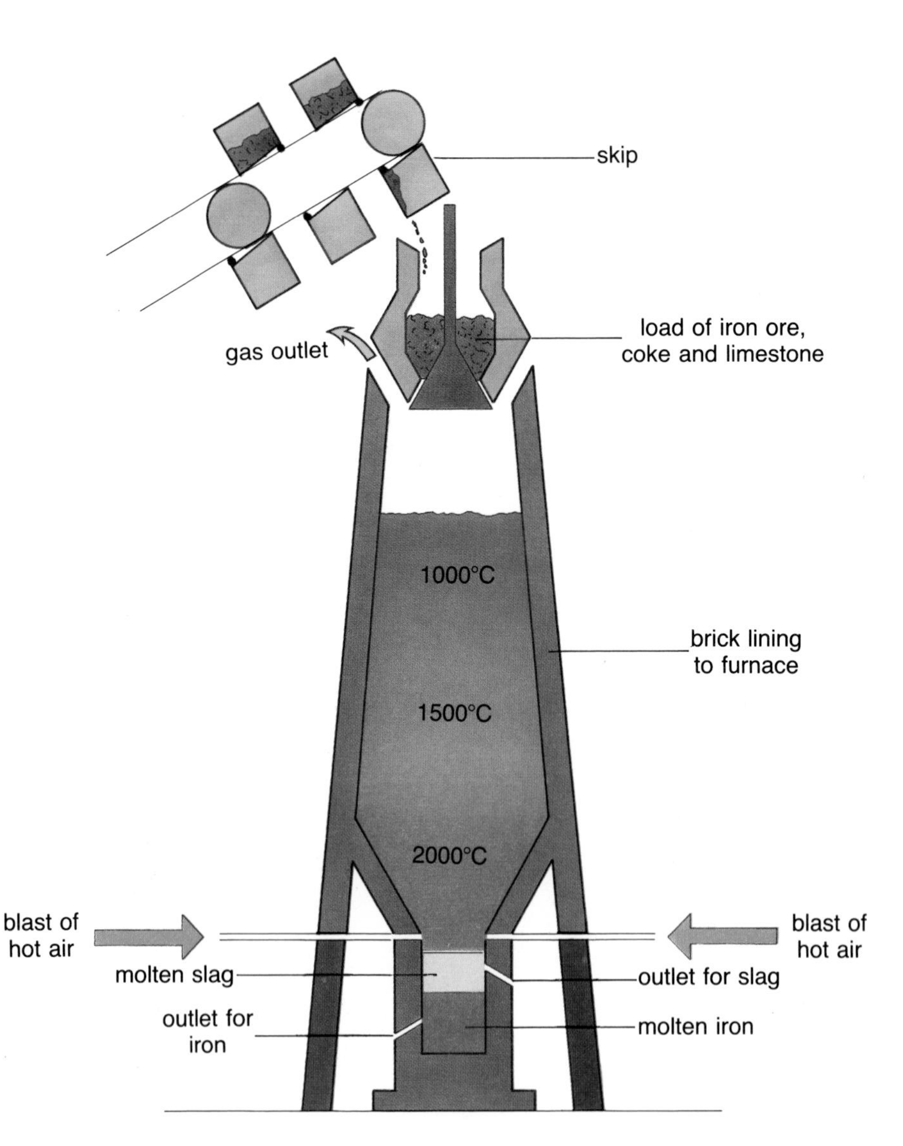

Figure 1
A blast furnace

This reaction gives out heat and the temperature in the furnace gets as high as 2000° C. As the carbon monoxide rises up the furnace, it reduces the iron ore (iron oxide) to iron:

$$Fe_2O_3 + 3CO \rightarrow 2Fe + 3CO_2$$

The high temperatures inside the furnace melt the iron. The molten iron runs to the bottom of the furnace and where it is tapped off from time to time. The molten iron can be used immediately to make steel or it can be poured into moulds to solidify. The large chunks of iron which form are called 'pigs' and the name 'pig-iron' is used for the metal.

Why is limestone used in the furnace?

Iron ore is usually found mixed with impurities like earth and sand (silicon dioxide, SiO_2). Limestone helps to remove these impurities. The limestone decomposes at the high temperatures inside the furnace. It forms calcium oxide and carbon dioxide.

$$CaCO_3 \rightarrow CaO + CO_2$$

The calcium oxide then reacts with sand (SiO_2) and other substances in the impurities to form 'slag':

$$CaO + SiO_2 \rightarrow CaSiO_3$$

The 'slag' falls to the bottom of the furnace and floats on the molten iron. It is tapped off from the furnace at a different level from the molten iron. The 'slag' is used for building materials and cement manufacture.

Blast furnaces work 24 hours a day. Raw materials are continually added at the top of the furnace and hot air is blasted in at the bottom. At the same time, molten slag and molten iron are tapped off from time to time as they collect. This process goes on all the time for about two years. After this time, the furnace has to be closed down so that the lining can be repaired.

Molten iron being tapped from a furnace

Questions

1 Why is the furnace used to make iron called a *blast* furnace?
2 (a) What materials are added to a blast furnace?
(b) What materials come out of a blast furnace?
3 Why is limestone added to a blast furnace?
4 What are the main uses of iron?
5 Summarise the processes used to manufacture iron in a blast furnace. Give equations for the reactions which occur.
6 Why are blast furnaces usually built near coal fields?

9 Extracting Metals

Mercury can be obtained from cinnabar (impure mercury sulphide) by simply heating the ore

Extracting metals from their ores involves 3 stages: mining and concentrating the ore; reducing the ore to the metal; purifying the metal.

Mining and concentrating the ore

Very often, the ore must be separated from soil and other impurities before it is processed.

Reducing the ore to the metal

The method used for a particular metal depends on:
(i) the position of the metal in the reactivity series
(ii) the cost of the process.

- **Heating the ore alone.** This is the cheapest way to extract a metal, but it only works with the compounds of *metals at the bottom of the reactivity series*. For example, mercury is extracted from cinnabar (HgS) by heating it in air:

$$HgS + O_2 \rightarrow Hg + SO_2$$

- **Reduction with carbon and carbon monoxide.** The *metals in the middle of the reactivity series* such as zinc, iron and lead cannot be obtained simply by heating their ores. They are usually obtained by reducing their oxides using carbon (coke) or carbon monoxide. In the last unit we studied the extraction of iron by reduction of iron(III) oxide using carbon monoxide.

Sometimes, these metals exist as carbonate or sulphide ores. These ores must be converted to oxides before reduction. Carbonates can be converted to oxides just by heating them; sulphides by heating in air:

$$ZnCO_3 \xrightarrow{heat} ZnO + CO_2$$

$$2ZnS + 3O_2 \xrightarrow{heat} 2ZnO + 2SO_2$$

- **Electrolysis of molten compounds.** *Metals at the top of the reactivity series*, like sodium, magnesium and aluminium, cannot be obtained by reduction of their oxides with carbon or carbon monoxide. This is because the temperature needed to reduce the oxides is too high. These metals cannot be obtained by electrolysis of their aqueous solutions because hydrogen from the water is produced at the cathode instead of the metal. The only way to extract these reactive metals is by electrolysing their molten (fused) compounds. Potassium, sodium, calcium and magnesium are obtained by electrolysis of their molten chlorides. Aluminium is obtained by electrolysis of molten Al_2O_3 (section D, unit 9). During electrolysis the metal ions are discharged at the cathode, forming the metal:

Cathode (−) $Al^{3+} + 3e^- \rightarrow Al$

Impurity	% in pig-iron	% in mild steel
Carbon	3–5	0.15
Silicon	1–2	0.03
Sulphur	0.05–0.10	0.05
Phosphorus	0.05–1.5	0.05
Manganese	0.5–1.0	0.50

The main impurities in pig-iron and in mild steel

Molten iron being poured into a basic oxygen furnace

Purifying the metal

Reducing copper ores produces copper with 2–3% of impurities. Sheets of this metal are purified by electrolysis with copper sulphate solution (see section D, unit 9). Pig-iron contains about 8% of impurities (see the table). These impurities make pig-iron very hard and brittle compared to iron and steel. In order to make steel from pig-iron, most of the carbon, silicon, sulphur and phosphorus must be removed. In the basic oxygen furnace (figure 1), this is done by blowing oxygen onto the hot, molten pig-iron. The oxygen converts carbon and sulphur to carbon dioxide and sulphur dioxide which escape as gases. Silicon and phosphorus are oxidized to solid oxides (P_2O_5 and SiO_2). These combine with lime in the furnace to form slag:

$$\underset{\text{lime}}{3CaO + P_2O_5} \rightarrow \underset{\text{calcium phosphate}}{Ca_3(PO_4)_2}$$

$$\underset{\text{lime}}{CaO + SiO_2} \rightarrow \underset{\text{calcium silicate}}{CaSiO_3}$$

slag

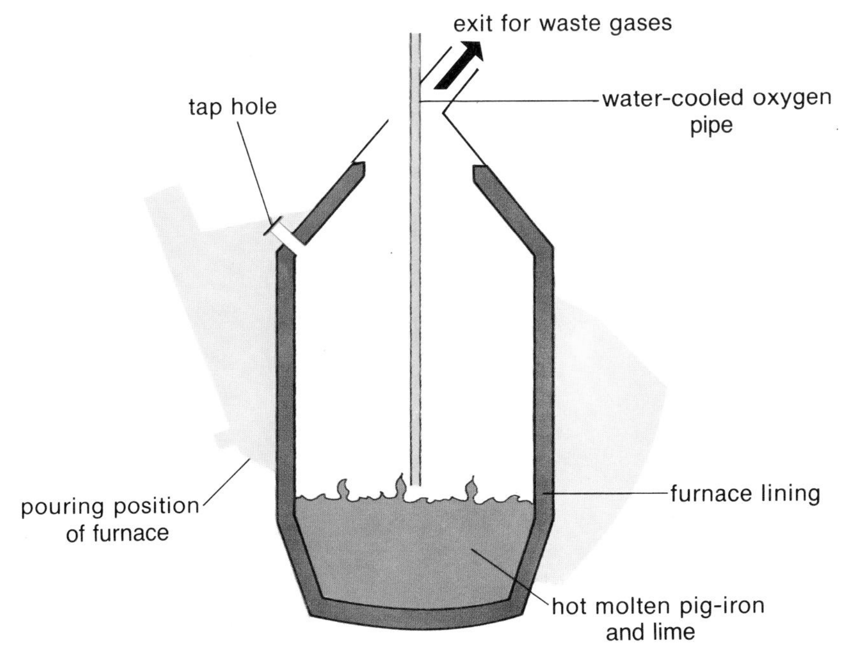

Figure 1
The basic oxygen converter

Questions

1 (a) List the three stages involved in extracting a metal from its ore.
(b) Describe what happens in these three stages when steel is manufactured from iron ore.

2 How does the method used to obtain a metal from its ore depend upon the position of the metal in the reactivity series?

3 Why is iron used in much larger quantities than any other metal?

4 Suggest a method of extracting
(i) silver from silver sulphide (Ag_2S);
(ii) zinc from zinc sulphide (ZnS);
(iii) magnesium from magnesium chloride ($MgCl_2$).

10 Displacing Metals

In the first unit of this section, we studied the reactions of metals with acids. Metals above copper in the reactivity series reacted to form a salt and hydrogen. For example,

$$Zn + H_2SO_4 \rightarrow ZnSO_4 + H_2$$

In this case, zinc has displaced hydrogen from the acid. We can write the equation in terms of ions as

$$Zn(s) + \underbrace{2H^+(aq) + SO_4^{2-}(aq)}_{\text{dilute sulphuric acid}} \rightarrow \underbrace{Zn^{2+}(aq) + SO_4^{2-}(aq)}_{\text{zinc sulphate solution}} + H_2(g)$$

The sulphate ions take no part in this reaction. We can cancel them from both sides of the equation and write

$$Zn(s) + 2H^+(aq) \rightarrow Zn^{2+}(aq) + H_2(g)$$

Now, if zinc can displace hydrogen from $H^+(aq)$ ions in acid, then it may be able to displace metals from solutions of their salts. Figure 1 shows what happens when strips of zinc are placed in solutions of various metal ions. Notice that zinc displaces lead from lead nitrate solution, copper from copper sulphate solution and silver from silver nitrate solution. But, zinc does not displace magnesium from magnesium nitrate solution. The table summarises the results of four other experiments in which magnesium, iron, lead and copper are used in place of zinc.

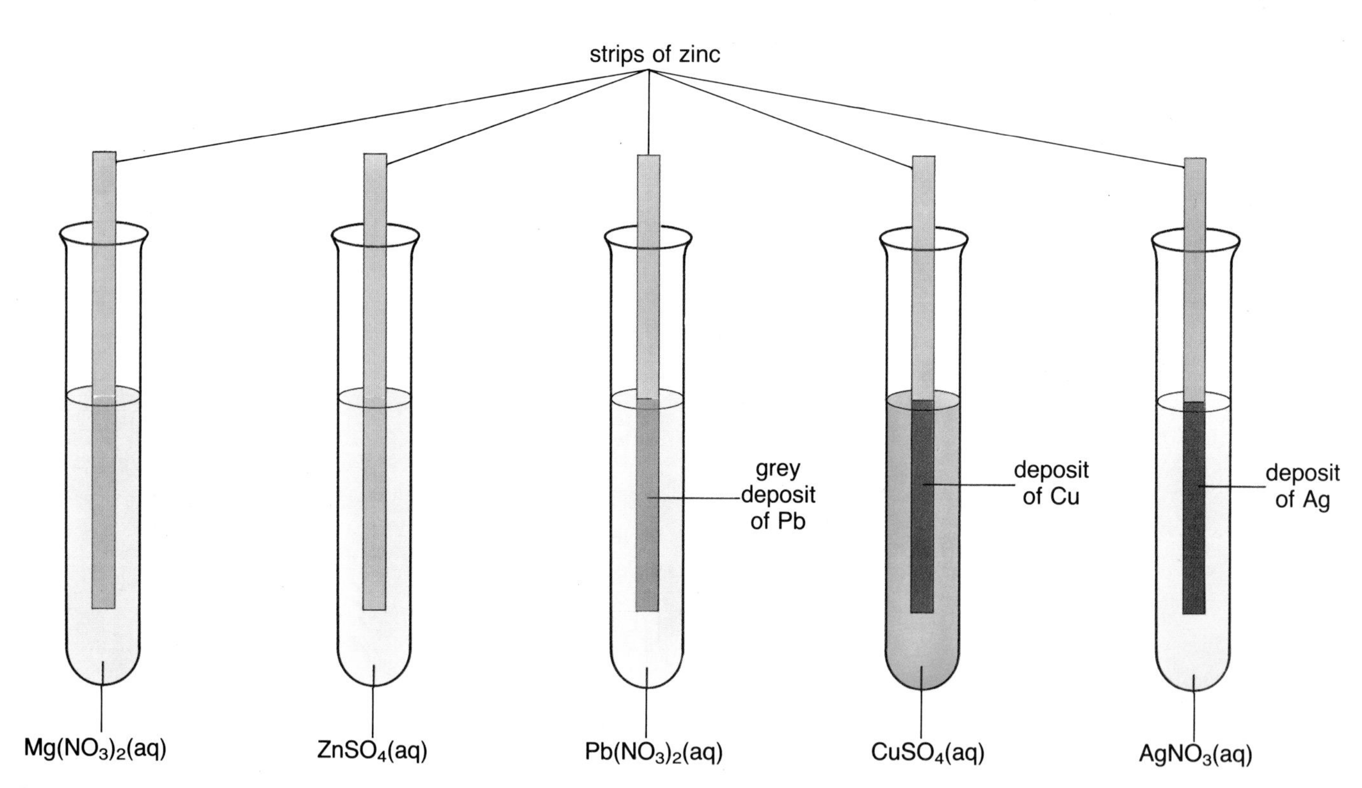

Figure 1

When zinc is placed in copper sulphate solution, the zinc becomes coated with a red brown deposit of copper. At the same time, the blue colour of the solution fades.

Metal used	Results with an aqueous solution of				
	$Mg(NO_3)_2$	$ZnSO_4$	$Pb(NO_3)_2$	$CuSO_4$	$AgNO_3$
Mg		dark grey deposit of zinc	grey deposit of lead	red-brown deposit of copper	black deposit of silver
Zn			grey deposit of lead	red-brown deposit of copper	black deposit of silver
Fe	No apparent reaction		grey deposit of lead	red-brown deposit of copper	black deposit of silver
Pb				red-brown deposit of Cu forms slowly	grey-black deposit of silver
Cu					black deposit of silver

The displacement reactions of some metals

Copper ions have been displaced from the solution as copper atoms. These have been deposited on the zinc and zinc ions from the zinc metal have replaced the copper ions in the solution. The equation for the reaction is

$$\underset{\text{zinc}}{Zn(s)} + \underbrace{Cu^{2+}(aq) + SO_4{}^{2-}(aq)}_{\text{copper sulphate solution}} \rightarrow \underset{\text{zinc ions}}{Zn^{2+}(aq)} + SO_4{}^{2-}(aq) + \underset{\text{copper}}{Cu(s)}$$

Notice two things in the table:

1 The deposits form because the metal added displaces the second metal from a solution of its ions.

2 The metals and their solutions are written in the order of the reactivity series. *One metal only displaces ions of a metal below it in the reactivity series.* So, zinc can displace lead, copper and silver but not magnesium. Lead can displace copper and silver, but not magnesium and zinc.

These experiments show that magnesium is more reactive than the other metals in the table. Magnesium reacts and forms its ions whilst the other metals are displaced from solution.

$$Mg(s) + Pb^{2+}(aq) \rightarrow Mg^{2+}(aq) + Pb(s)$$

Thus, magnesium is higher in the reactivity series than the other metals in the table. In the same way, zinc is more reactive than lead, copper and silver but less reactive than magnesium. Thus, zinc is above lead, copper and silver in the reactivity series but below magnesium. These results provide further evidence for the order of metals in the reactivity series. They show that metals become less reactive down the reactivity series (figure 2).

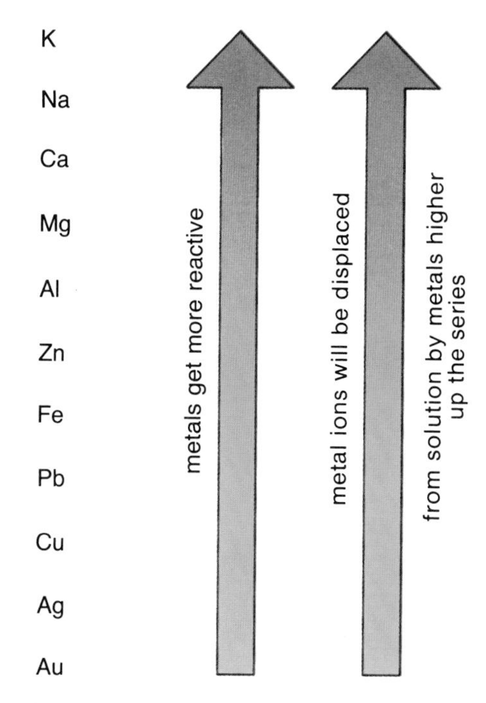

Figure 2

Questions

1 (a) What happens when an iron nail is placed in zinc sulphate solution?
(b) What happens when an iron nail is placed in copper sulphate solution?
(c) Write equations for any reactions in (a) and (b).

2 Iron (steel) is sometimes protected from corrosion by a coating of zinc to form galvanized iron. Use the reactivity series to explain why the zinc protects the iron even when the zinc layer gets scratched and the iron is exposed.

3 Say whether reactions occur in (i) to (v) below. Write equations for any reactions which occur:
(i) *magnesium and aqueous silver nitrate;*
(ii) *silver and aqueous sodium chloride;*
(iii) *copper and aqueous lead nitrate;*
(iv) *aluminium and aqueous iron(II) sulphate;*
(v) *iron and aqueous lead nitrate.*

11 Metal Compounds

Some of the reactions of metal compounds can be linked to the reactivity series like the reactions of metals themselves. We have noticed this already in the reduction of metal oxides with hydrogen and carbon.

Heating metal compounds

The effect of heat on metal compounds can also be related to the reactivity series. The equations below summarise what happens when some copper compounds are heated.

$$Cu(OH)_2(s) \rightarrow CuO(s) + H_2O(g)$$
$$CuCO_3(s) \rightarrow CuO(s) + CO_2(g)$$
$$Cu(NO_3)_2(s) \rightarrow CuO(s) + 2NO_2(g) + \tfrac{1}{2}O_2(g)$$

Notice how similar the equations are. The metal compounds decompose to a solid metal oxide and a gaseous non-metal oxide:

$$\text{metal compound(s)} \xrightarrow{heat} \text{metal oxide(s)} + \text{non-metal oxide(g)}$$

Other common metal compounds—sulphates, sulphides and chlorides, are more stable than hydroxides, carbonates and nitrates, and they do not decompose so readily on heating.

The action of heat on metal hydroxides, carbonates and nitrates is summarised in the table. Notice how the metal compounds fall into three groups.

Metal	Action of heat on hydroxide	Action of heat on carbonate	Action of heat on nitrate
K Na	stable	stable	decompose to nitrite + O_2 $NaNO_3 \rightarrow NaNO_2 + \tfrac{1}{2}O_2$
Ca Mg Al Zn Fe Pb Cu	decompose to oxide + H_2O e.g. $Ca(OH)_2 \rightarrow CaO + H_2O$	decompose to oxide + CO_2 e.g. $CaCO_3 \rightarrow CaO + CO_2$	decompose to oxide + NO_2 + O_2 e.g. $Ca(NO_3)_2 \rightarrow CaO + 2NO_2 + \tfrac{1}{2}O_2$
Ag Au	hydroxides are too unstable to exist	carbonates are too unstable to exist	decompose to metal + NO_2 + O_2 $AgNO_3 \rightarrow Ag + NO_2 + \tfrac{1}{2}O_2$

The action of heat on metal compounds

1 The most reactive metals, potassium and sodium, have the most stable compounds. Only the nitrates of these metals decompose on heating with a Bunsen (which reaches about 900° C).

2 Metals in the middle of the reactivity series from calcium to copper form less-stable compounds. Their hydroxides, carbonates and nitrates decompose to give the metal oxide on heating.

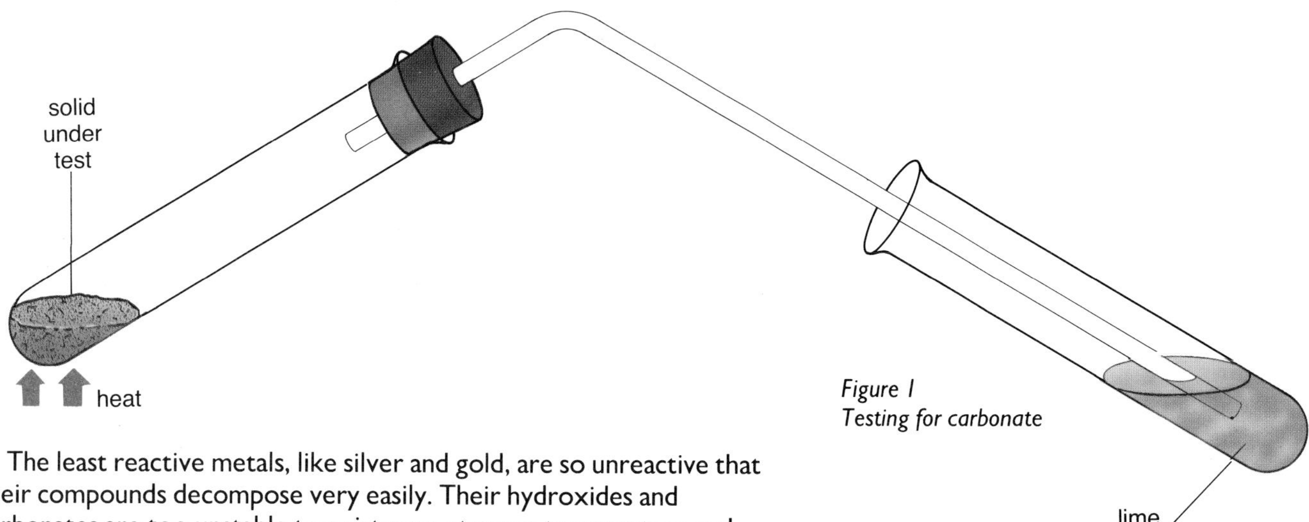

Figure 1
Testing for carbonate

3 The least reactive metals, like silver and gold, are so unreactive that their compounds decompose very easily. Their hydroxides and carbonates are too unstable to exist even at room temperature and their nitrates decompose to the metal on heating.

> *In general, the lower a metal is in the reactivity series, the easier it is to decompose its compounds.*

Identifying anions

The action of heat on metal compounds can be used to identify carbonates (CO_3^{2-}) and nitrates (NO_3^-).

- **Test for carbonate.** Heat the substance and pass any gases produced on heating through lime water (figure 1). If the solid is a carbonate, carbon dioxide which turns lime water milky may be produced.

  In order to make sure that it is a carbonate, add dilute nitric acid to the solid. Carbon dioxide should be produced:

$$CO_3^{2-}(s) + 2H^+(aq) \rightarrow CO_2 + H_2O$$

- **Test for nitrate.** Most nitrates give brown fumes of nitrogen dioxide (NO_2) on heating. This is good evidence for a nitrate. In order to check that it is a nitrate, add fresh iron(II) sulphate solution to an aqueous solution of the substance. Then, **carefully (wearing eye protection)** pour concentrated sulphuric acid down the inside of the test tube (figure 2). A brown ring should form where the acid and the aqueous layer meet.

- **Test for chloride.** Add dilute nitric acid to a solution of the substance. Then add silver nitrate solution. If the substance contains chloride (Cl^-), a white precipitate of silver chloride forms, which turns purple-grey in sunlight:

$$Ag^+(aq) + Cl^-(aq) \rightarrow AgCl(s)$$

- **Test for sulphate.** Add dilute nitric acid to a solution of the substance. Then add barium nitrate solution. If the substance contains sulphate (SO_4^{2-}), a thick white precipitate of barium sulphate ($BaSO_4$) forms:

$$Ba^{2+}(aq) + SO_4^{2-}(aq) \rightarrow BaSO_4(s)$$

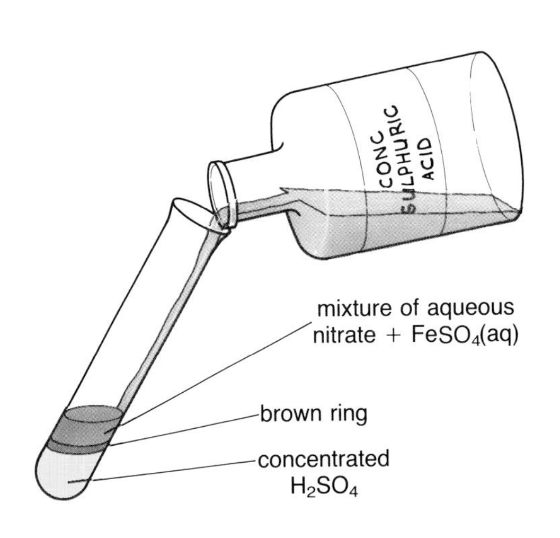

Figure 2
Testing for nitrate

Questions

1 How is the decomposition of metal carbonates related to the reactivity series?
2 Write equations for the action of heat on the following compounds. If decomposition does not occur, say so: $Zn(OH)_2$; KOH; Ag_2O; Rb_2CO_3; $PbCO_3$; $Mg(NO_3)_2$.
3 What are the tests for (i) carbonate; (ii) nitrate; (iii) chloride; (iv) sulphate?
4 Two fertilizers are labelled 'sulphate of potash' and 'nitrate of potash'. How would you check which was which?

12 Identifying Cations

We can identify different substances using tests for the atoms and ions in them. The best way to identify a substance is to find a property or a reaction which is shown by that substance only and is easy to see.

Two of the best tests for cations use flame tests and sodium hydroxide solution.

Flame tests

Cation present in substance	Flame colour
K^+	lilac
Na^+	yellow
Ca^{2+}	red
Cu^{2+}	blue/green

Table 1: the flame colours from some cations

When substances are heated strongly, the electrons in their particles absorb extra energy. We say that the electrons are **excited**. Very soon they give out this excess energy as light and they become stable again. Different cations are found to emit different colours of light (table 1). See if you can get a good flame test for a potassium or a copper compound using the following method. **(Take care: wear eye protection.)**

Dip a nichrome wire in concentrated hydrochloric acid and then heat it in a roaring Bunsen until it gives no colour to the flame. The wire is now clean. Dip it in concentrated hydrochloric acid again, and then in the substance to be tested. Heat the wire in the Bunsen and note the flame colour.

Tests using sodium hydroxide solution

The hydroxides of all metals (except those in Group I) are insoluble. So, these insoluble hydroxides form as solids (precipitates) when aqueous sodium hydroxide, containing hydroxide ions (OH^-), is added to a solution of metal ions. For example:

$$Cu^{2+}(aq) + 2OH^-(aq) \rightarrow \underset{\text{blue precipitate}}{Cu(OH)_2(s)}$$

Cation in solution	3 drops of NaOH(aq) added to 3 cm^3 of solution of cation	10 cm^3 of NaOH(aq) added to 3 cm^3 of solution of cation
K^+	No precipitate	No precipitate
Na^+	No precipitate	No precipitate
Ca^{2+}	A white precipitate of $Ca(OH)_2$ forms $Ca^{2+} + 2OH^- \rightarrow Ca(OH)_2$	White precipitate remains
Mg^{2+}	A white precipitate of $Mg(OH)_2$ forms $Mg^{2+} + 2OH^- \rightarrow Mg(OH)_2$	White precipitate remains
Al^{3+}	A white precipitate of $Al(OH)_3$ forms $Al^{3+} + 3OH^- \rightarrow Al(OH)_3$	White precipitate dissolves to give a colourless solution
Zn^{2+}	A white precipitate of $Zn(OH)_2$ forms $Zn^{2+} + 2OH^- \rightarrow Zn(OH)_2$	White precipitate dissolves to give a colourless solution
Fe^{2+}	A green precipitate of $Fe(OH)_2$ forms $Fe^{2+} + 2OH^- \rightarrow Fe(OH)_2$	Green precipitate remains
Fe^{3+}	A brown precipitate of $Fe(OH)_3$ forms $Fe^{3+} + 3OH^- \rightarrow Fe(OH)_3$	Brown precipitate remains
Cu^{2+}	A blue precipitate of $Cu(OH)_2$ forms $Cu^{2+} + 2OH^- \rightarrow Cu(OH)_2$	Blue precipitate remains

Table 2: testing for cations with sodium hydroxide solution

Carrying out a flame test for potassium

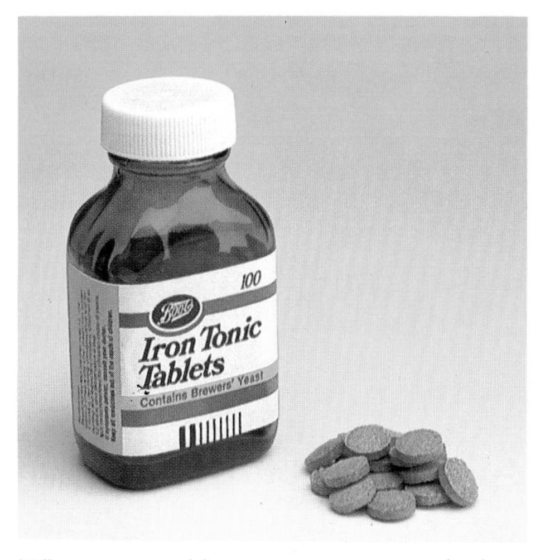

What tests would you carry out to see whether iron tablets contain Fe^{2+} or Fe^{3+} ions?

Table 2 shows what happens when
(i) a little sodium hydroxide solution, NaOH (aq), and then
(ii) excess sodium hydroxide solution is added to solutions of some common cations.

Look carefully at table 2.

1 Which cations give no precipitate with sodium hydroxide solution? Could you tell the difference between these cations using flame tests?
2 Which cations give a white precipitate which does *not* dissolve in excess sodium hydroxide solution? Could you tell the difference between these cations using flame tests?
3 Which cations give a coloured (non-white) precipitate with sodium hydroxide solution? Could you tell the difference between these cations from the colour of their hydroxides?
4 Which cations give a white precipitate which dissolves in excess sodium hydroxide solution? These metal ions can be identified by heating their hydroxides to dryness:

$$2Al(OH)_3 \xrightarrow{heat} \underset{\text{white}}{Al_2O_3} + 3H_2O$$

$$Zn(OH)_2 \xrightarrow{heat} \underset{\text{yellow when hot}}{ZnO} + H_2O$$

Notice how several cations can be identified from the colour and solubility of their hydroxides.

Questions

1 (a) How would you carry out a flame test on a sample of chalk?
(b) What colour should the flame test give?
(c) Why must eye protection be worn during flame testing?
(d) What causes flame colours from certain cations?

2 A garden pesticide is thought to contain copper sulphate. Describe two tests that you would do to find out whether it contains Cu^{2+} ions.

3 A metal has become coated with a layer of oxide. How would you check whether the metal is aluminium or zinc?

Section F: Study Questions

1 (a) From the reactivity series choose
(i) an element that occurs uncombined in the Earth's crust (1)
(ii) an element that reacts vigorously with cold water (1)
(iii) an element that is used in flares. (1)
(b) Tin (Sn) has the same valency as lead and is slightly more reactive. Predict what *you would see* when tin foil is placed in dilute hydrochloric acid. (2)
(c) A gas may be made by passing steam over heated zinc using the apparatus shown below:

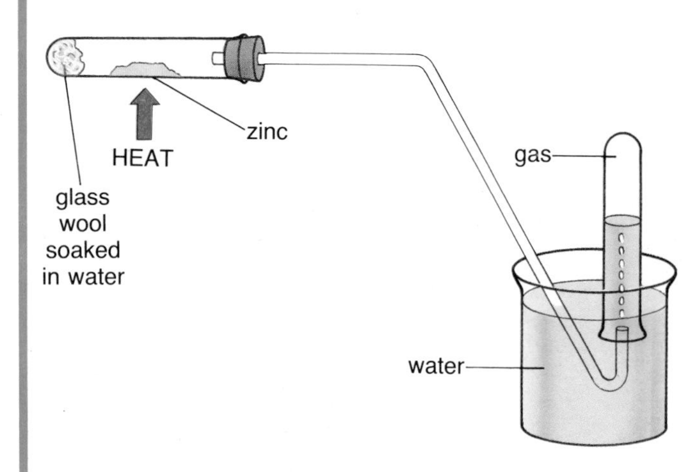

(i) Name the gas collected. (1)
(ii) Write an equation for the reaction. (1)
(iii) Why should you wear safety goggles when carrying out this experiment? (2)
(iv) Which substance has acted as a reducing agent in this reaction? (1)
(v) Name one metal that you would not react with steam in this way. Explain your choice. (2)
(d) Chromium (Cr) metal may be prepared by heating chromium(III) oxide with aluminium powder.
(i) What does this tell you about the relative positions of chromium and aluminium in the reactivity series? (1)
(ii) Write an equation for the reaction. (2)
SEG

2 Complete the following table. (Sodium has been done as an example.)

Metal	Symbol	Name of an ore of the metal
Sodium	Na	Rocksalt
Aluminium		
	Fe	

[4]
SEG

3 Below is a diagram of a blast furnace in which iron is extracted from iron ore.

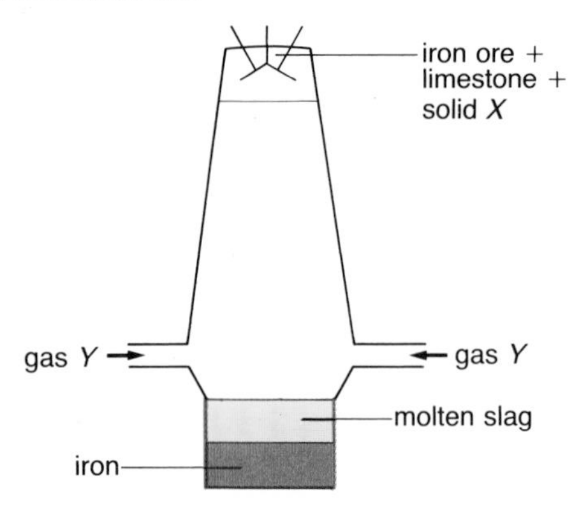

(a) Give the name of
(i) solid *X*
(ii) gas *Y*. (2)
(b) Why does the molten slag *float on top of the molten iron?* (1)
(c) The inside of the furnace is hot. What chemical reaction produces this heat? (1)
(d) Most iron is made into steel. What is the difference between iron and steel? (1)
(e) Name *one* metal other than iron used in the alloy stainless steel. (1)
(f) (i) What is the advantage of using stainless steel instead of iron? (1)
(ii) Name *one* object made of stainless steel. (1)
(g) Oxygen is used to convert iron into steel. What is the oxygen used for? (1)
(h) (i) From what raw material is the oxygen obtained? (1)
(ii) What method is used to obtain the oxygen in h(i)? (1)
(iii) Name *one* alloy containing *no* iron. (1)
Total [12] **LEAG**

4 The labels have become unreadable on three bottles that were known to contain
ammonium chloride
calcium carbonate
iron(II) sulphate.
What *chemical* tests would you carry out to enable you to re-label the bottles correctly? Your answers must include at least *one positive, chemical* test for each substance that would not be given by either of the other two. [2]
LEAG

5 (a) Name *one* natural form of calcium carbonate. (1)
(b) Describe tests you might use to prove that a piece of rock contained
(i) carbonate ions (3)
(ii) calcium ions. (3)
(c) Strontium (Sr) is in the same group of the periodic table as calcium. Write the formula of strontium nitrate. (1)
SEG

SECTION G

Acids, Bases and Salts

Oranges and other citrus fruits contain citric acid. This gives them a sharp taste

1 Introducing Acids

The sharp taste of lemon juice and vinegar is caused by acids. It is easy to identify some acids by their sour taste. But it would be very foolish to rely on taste for all acids. Think of the dangers of tasting concentrated sulphuric acid or nitric acid.

A more sensible test for acids is to use indicators like litmus and universal indicator. Acidic substances dissolve in water to produce solutions which turn litmus red and give a yellow, orange or red colour with universal indicator (figure 1). Acidic solutions have a pH below 7. Acids are important in everyday life and in the chemical industry. They are used to make our clothes, our food and the medicines that protect us from infection and disease.

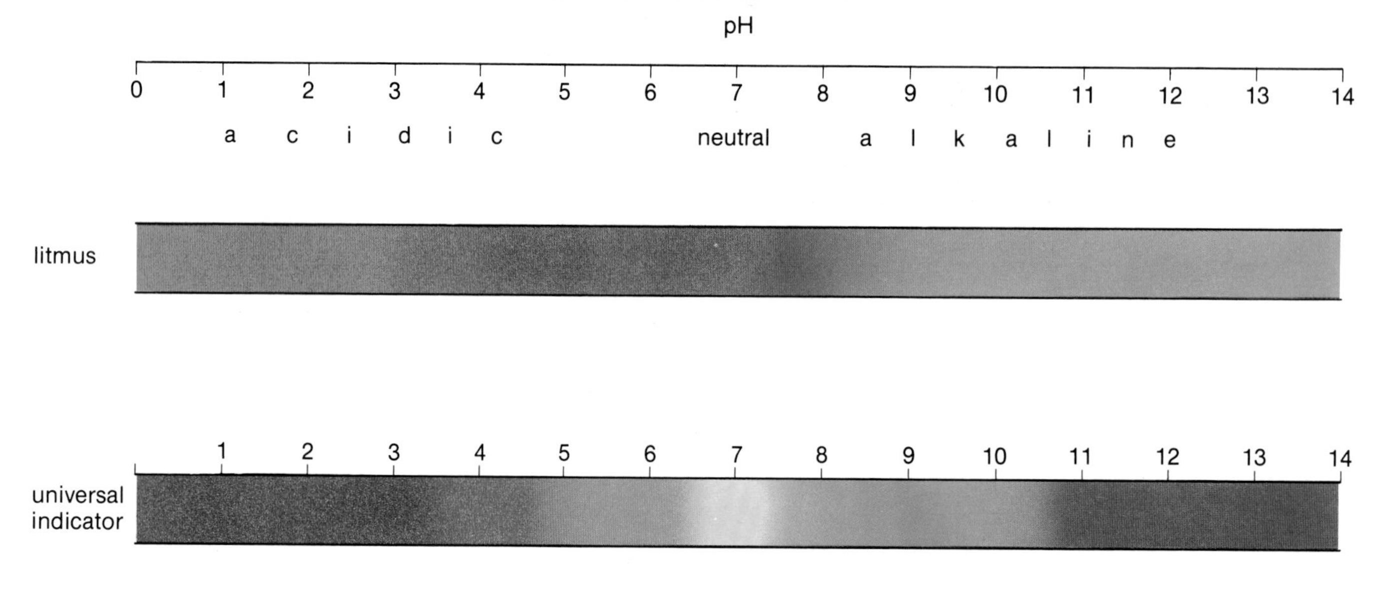

Figure 1
The colours of litmus and universal indicator with solutions of different pH

What protective clothing is this technician wearing to protect herself from the acid she is pipetting?

Making acids

- **From non-metal oxides and water.** When non-metal oxides are added to water, acids are produced. For example:

$$\text{carbon dioxide} + \text{water} \rightarrow \text{carbonic acid}$$
$$CO_2 + H_2O \rightarrow H_2CO_3$$

$$\text{sulphur dioxide} + \text{water} \rightarrow \text{sulphurous acid}$$
$$SO_2 + H_2O \rightarrow H_2SO_3$$

Sulphuric acid is manufactured via sulphur dioxide. The raw materials are sulphur, air (oxygen) and water. Most of the sulphur is shipped to the UK from North America.

There are three stages in the manufacture of sulphuric acid.

1 First, sulphur dioxide is made by burning sulphur in air (oxygen).

$$S + O_2 \rightarrow SO_2$$

2 The sulphur dioxide is then converted to sulphur trioxide. This is called **the contact process**. The sulphur dioxide is mixed with air (oxygen) and passed over a catalyst (vanadium(V) oxide) at 450°C.

$$\text{sulphur dioxide} + \text{oxygen} \rightarrow \text{sulphur trioxide}$$
$$2SO_2 + O_2 \rightarrow 2SO_3$$

A photograph of a sulphuric acid plant

3 Finally, the sulphur trioxide is converted to sulphuric acid. Sulphur trioxide does not dissolve very easily in water. So, sulphur trioxide is dissolved in 98% sulphuric acid. At the same time, water is added to keep up the acid concentration.

$$SO_3 + H_2O \rightarrow H_2SO_4$$

- **By combining hydrogen with reactive elements.** The most important acid to be manufactured in this way is hydrochloric acid. Hydrogen chloride is first made by burning hydrogen in chlorine.

$$H_2 + Cl_2 \rightarrow 2HCl$$

The hydrogen chloride is then passed up a tower packed with stones (or some other unreactive substance) over which water trickles. The gas dissolves in the water to form hydrochloric acid. At one time, hydrochloric acid was made from salt, and sulphuric acid was made from iron (II) sulphate. Because of this, they were called **mineral acids**. Nitric acid is also a mineral acid. It was once made from potassium nitrate (salt petre).

Sulphur being mined in North America

Questions

1 How would you show that a substance was an acid?
2 Why are acids important?
3 What two general methods are used to make acids?
4 This question is about the manufacture of sulphuric acid.
(a) Write equations for the three stages in the manufacture of sulphuric acid.
(b) What is the contact process?
(c) Suggest two reasons why a catalyst is used in the contact process.
(d) 450°C is an optimum temperature for the contact process. What do you think the term 'optimum temperature' means?
(e) Sulphur trioxide from the contact process is not reacted directly with water to make sulphuric acid. Explain why.
(f) Most of the sulphuric acid plants in the UK are situated near large ports. Suggest reasons for this.

2 Acids in Everyday Life

Acid rain can have severe effects on trees

Acids in the air—acid rain

When fuels burn, sulphur dioxide and carbon dioxide are produced. Because of this, city air may contain five times as much sulphur dioxide as air in the country. Sulphur dioxide and carbon dioxide react with water vapour and rain in the air to form sulphurous acid and carbonic acid.

$$SO_2 + H_2O \rightarrow H_2SO_3 \text{ (sulphurous acid)}$$

$$CO_2 + H_2O \rightarrow H_2CO_3 \text{ (carbonic acid)}$$

These acids make the rain acidic and this has led to the term **'acid rain'**. (See section B, unit 4.)

Acid rain in some areas can cause severe damage to stone work such as this statue

Acids in the body

The stomach wall produces hydrochloric acid, causing a pH of about 2 in the stomach. These acidic conditions in the stomach help the breakdown of foods, particularly proteins and carbohydrates. Proteins are broken down into smaller molecules called peptides and amino acids. Carbohydrates, like starch, are broken down into smaller molecules, like glucose. These small molecules can pass through the lining of the small intestine into the bloodstream. The blood carries these small molecules round the body to where they are needed.

The pH of the blood is 7.4 (slightly alkaline). Most reactions in our bodies can take place only within a narrow range of pH. Changes in the pH in your stomach might give you indigestion but a pH change of just 0.5 in the blood would probably kill you. In order to prevent changes in pH, our bodies contain substances that counteract the effects of acids and alkalis.

Acids in the home

Many foods and drinks contain acids. Citrus fruits—oranges, lemons, pineapples, grapefruit—contain citric acid. Tomato sauce, brown sauce and mint sauce get their sharp taste from vinegar which contains acetic acid.

One of the simplest and cheapest drinks is soda water. This is made by dissolving carbon dioxide in water under pressure. Some of the carbon dioxide reacts with the water to form carbonic acid. When the soda water is poured from the bottle it fizzes because the pressure is lower and carbon dioxide forms as bubbles in the drink. Other fizzy drinks, like Coke and lemonade, also contain carbonic acid.

Sauces usually contain acetic acid from vinegar

Acids in the soil

The pH of soil can vary from about 4 to 8, but most soils have a pH of between 6.5 and 7.5. In chalk or limestone areas, the soil is usually alkaline, but in moorland, sandstone and forested areas it is generally acidic. Peat bogs and clay soils are normally acidic. For general gardening and farming purposes, the best results are obtained from a neutral or slightly acidic soil of pH 6.5 to 7.0. Only a few plants, including rhododendrons and azaleas, can grow well in soils which are acidic. In areas where the soil is too acidic, it can be improved by treatment with powdered limestone (calcium carbonate) or slaked lime (calcium hydroxide) which react with the acids in it and raise the pH to the right value. Substances like these which neutralize acids are called **bases** (see unit 7 of this section).

Liming the soil

Rhododendrons grow well in acid soil, unlike most plants

Questions

1 How would you check the pH of soil samples?
2 Why do you think that forested and peaty areas usually have acid soils?
3 Why is rain water acidic?
4 What is the advantage of acidic conditions in the stomach?
5 Make a list of a variety of foodstuffs which are acidic. Try to find out what acids they contain.
6 What sort of substances are used to cure indigestion?
7 Why are some mineral waters both fizzy and acidic?

3 Properties of Acids

Car batteries contain sulphuric acid

We know a good deal about acids already.

1 Acids are soluble in water. They produce solutions with a pH below 7.

2 Acids give characteristic colours with indicators. They give a red colour with litmus and an orange or red colour with universal indicator.

3 Acids react with metals above copper in the reactivity series to form a salt and hydrogen.

metal + acid → salt + hydrogen

$$\text{e.g. } Mg + H_2SO_4 \rightarrow MgSO_4 + H_2$$

Nearly all salts contain a metal and at least one non-metal. They are ionic compounds with a cation (e.g. Cu^{2+}, Mg^{2+}) and an anion (Cl^-, SO_4^{2-}). Sodium chloride (Na^+Cl^-) is known as common salt.

4 Acids react with metal oxides to form a salt and water.

metal oxide + acid → salt + water

For example, when black copper oxide is added to warm dilute sulphuric acid, the black solid disappears and a blue solution of copper sulphate forms.

$$CuO + H_2SO_4 \rightarrow CuSO_4 + H_2O$$

5 Acids react with carbonates to give a salt plus carbon dioxide and water.

carbonate + acid → salt + $CO_2 + H_2O$

For example, when sodium carbonate is added to hydrochloric acid, sodium chloride forms and bubbles of carbon dioxide are produced.

$$Na_2CO_3 + 2HCl \rightarrow 2NaCl + CO_2 + H_2O$$

What citrus fruits does this photograph show? What acid is present in citrus fruit?

6 Acid solutions conduct electricity and are decomposed by it. This shows that solutions of acids consist of ions. All acids produce hydrogen at the cathode during electrolysis. This suggests that all acids contain H^+ ions (table 1).

Name of acid	Formula	Ions produced from acid
Acetic acid	CH_3COOH	$H^+ + CH_3COO^-$
Carbonic acid	H_2CO_3	$2H^+ + CO_3^{2-}$
Hydrochloric acid	HCl	$H^+ + Cl^-$
Nitric acid	HNO_3	$H^+ + NO_3^-$
Sulphuric acid	H_2SO_4	$2H^+ + SO_4^{2-}$
Sulphurous acid	H_2SO_3	$2H^+ + SO_3^{2-}$

Table 1: some acids and the ions which they form

The part of water in acidity

All these properties of acids apply to solutions of acids in water. But what happens when water is not present?

Dry hydrogen chloride (HCl) and pure *dry* acetic acid have no effect on *dry* litmus paper or magnesium. **So, substances which we call acids do not behave as acids in the absence of water.** When a little water is added, they show acidic properties straight away. Blue litmus paper turns red and bubbles of hydrogen are produced with magnesium.

To study the part that water plays, we can compare a solution of hydrogen chloride in water (hydrochloric acid) with a solution of hydrogen chloride in toluene (a solvent with properties like petrol). Table 2 shows what happens. Look carefully at table 2. You will see that hydrochloric acid in toluene has no acidic properties. It does not conduct electricity and therefore it contains no ions.

Test	**Solution of HCl in water**	**Solution of HCl in toluene**
Colour of *dry* blue litmus paper in solution	Red	Blue
Add *dry* magnesium ribbon to solution	Bubbles of hydrogen are produced	No bubbles form
Add *dry* anhydrous sodium carbonate to the solution	Bubbles of carbon dioxide are produced	No bubbles form
Does the solution conduct electricity?	Yes	No

Table 2: comparing solutions of HCl in water and HCl in toluene

When hydrochloric acid in water is electrolysed, it produces hydrogen at the cathode and chlorine at the anode. This shows that it contains H^+ ions and Cl^- ions. Chemists now know that it is the H^+ ions which cause the typical properties of acids with indicators, metals, metal oxides and carbonates.

Because of this,

> *acids are defined as substances which can donate (give) H^+ ions.*

Thus, hydrogen chloride splits up into aqueous H^+ and Cl^- ions when it dissolves in water.

$$HCl(g) \xrightarrow{water} \underbrace{H^+(aq) + Cl^-(aq)}_{\text{hydrochloric acid}}$$

So, we see that

> *water reacts with substances to produce the $H^+(aq)$ ions which cause acid properties.*

Questions

1 Make a table to summarise the properties and chemical reactions of acids.

2 (a) What part does water play in acidity?
(b) Why does dry HCl gas not act like an acid?
(c) What is an acid?

3 When a thermometer bulb dipped in water is placed in a gas jar of dry hydrogen chloride, the temperature rises by 10°C. When a thermometer bulb dipped in toluene is placed in dry hydrochloric acid, there is no temperature rise. Explain these differences.

4 Write word equations and balanced equations for the following reactions:
sodium oxide + sulphuric acid
zinc oxide + hydrochloric acid
copper oxide + nitric acid
iron + sulphuric acid
aluminium + hydrochloric acid
copper carbonate + hydrochloric acid

5 Why can vinegar be used to descale kettles?

Concentrated and dilute acids

Concentrated acids contain a lot of acid in a small amount of water. (Concentrated sulphuric acid has 98% sulphuric acid and only 2% water.) Dilute acids have a small amount of acid in a lot of water. Concentration tells us how much substance is dissolved in a certain volume of solution. It is usually given as the number of grams of solute per dm^3, or moles of solute per dm^3. Solutions with a concentration of 1.0 mole per dm^3 are often described as 1.0 M (one molar). Concentrated acids may contain 12 moles per dm^3 (12.0 M, twelve molar). Dilute acids usually have concentrations of 2.0 M or less.

Strong and weak acids

When different acids with the *same* concentration in moles per dm^3 are added to the *same* indicator, they sometimes give different pHs. Figure 1 shows the results when acids of the same concentration were tested with universal indicator. The different pH values mean that some acids produce H^+ ions more readily than others. The results in figure 1 show that nitric acid, hydrochloric acid and sulphuric acid produce more H^+ ions than the other acids. We say that they **dissociate** (split up) into ions more easily. We call them **strong acids** and the others **weak acids**.

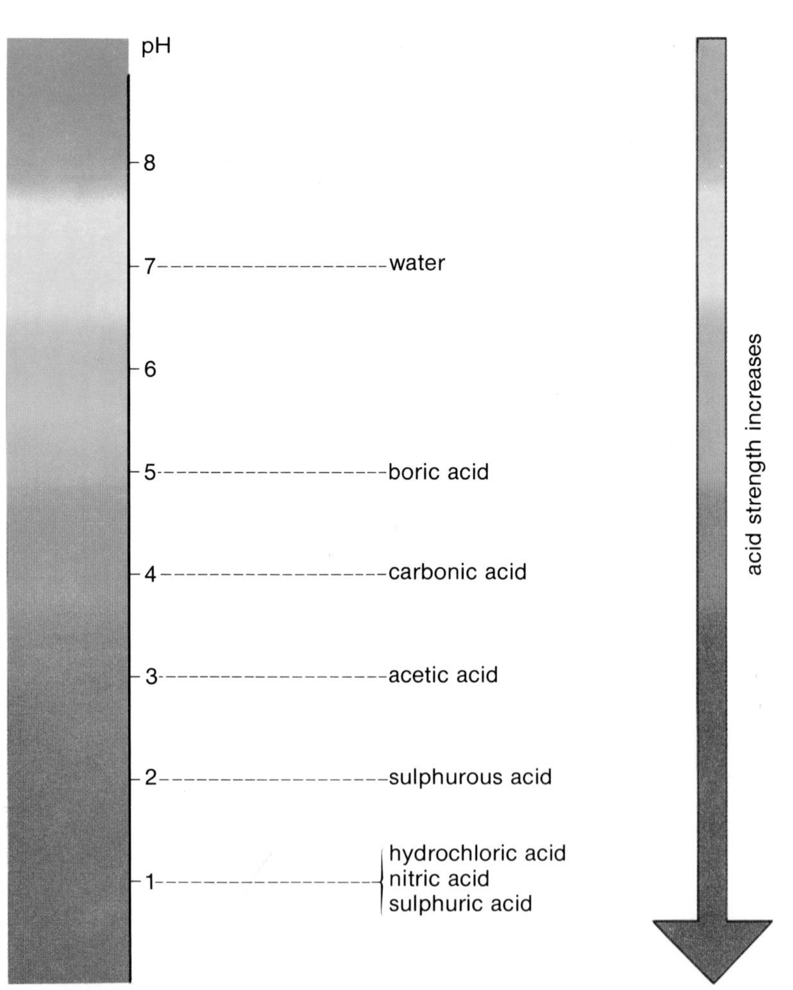

Figure 1 (opposite)
The pH of solutions of various acids (All solutions have a concentration of 0.1 mole per dm^3)

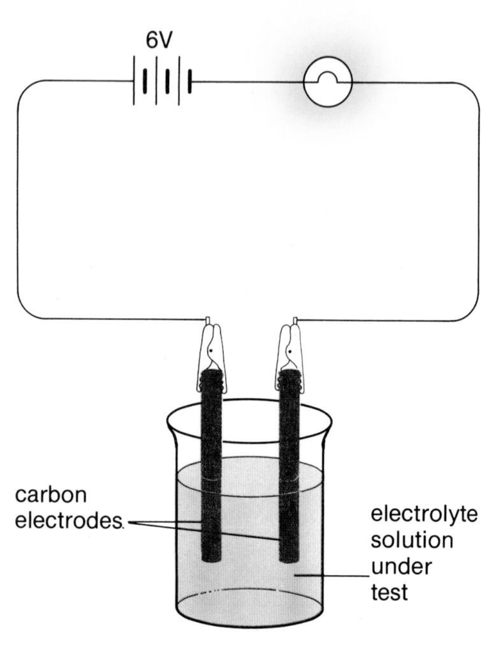

Figure 2
Testing the strengths of different electrolytes

> *In aqueous solution:*
> - *strong acids are completely dissociated into ions,*
> - *weak acids are only partly dissociated into ions.*

We can show the difference between strong and weak acids in the way we write equations for their dissociation.

$$HCl\,(aq) \rightarrow H^+(aq) + Cl^-\,(aq)$$

$$CH_3COOH\,(aq) \rightleftharpoons H^+(aq) + CH_3COO^-\,(aq)$$

(acetic acid)

A single arrow, as for hydrochloric acid, shows that *all* the HCl has formed H^+ and Cl^- ions in aqueous solution. The arrows in opposite directions for acetic acid show that some of the CH_3COOH molecules have formed H^+ and CH_3COO^- ions. Most of the acetic acid remains *undissociated* as CH_3COOH molecules dissolved in water.

Notice the difference between the terms 'concentration' and 'strength'. **Concentration** tells us how much solute is dissolved in the solution, and we use the words 'concentrated' and 'dilute'. **Strength** tells us how much of the acid is dissociated into ions, and we use the words 'strong' and 'weak'.

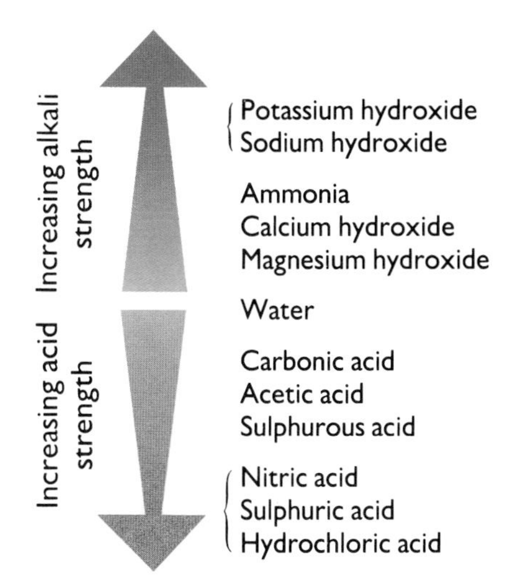

Table 2: relative strengths of acids and alkalis

Strong and weak electrolytes

Compounds, like acids, which conduct electricity are called electrolytes (section D, unit 3). Strong electrolytes dissociate into ions more readily than weak electrolytes. Thus, strong electrolytes conduct electricity better than weak electrolytes. Figure 2 shows how we can test this electrical conductivity. Some results are given in table 1. These results show that acids and alkalis (unit 7 of this section) are not all equal in strength.

Acids and alkalis can be arranged in order of their relative strengths. The easiest way to do this is to use universal indicator to find the pH of their solutions. The lower the pH of an acid, the stronger it is. The higher the pH of an alkali, the stronger it is. Table 2 shows the relative strengths of some acids and alkalis. Pure water, which is neutral, divides the acids from the alkalis.

Aqueous solution	Type of compound	Bulb brightness	Strong or weak electrolyte
$CuSO_4$	Salt	Very bright	Strong
NaCl	Salt	Very bright	Strong
HCl	Acid	Very bright	Strong
CH_3COOH	Acid	Dim	Weak
H_2SO_4			
NaOH	Alkali	Very bright	Strong
KOH			
NH_3	Alkali	Dim	Weak

Table 1: the electrical conductivity of some compounds

Questions

1 Explain the following terms: *dissociate; strong acid; weak acid; strong electrolyte.*

2 Explain the terms *concentration* and *strength* as applied to acids.

3 Classify vinegar and battery acid (approximately 4 M sulphuric acid) as (i) concentrated or dilute; (ii) strong or weak.

4 Look at figure 2 and the results in table 1.

(a) Why do strong electrolytes give a brighter bulb than weak electrolytes?

(b) Why should the solutions tested have the same concentration?

(c) Suggest results in table 1 for sulphuric acid and potassium hydroxide.

(d) Are salts strong or weak electrolytes?

5 Titrating Acid and Base

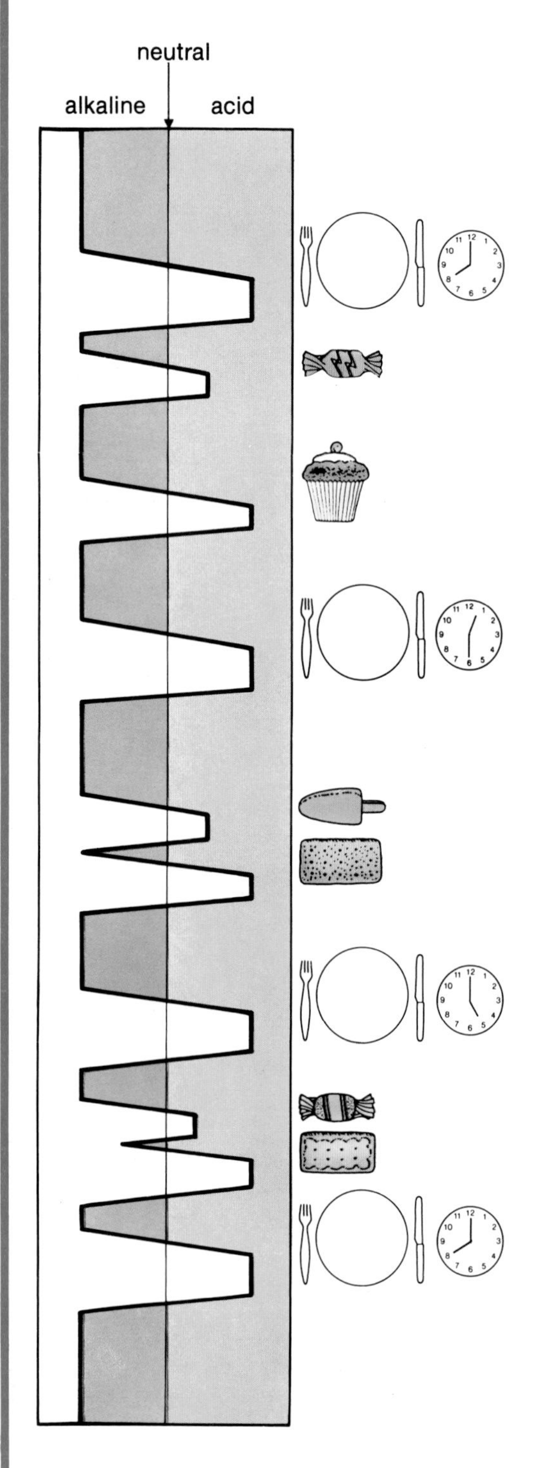

Figure 1
This shows how the pH in your mouth changes during the day. Notice how the pH becomes acidic during and just after meals when sugar and other chemicals in the food are being broken down into acids. These acids cause tooth decay

Certain substances can neutralize acids. Farmers use lime to 'cure' acid soils. We use toothpaste to neutralize the acids produced as food breaks down (figure 1). These substances which neutralize acids are called bases. When hydrogen chloride dissolves in water, it reacts to form H^+ (aq) ions.

$$HCl(g) \xrightarrow{water} \underset{\text{hydrochloric acid}}{H^+(aq) + Cl^-(aq)}$$

HCl(aq) acts as an acid because it produces H^+ ions, like all acids. Bases react with these H^+ ions and neutralize the acidity. Bases (unit 7 of this section) are therefore the chemical opposites of acids. **Acids give up H^+ ions whereas bases take H^+ ions.**

How much acid and base react?

When acids are neutralized by bases, the pH of the solution changes. We can use this pH change to find how much base (e.g. sodium hydroxide solution) reacts with the acid (e.g. hydrochloric acid).

Measure 25 cm^3 of sodium hydroxide solution containing 1.0 mole per dm^3 (1.0 M NaOH(aq)) into a conical flask using a pipette (figure 2). Add 5–10 drops of universal indicator and note the colour. Add 5 cm^3 of 1.0 M hydrochloric acid from a burette (figure 3), mix well and record the colour again. Record the colour also when 10, 15 and 20 cm^3 of hydrochloric acid have been added. Now add 1 cm^3 of hydrochloric acid and record the colour again. Repeat the addition of 1 cm^3 nine more times and note the colour each time.

The table shows the results that you should get. Notice that the indicator shows a neutral pH colour (yellow) when 25 cm^3 of hydrochloric acid have been added. So, 25 cm^3 of 1.0 M hydrochloric acid just neutralize 25 cm^3 of 1.0 M sodium hydroxide.

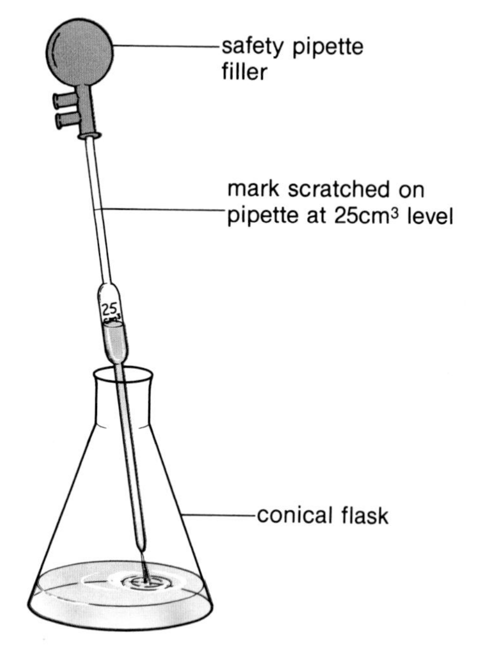

Figure 2
Measure 25 cm^3 of 1.0 M NaOH (aq) using a pipette

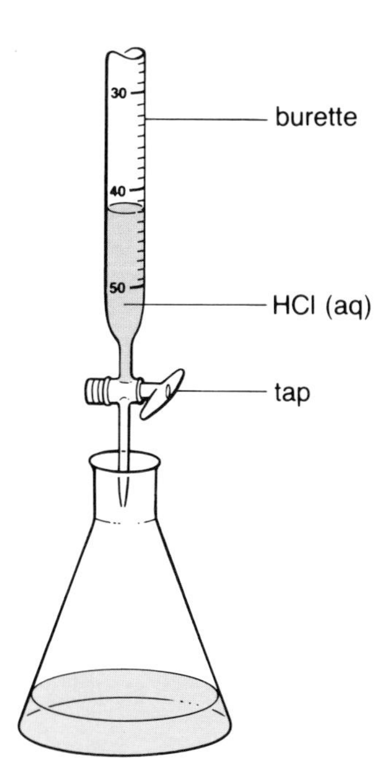

Figure 3
Add 1.0 M HCl from a burette

Volume of 1.0 M HCl(aq) added/cm^3	Colour of indicator
0	Purple
5	Purple/blue
10	Blue
15	Blue
20	Blue
21	Blue
22	Blue
23	Blue
24	Blue/green
25	Yellow
26	Pink
27	Pink
28	Pink
29	Pink
30	Pink

This method of adding one solution from a burette to another solution in order to find out how much of the two solutions will *just* react with each other is called a **titration**. When the two solutions just react and neither is in excess, we have found the **neutral point** or **end point** of the titration. From the table, we know that:

25 cm^3 of 1.0 M HCl just react with 25 cm^3 of 1.0 M NaOH.

Now, 1000 cm^3 of a 1.0 M solution contains 1 mole, so

$$25\ cm^3 \text{ of } 1.0\ M\ HCl \text{ contains } \frac{25}{1000} \times 1 = 0.025 \text{ moles of HCl and}$$

$$25\ cm^3 \text{ of } 1.0\ M\ NaOH \text{ contains } \frac{25}{1000} \times 1 = 0.025 \text{ moles of NaOH}$$

$\therefore$ 0.025 moles of HCl react with 0.025 moles of NaOH

$\Rightarrow$ 1 mole of HCl reacts with 1 mole of NaOH.

We can write the left-hand side of the equation as:

$$HCl(aq) + NaOH(aq) \rightarrow$$

If all the water from the end-point solution is evaporated, then the only product which remains is sodium chloride, NaCl(s). The complete equation is therefore:

$$HCl(aq) + NaOH(aq) \rightarrow NaCl(aq) + H_2O(l)$$

$\therefore$ 1 mole of HCl reacts with 1 mole of NaOH to give 1 mole of NaCl and 1 mole of H_2O.

The acid is *neutralized* by the base.

So, this type of reaction: acid + base $\rightarrow$ salt + water is called **neutralization**. Substances such as sodium chloride, which are obtained by replacing the H^+ ions in acids by metal ions, are called **salts**. For example, copper sulphate ($Cu^{2+}SO_4^{2-}$) is obtained by replacing the H^+ ions in sulphuric acid (H_2SO_4) by Cu^{2+} ions.

Questions

1 Define the following:
acid; base; salt; neutralization.

2 Explain the following:
titration; neutral point.

3 What is (i) an indicator; (ii) a pipette; (iii) a burette?

4 Complete the following word equations and then write balanced equations with formulas.

(a) hydrochloric acid + potassium hydroxide $\rightarrow$?

(b) nitric acid + sodium hydroxide $\rightarrow$?

(c) hydrochloric acid + calcium hydroxide $\rightarrow$?

5 30 cm^3 of 2 M NaOH just react with 10 cm^3 of 3 M H_2SO_4.

(a) How many moles of NaOH react?

(b) How many moles of H_2SO_4 react?

(c) How many moles of NaOH react with 1 mole of H_2SO_4?

(d) Write an equation for the reaction.

6 Neutralization

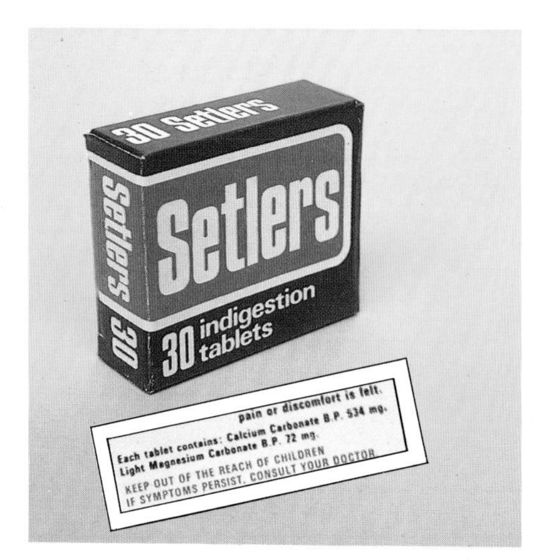

What substances in 'Setlers' help to neutralise acids in the stomach?

Adverts for indigestion cures usually talk of 'acid stomach' and 'acid indigestion'. Medicines which ease stomach ache (such as Milk of Magnesia, Setlers and Rennies) are called antacids (anti-acids) because they contain bases which neutralize excess acid in the stomach.

Neutralization is also important in treating splashes of acid or alkali on skin or clothes. The stings from many plants and animals contain acids or bases so they can be treated by neutralization. For example, nettle stings and wasp stings contain bases like *histamine*. They are treated with acidic substances called *anti-histamines*. Unlike wasp stings, those caused by bees are acidic and should be treated with bases.

Neutralization is used to remove carbon dioxide from the air in air-conditioned buildings. Carbon dioxide is an acidic oxide. So, the air is passed over soda lime, a mixture of two bases (sodium hydroxide and calcium hydroxide). These react with the carbon dioxide and neutralize the air.

One of the most important applications of neutralization is in making fertilizers like ammonium sulphate ($(NH_4)_2SO_4$) and ammonium nitrate (NH_4NO_3). Ammonium sulphate is manufactured by neutralizing sulphuric acid (H_2SO_4) with ammonia (NH_3). Ammonium nitrate is manufactured by neutralizing nitric acid (HNO_3) with ammonia (NH_3).

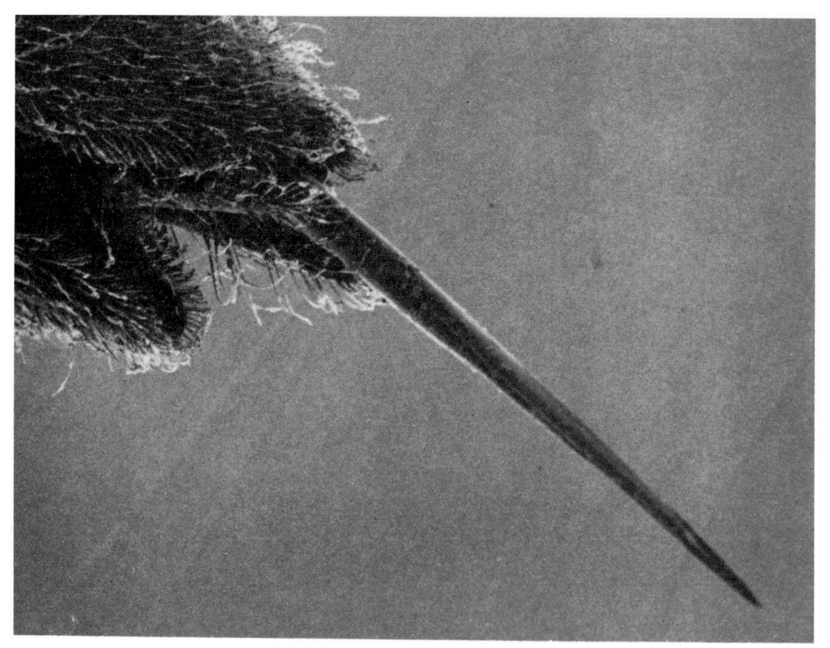

This needle-like extension from the end of a wasp's abdomen is used to sting its victims with bases, like histamine. Wasp stings are therefore neturalised by acidic substances such as anti-histamines.

Explaining neutralization in terms of ions

Hydrochloric acid contains equal numbers of $H^+(aq)$ and $Cl^-(aq)$ ions. So we can write $H^+(aq) + Cl^-(aq)$ instead of $HCl(aq)$ in equations. The '+' sign between the ions shows that they are quite separate from one another, moving around freely in the solution. Similarly, sodium hydroxide ($NaOH(aq)$) and sodium chloride ($NaCl(aq)$) consist of ions. They can be written as $Na^+(aq) + OH^-(aq)$ and $Na^+(aq) + Cl^-(aq)$. Water does not consist of ions so it must be written as H_2O.

A bag of Nitram fertilizer. Nitram is ammonium nitrate

We can now rewrite our initial equation:

$$HCl(aq) + NaOH(aq) \rightarrow NaCl(aq) + H_2O(l)$$

in terms of ions as:

$$H^+(aq) + Cl^-(aq) + Na^+(aq) + OH^-(aq) \rightarrow Na^+(aq) + Cl^-(aq) + H_2O(l)$$

Notice that $Na^+(aq)$ and $Cl^-(aq)$ appear on both sides of the last equation. These two ions do not really react. They are just the same after the reaction as they were before. If we cancel $Na^+(aq)$ and $Cl^-(aq)$ on both sides of the equation we get

$$H^+(aq) + OH^-(aq) \rightarrow H_2O(l)$$

This final equation shows clearly what happens during the reaction. H^+ ions in the acid have been neutralized by OH^- ions in the base, while Na^+ and Cl^- ions have not changed. Na^+ and Cl^- have just 'stood by' and watched the reaction like spectators at a sports fixture. Ions like Na^+ and Cl^-, which are present in the mixture but take no part in the reaction, are called **spectator ions**.

Naming salts

The name of a salt comes from the ions which it contains: the positive metal ion from the base and the negative ion from the acid. Some examples are shown in the table. Notice that acids with names ending in '–ic' form salts with names ending in '–ate'. Acids with names ending in '–ous' form salts with names ending in '–ite'. Hydrochloric acid is an exception: its salts are called chlorides.

Base	**Name of salt**
Magnesium oxide	Magnesium . . .
Potassium hydroxide	Potassium . . .
Zinc oxide	Zinc . . .
Acid	**Name of salt**
Sulphuric	. . . sulphate
Sulphurous	. . . sulphite
Nitric	. . . nitrate
Nitrous	. . . nitrite
Carbonic	. . . carbonate
Hydrochloric	. . . chloride

The name of a salt is obtained from the names of the base and acid from which it is formed

Questions

1 (a) What causes indigestion?
(b) How do 'Rennies' cure indigestion?

2 (a) Write a word equation for the reaction between nitric acid and potassium hydroxide.
(b) Write a balanced equation for the reaction given in (a) using formulas.
(c) Rewrite the last equation in terms of ions.
(d) Explain the reaction in terms of ions.

3 (a) Why is ammonium nitrate (NH_4NO_3) used as a fertilizer?
(b) What substances are used to manufacture ammonium nitrate?
(c) Write an equation for the reaction to make ammonium nitrate.

4 Complete the following word equations:
Magnesium oxide + sulphuric acid → ? + ? ;
Potassium hydroxide + ? → ? nitrate + water ;
? + hydrochloric acid → zinc ? + ?

7 Bases and Alkalis

Bases are the chemical opposites to acids. Bases take H^+ ions whereas acids donate them.

The largest group of bases are metal oxides and metal hydroxides such as sodium oxide, zinc oxide, copper oxide, sodium hydroxide, zinc hydroxide and copper hydroxide. They react with acids to form a salt and water.

For example:

$ZnO(s)$	+	$H_2SO_4(aq)$	$\rightarrow$	$ZnSO_4(aq)$	+	$H_2O(l)$
zinc oxide		sulphuric acid		zinc sulphate		water
$Zn(OH)_2(s)$	+	$H_2SO_4(aq)$	$\rightarrow$	$ZnSO_4(aq)$	+	$2H_2O(l)$
zinc hydroxide		sulphuric acid		zinc sulphate		water

These bases contain either oxide ions (O^{2-}) or hydroxide ions (OH^-) which react with H^+ ions in acids to form water.

$$2H^+ + O^{2-} \rightarrow H_2O$$
$$H^+ + OH^- \rightarrow H_2O$$

A special class of bases is called **alkalis. Alkalis are bases which are soluble in water**.

The commonest alkalis are sodium hydroxide ($NaOH$), calcium hydroxide ($Ca(OH)_2$) and ammonia (NH_3). Calcium hydroxide is much less soluble than sodium hydroxide. A solution of calcium hydroxide in water is often called 'lime water'. Sodium oxide (Na_2O), potassium oxide (K_2O) and calcium oxide (CaO) react with water to form their hydroxides. So, the reaction of these three metal oxides with water produces alkalis. For example.

$$Na_2O(s) + H_2O(l) \rightarrow 2NaOH(aq)$$
$$CaO(s) + H_2O(l) \rightarrow Ca(OH)_2(aq)$$

Most other metal oxides and hydroxides are insoluble in water. These insoluble metal oxides and hydroxides are bases but *not* alkalis. The relationship between bases and alkalis is shown in a Venn diagram in figure 1.

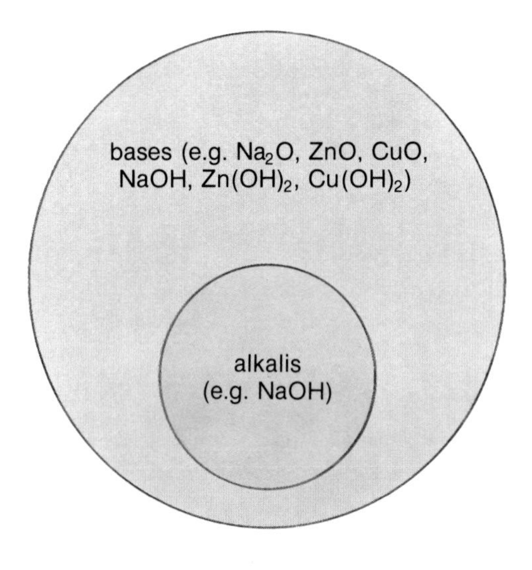

Figure 1
A Venn diagram showing the relationship between bases and alkalis

Alkalis in industry

The most important industrial alkalis are sodium hydroxide (caustic soda) and calcium hydroxide (slaked lime). Calcium hydroxide is made by adding water to lime (calcium oxide). It is used in cement and in the manufacture of sodium hydroxide and bleaching powder. Large amounts of sodium hydroxide are used to make soap, paper, rayon and other cellulose fibres.

Paper, rayon and cellulose fibres are all made from wood. The wood is made into pulp and soaked in sodium hydroxide solution. This removes gums and resins and leaves the natural fibres of cellulose. The cellulose fibres are then squashed into thin white sheets which look like blotting paper. This purified pulp can now be used to make paper, rayon or cellulose acetate (Tricel).

Soaps and soap powders are made by boiling fats and oils with sodium hydroxide in large vats (figure 2).

$$\begin{array}{l} \wedge\!\!\wedge\!\!\wedge\!\!\wedge\!\!\wedge\!-CH_2 \\ \quad\quad\quad\quad\quad | \\ \wedge\!\!\wedge\!\!\wedge\!\!\wedge\!\!\wedge\!-CH \\ \quad\quad\quad\quad\quad | \\ \wedge\!\!\wedge\!\!\wedge\!\!\wedge\!\!\wedge\!-CH_2 \end{array} + 3NaOH \longrightarrow 3Na-\wedge\!\!\wedge\!\!\wedge\!\!\wedge\!\!\wedge + \begin{array}{c} CH_2OH \\ | \\ CHOH \\ | \\ CH_2OH \end{array}$$

oil or fat molecule — soap — glycerine (glycerol)

($\wedge\!\!\wedge\!\!\wedge\!\!\wedge\!\!\wedge\!-$ $\equiv$ long chain of carbon, hydrogen and oxygen atoms)

Figure 2

In the manufacture of paper, wood pulp (cellulose) is purified by soaking in concentrated sodium hydroxide solution

Rayon is made by forcing a solution containing cellulose (wood pulp) through tiny holes into dilute acid to form filaments of yarn

Questions

1 Define the words:
base; acid; alkali; salt.

2 Give the names and formulas of
(i) three bases which are not alkalis;
(ii) three alkalis; (iii) three acids;
(iv) three salts.

3 Write equations for the following reactions:
lime (calcium oxide) with water;
ammonia with water;
sodium hydroxide with nitric acid;
sodium hydroxide with sulphuric acid.

4 What are the main industrial uses of alkalis?

5 (a) Why is it important to recycle paper?
(b) What are the main stages in recycling paper?
(c) Paper can be made from rags. Explain why this is so.

8 Properties of Alkalis

All alkalis, except ammonia solution are the hydroxides of reactive metals. They have similar properties because they all contain hydroxide ions (OH^-). The commonest alkalis are calcium hydroxide, sodium hydroxide and ammonia.

- *Alkalis are soluble in water.*
- *Alkalis are electrolytes.*

Alkalis dissolve in water giving positive ions and hydroxide ions (OH^-). These ions allow the solution to conduct electricity. Experiments (see unit 4) show that sodium hydroxide, potassium hydroxide and calcium hydroxide are strong electrolytes and strong alkalis. Ammonia solution is a weak electrolyte and therefore a weak alkali. This means that sodium hydroxide, potassium hydroxide and calcium hydroxide are fully dissociated (split up) into ions in solution, but ammonia solution is only partly dissociated.

$$NaOH(aq) \rightarrow Na^+(aq) + OH^-(aq)$$
$$NH_3(aq) + H_2O(l) \rightleftharpoons NH_4^+(aq) + OH^-(aq)$$

1 *Alkalis give characteristic colours with indicators.*
Alkalis form solutions with a pH above 7. They turn litmus blue and give a green, blue or purple colour with universal indicator.

2 *Alkalis react with acids to make a salt and water.*
These reactions are examples of neutralizations. For example,

$$\underset{\text{nitric acid}}{2HNO_3(aq)} + \underset{\text{calcium hydroxide}}{Ca(OH)_2} \rightarrow \underset{\text{calcium nitrate}}{Ca(NO_3)_2} + \underset{\text{water}}{2H_2O}$$

3 *Alkalis react with metal ions.*

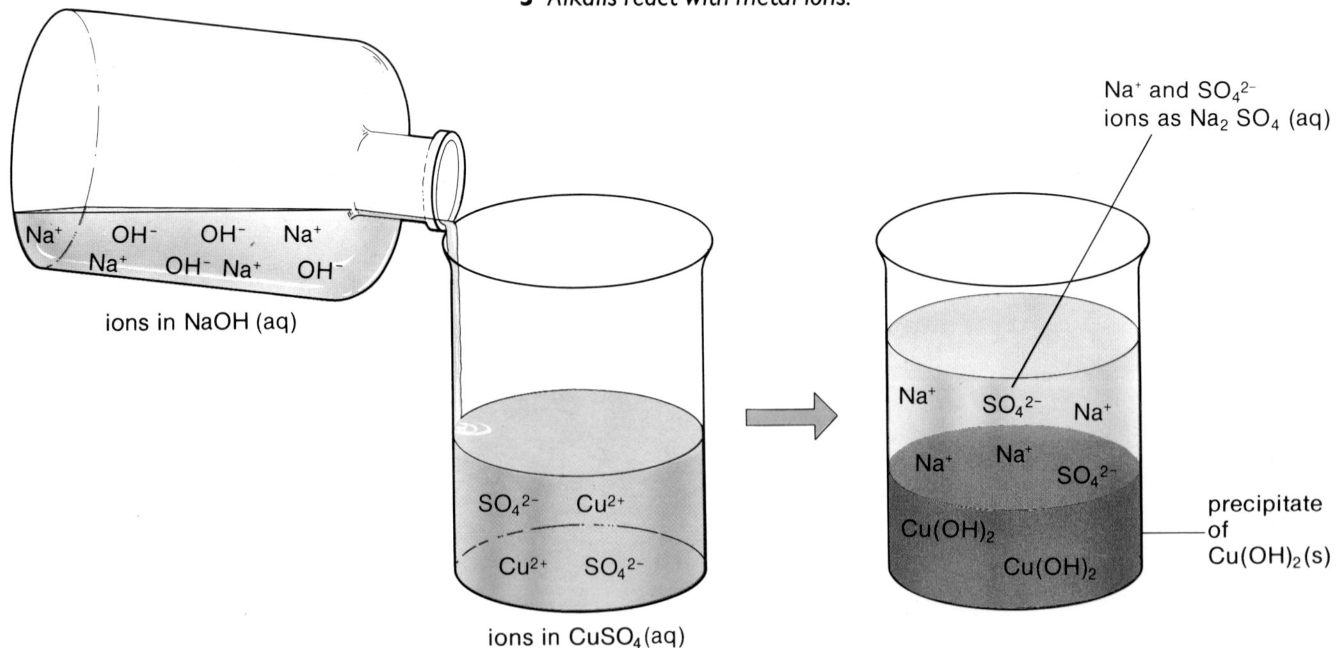

Figure 1
The precipitation of copper hydroxide

When an alkali is added to a solution containing metal ions (other than K^+ or Na^+), a solid forms and falls to the bottom. The solid is called a **precipitate**. Figure 1 shows what happens when sodium hydroxide solution, NaOH(aq), is added to copper sulphate solution, $CuSO_4$(aq). Sodium hydroxide and copper sulphate are both soluble. When the two solutions are mixed, two new combinations of ions are possible: sodium ions with sulphate ions, and copper ions with hydroxide ions. Sodium sulphate is soluble, but copper hydroxide is insoluble. So, a blue precipitate of copper hydroxide forms. We can write an equation for the reaction as:

$$\underbrace{Cu^{2+}(aq) + SO_4^{2-}(aq)}_{\text{copper sulphate solution}} + \underbrace{2Na^+(aq) + 2OH^-(aq)}_{\text{sodium hydroxide solution}}$$

$$\rightarrow \underbrace{Cu(OH)_2(s)}_{\text{copper hydroxide precipitate}} + \underbrace{2Na^+(aq) + SO_4^{2-}(aq)}_{\text{sodium sulphate solution}}$$

Notice that sodium ions and sulphate ions are left in solution at the end of the reaction. They have taken no part in the reaction. They are spectator ions. If we cancel these spectator ions in the last equation, then we get a simpler equation which shows only those ions which react:

$$Cu^{2+}(aq) + 2OH^-(aq) \rightarrow Cu(OH)_2(s)$$

The colour of insoluble metal hydroxides is sometimes used to identify different metal ions. This is studied further in section F, unit 12.

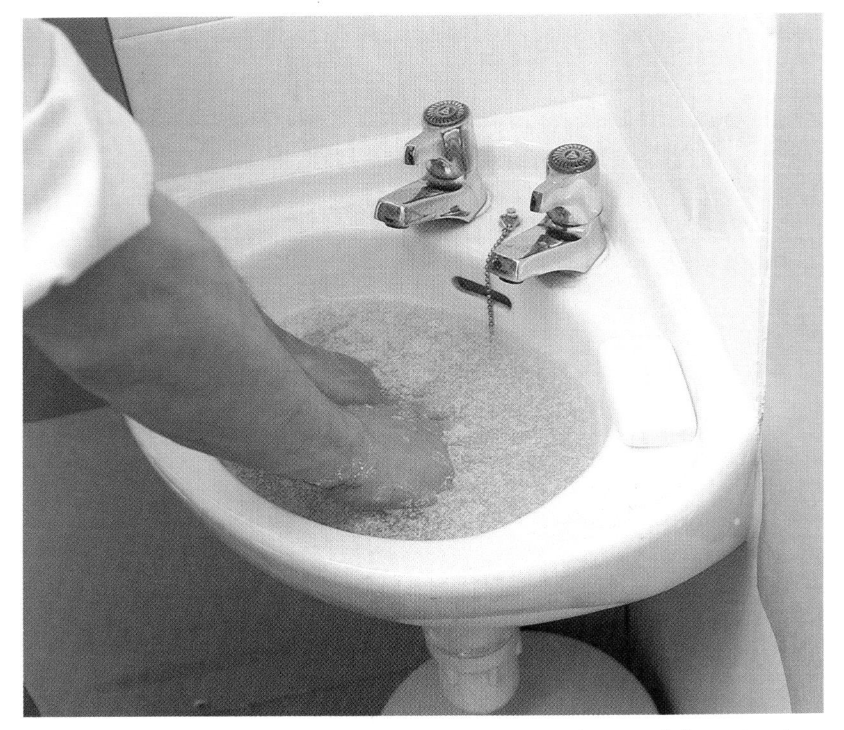

Scum is a precipitate. It forms when soap is used with hard water. Calcium ions in the hard water react with complex (stearate) ions in the soap to form scum—insoluble calcium stearate

Questions

1 Explain the following:
alkali; electrolyte; indicator; precipitate.

2 Summarise the important properties of alkalis.

3 Answer *true* or *false* to parts A to F:

Alkalis
A are always soluble in water.
B are reduced by acids.
C are generally solids.
D react exothermically with acids.
E always contain metal ions.
F are all bases.

4 (a) How would you investigate whether sodium hydroxide and ammonium hydroxide are strong or weak alkalis? (See unit 4 of this section.)
(b) What results would you expect?

9 Salts

Large cubic crystals of galena (lead sulphide)

Salts are formed when acids react with metals or bases. Most salts contain a positive metal ion and a negative non-metal or radical ion. Salts are ionic compounds:

1 they have high melting points and boiling points,

2 they are electrolytes,

3 they are often soluble in water.

The best known salt is sodium chloride, NaCl, which is often called common salt. Many ores and minerals are composed of salts. These include chalk and limestone (calcium carbonate), gypsum (calcium sulphate) and iron pyrites (a mixture of copper sulphide and iron sulphide).

Salt crystals, like those of sodium chloride, are often formed by crystallization from aqueous solution. When this happens, water molecules sometimes form part of the crystal structure. This occurs in Epsom salts ($MgSO_4.7H_2O$), gypsum ($CaSO_4.2H_2O$) and washing soda ($Na_2CO_3.10H_2O$). The water which forms part of the crystal structure is called **water of crystallization**. Salts containing water of crystallization are called **hydrates** or hydrated salts.

Soluble and insoluble salts

If you are using a salt or making a salt, it is important to know whether it is soluble or insoluble.

Table 1 shows the solubilities of various salts in water at 20°C. Notice the wide range in solubilities from potassium nitrite (300 g per 100 g water) to silver chloride (0.000 000 1 g per 100 g water). It is useful to divide salts into two categories—soluble and insoluble.

Purple cubic crystals of fluorite (calcium fluoride)

Salt	Formula	Solubility /g per 100 g water at 20°C
Barium chloride	$BaCl_2$	36.0
Barium sulphate	$BaSO_4$	0.000 24
Calcium chloride	$CaCl_2$	74.0
Calcium sulphate	$CaSO_4$	0.21
Copper(II) sulphate	$CuSO_4$	20.5
Copper(II) sulphide	CuS	0.000 03
Lead(II) sulphate	$PbSO_4$	0.004
Potassium chlorate	$KClO_3$	7.3
Potassium nitrite	KNO_2	300.0
Silver chloride	$AgCl$	0.000 000 1
Silver nitrate	$AgNO_3$	217.0
Sodium chloride	$NaCl$	36.0
Sodium nitrate	$NaNO_3$	87.0

Table 1: the solubilities of various salts

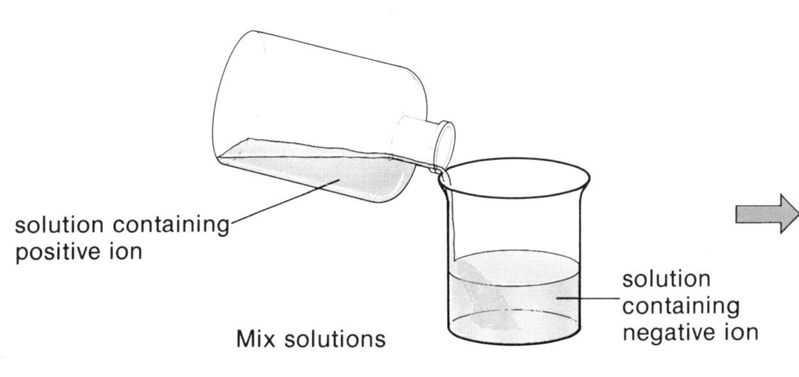

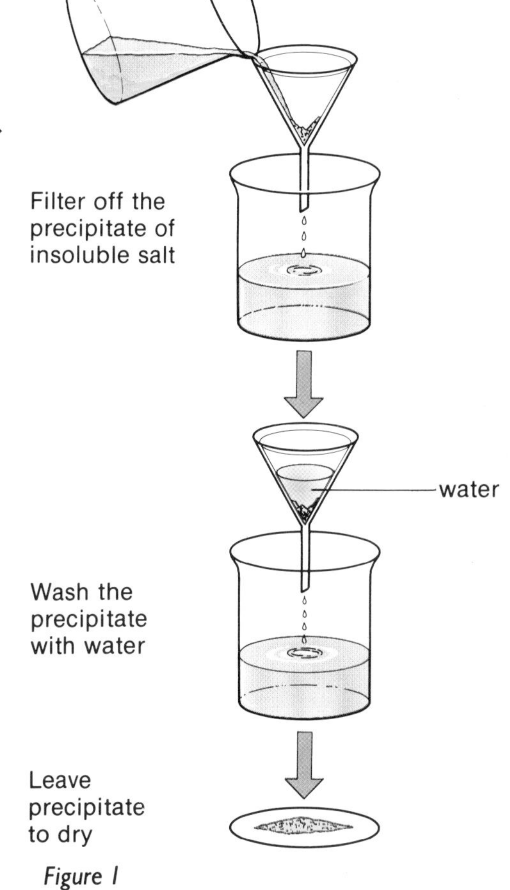

Figure 1
Preparing an insoluble salt

Salts with a solubility greater than 1 g per 100 g water are classed as soluble; salts with a solubility less than 1 g per 100 g water are classed as insoluble.

Table 2 gives a summary of the solubilities of common salts.

All	sodium, potassium, ammonium	salts are soluble	
All	nitrates	are soluble	
All	sulphates	are soluble except	Ag_2SO_4, $BaSO_4$, $CaSO_4$, $PbSO_4$
All	chlorides	are soluble except	$AgCl$, $PbCl_2$
All	carbonates, sulphides, sulphites	are **insoluble** except	those of Na^+, K^+ and NH_4^+

Table 2: solubilities of common salts

Preparing insoluble salts

The method used to prepare a salt depends on whether the salt is soluble or insoluble. Methods for soluble salts are described in the next unit. Insoluble salts, like lead chloride, silver chloride, calcium carbonate and barium sulphate, are prepared by making the salt as a precipitate.

Suppose you are making insoluble silver chloride, AgCl. You will need to mix a soluble Ag^+ salt and a soluble chloride.

1 Which Ag^+ salt is certain to be soluble? Look at table 2.
2 Which chloride is certain to be soluble? Look at table 2.

These questions show that you can precipitate any insoluble salt (say XY) by mixing solutions of NaY and XNO_3. Both NaY and XNO_3 are soluble, since all sodium salts and all nitrates are soluble. Figure 1 shows how an insoluble salt is precipitated and then purified. Make one solution containing the positive ion in the insoluble salt and another solution containing the negative ion. Mix the two solutions, filter off the insoluble salt, wash it with water and then leave it to dry at room temperature.

Questions

1 Explain the following:
hydrated; water of crystallization; precipitation; insoluble.

2 What units are used for
(i) concentration; (ii) solubility?

3 (a) Summarise the stages in preparing an insoluble salt.
(b) Describe how you would prepare a pure sample of insoluble barium sulphate.
(c) Write an equation for the reaction which occurs.

4 Make a table to show whether the following salts are soluble or insoluble:
$Pb(NO_3)_2$; Ag_2S; $CuCO_3$; K_2SO_3; NH_4Cl; $FeSO_4$.

5 Epsom Salts have the formula $MgSO_4.7H_2O$. What does this tell you about Epsom Salts?

10 Preparing Salts

When you are making a salt, the first question to ask is 'Is the salt soluble or insoluble?'. If the salt is insoluble, it is usually prepared by precipitation (see unit 9 of this section).

Soluble salts are usually prepared by reacting an acid with a metal, a base or a carbonate (see unit 3 of this section).

$$\text{metal} + \text{acid} \rightarrow \text{salt} + H_2$$
$$\text{base} + \text{acid} \rightarrow \text{salt} + H_2O$$
$$\text{carbonate} + \text{acid} \rightarrow \text{salt} + CO_2 + H_2O$$

- **Method 1: For metals, insoluble bases and insoluble carbonates.** Figure 1 shows the main stages in this method.

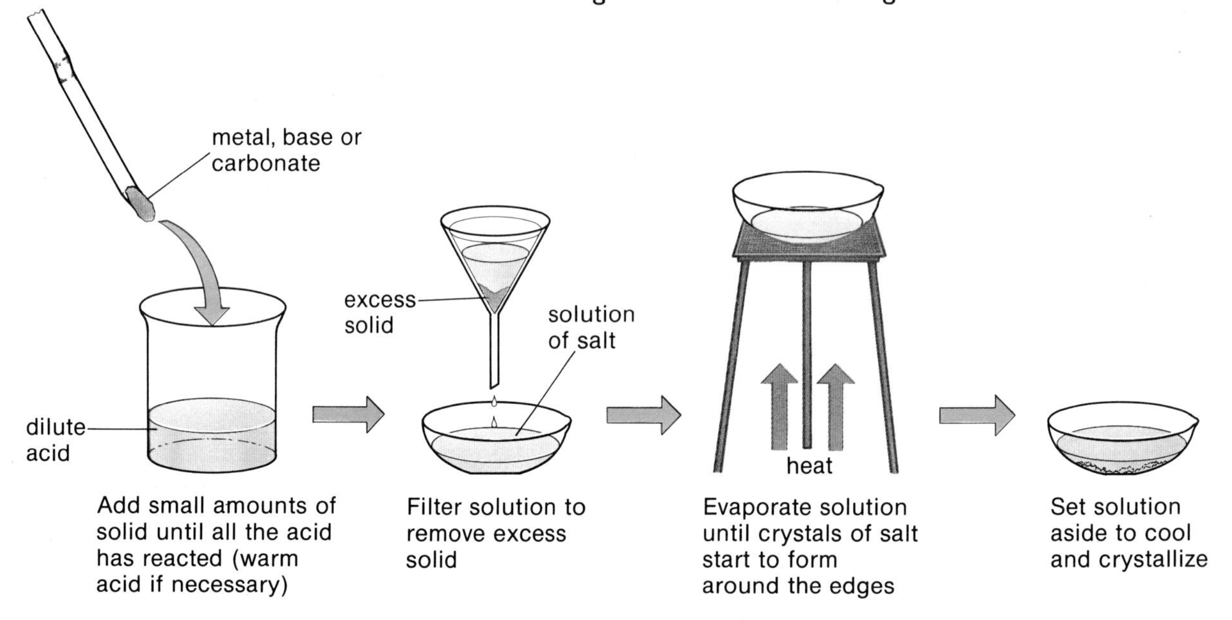

Figure 1

Zinc sulphate can be made by this method using sulphuric acid with either zinc, zinc oxide or zinc carbonate.

$$Zn + H_2SO_4 \rightarrow ZnSO_4 + H_2$$
$$ZnO + H_2SO_4 \rightarrow ZnSO_4 + H_2O$$
$$ZnCO_3 + H_2SO_4 \rightarrow ZnSO_4 + CO_2 + H_2O$$

- **Method 2: For soluble bases and carbonates.** In method 1, we can tell when the acid has been used up because unreacted metal, base or carbonate remains in the liquid as undissolved solid. But, if the solid is soluble (like sodium hydroxide or sodium carbonate), we cannot tell when the acid has been used up, because excess solid will dissolve even after the acid has been neutralized. To get round this, we must use an indicator to tell us when we have added just enough base or carbonate to neutralize the acid. Figure 2 shows the main stages involved.

Potassium chloride can be made by this method using hydrochloric acid with either potassium hydroxide or potassium carbonate.

$$KOH + HCl \rightarrow KCl + H_2O$$
$$K_2CO_3 + 2HCl \rightarrow 2KCl + CO_2 + H_2O$$

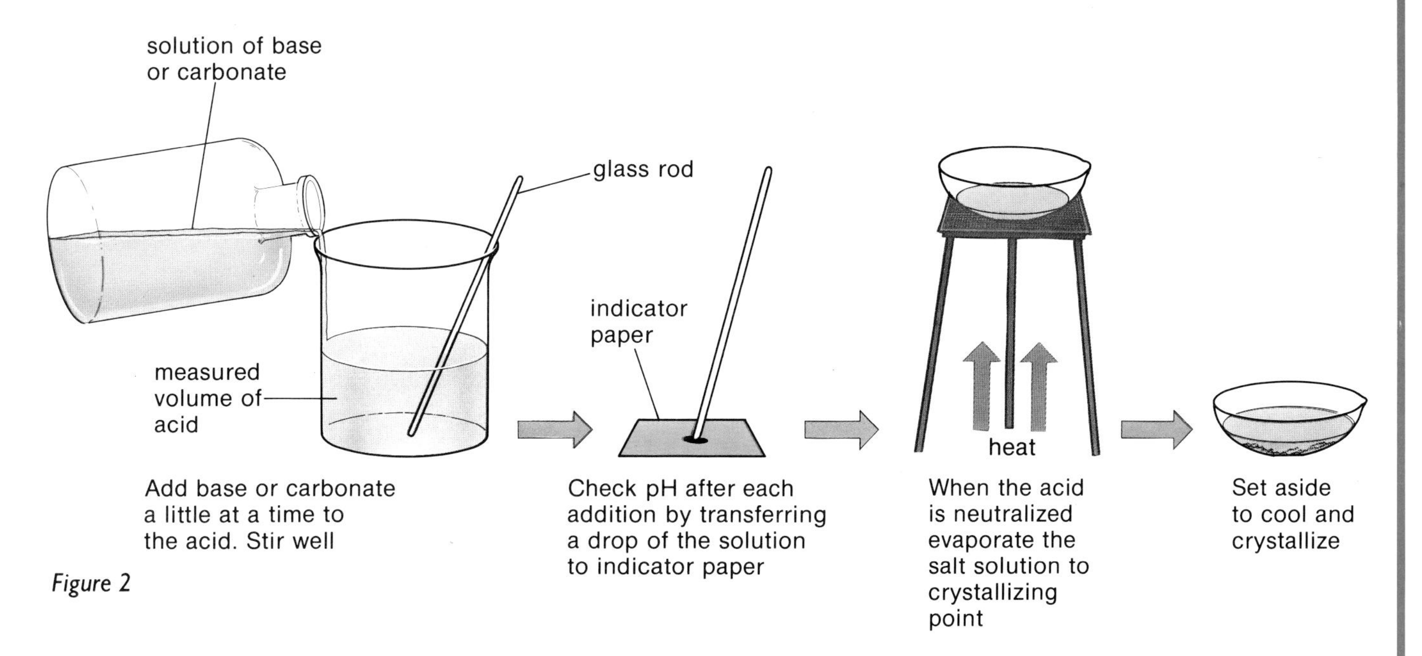

Figure 2

Method 2 is used to make the salts of sodium, potassium and ammonium, because the bases and carbonates containing sodium, potassium and ammonium are all soluble. Other soluble salts are usually made by method 1. Figure 3 shows a flowchart which can be used to decide how to prepare a particular salt.

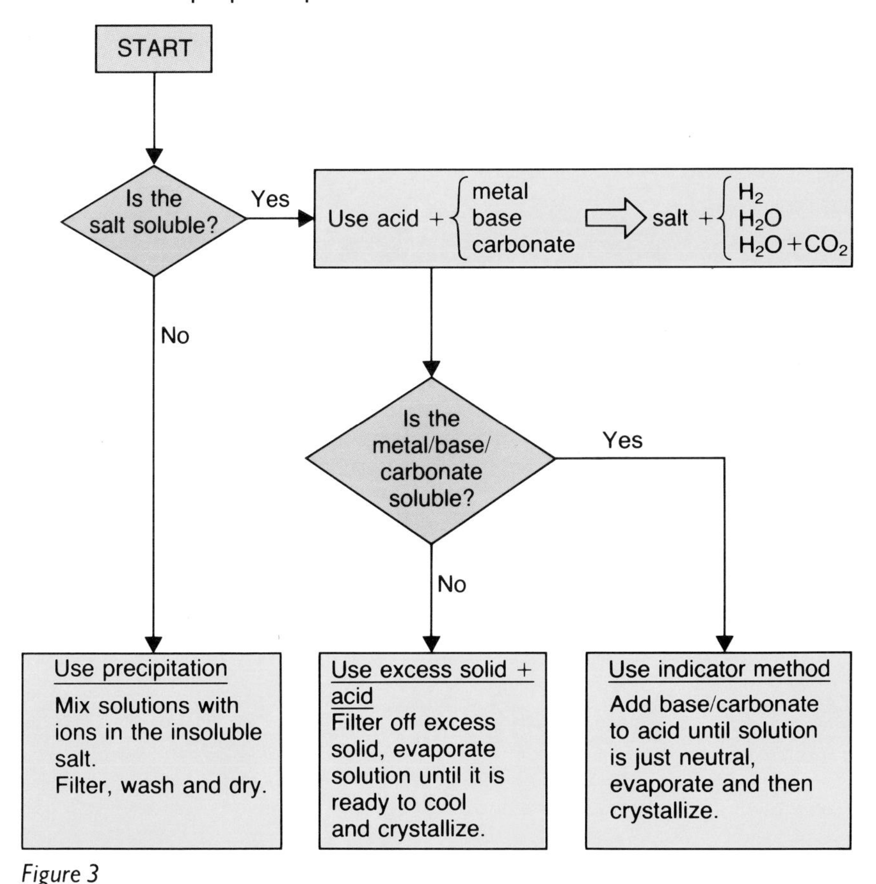

Figure 3
How to prepare a salt

Questions

1 Look at method 1 for preparing soluble salts.
  (a) Explain why this method will not work for metals below hydrogen in the activity series.
  (b) Why is this method not used with sodium?
  (c) How can you tell when all the acid is used up if the solid used is (i) zinc; (ii) copper oxide; (iii) copper carbonate?
  (d) Why is the salt produced not contaminated with (i) the acid used; (ii) the solid added?
  (e) Why is method 1 no good for insoluble salts?
  (f) Why is method 1 no good if the solid added dissolves in water?

2 Look at method 2 for preparing salts.
  (a) Why is the pH of the solution tested using indicator paper rather than putting indicator solution into the acid?
  (b) Describe how you would make sodium nitrate by this method.
  (c) Write a word equation and a balanced equation with formulas for the reaction in (b).

Section G: Study Questions

1 Question 1 concerns acids.

(a) Name *one* indicator that you could use to test for an acid.

(b) What would happen to the indicator that you named in part (a) when you placed it in an acidic solution?

(c) Name the acid that is used in a car battery.

(d) Name *one* substance, usually found around the house, that you could use to neutralise some spilt battery acid.

(e) Bath salts contain crystals of sodium carbonate. What gas would be given off if vinegar (ethanoic acid) were dropped onto some of these crystals? **LEAG**

2 The pH scale is used to indicate how acidic or alkaline a solution is. Here are some numbers from the pH scale:

1 3 7 9 14

(a) What kind of a solution would have a pH of 1?

(b) Which pH number from the above list would lime water have?

(c) Which number represents the pH of pure water? **LEAG**

3 Complete the table below, which describes the preparation of some salts.

REACTANTS	PRODUCTS
magnesium oxide +	→ magnesium sulphate +
+	→ zinc chloride + hydrogen

[4] **SEG**

4 Magnesium sulphate crystals ($MgSO_4.7H_2O$) can be made by adding excess magnesium oxide (MgO), which is insoluble in water, to dilute sulphuric acid.

(a) Why is the magnesium oxide added in excess? (1)

(b) The following apparatus could be used to separate the excess magnesium oxide from the solution. Label the diagram by putting the correct words in each of the spaces below. (4)

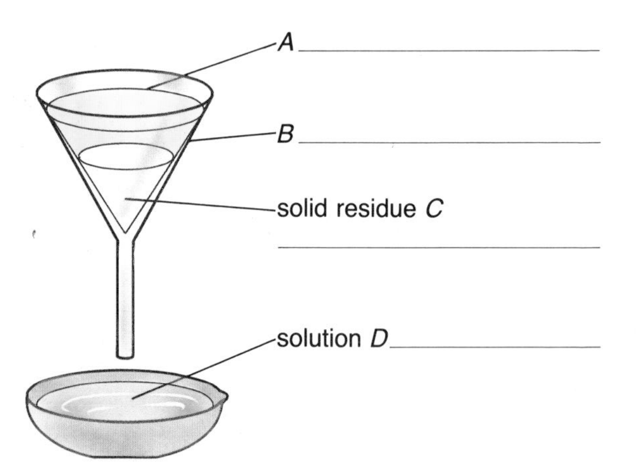

(c) After the excess magnesium oxide has been removed, the solution is partly evaporated and set aside. Some days later the sample is filtered again and the product is washed and finally dried.

(i) Why is the solution partly evaporated? (1)

(ii) Which substance is removed by the *second* filtering? (1)

(iii) How could the product be dried? (1)

(iv) What would happen to the product if it was then heated strongly? (1)

Total [9] **LEAG**

5 A class is investigating the chemical reactions of the following metals and some of their compounds:
iron, zinc, nickel, copper, lead.

(a) It is suspected that nickel lies between zinc and lead in the electrochemical (activity) series. If the class had available samples of the three metals and their soluble nitrate salts describe experiments to verify this. (7)

(b) Given that the reactions of zinc and nickel with sulphuric acid are similar and that the salts of these metals have similar solubilities in water, describe and explain class experiments to make a reasonably pure sample of nickel carbonate from nickel using the following two-stage process:
nickel → nickel sulphate → nickel carbonate (10)

(c) Explain, giving practical details, how the following aqueous solutions could be distinguished from each other:
(i) iron(II) sulphate and iron(III) sulphate
(ii) magnesium sulphate and zinc sulphate. (10)

Total [27] **WJEC**

6 Study the following reaction scheme:

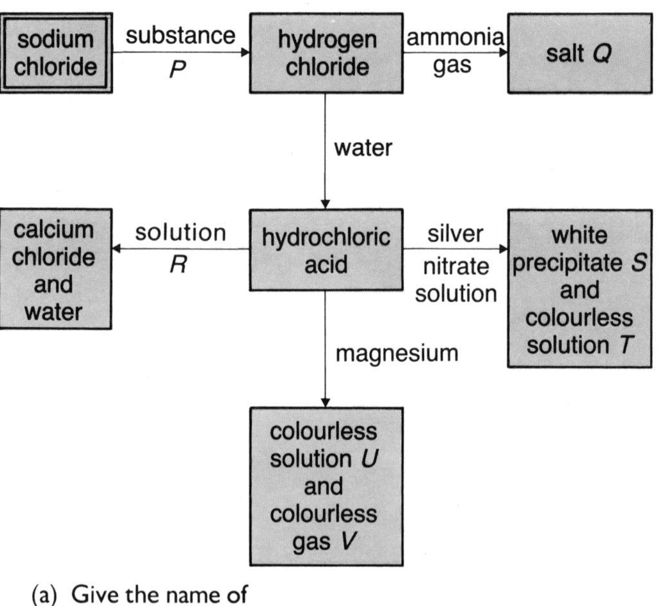

(a) Give the name of
substance *P* solution *T*
salt *Q* solution *U*
solution *R* gas *V*
precipitate *S* (7)

(b) (i) State the type of reaction taking place when hydrochloric acid reacts with solution *R*. (1)

(ii) Name one other compound that will give a white precipitate with silver nitrate solution. (1)

(iii) Describe a test by which you could identify gas *V*. (1)

Total [10] **NEA**

SECTION H

The Structure of Substances

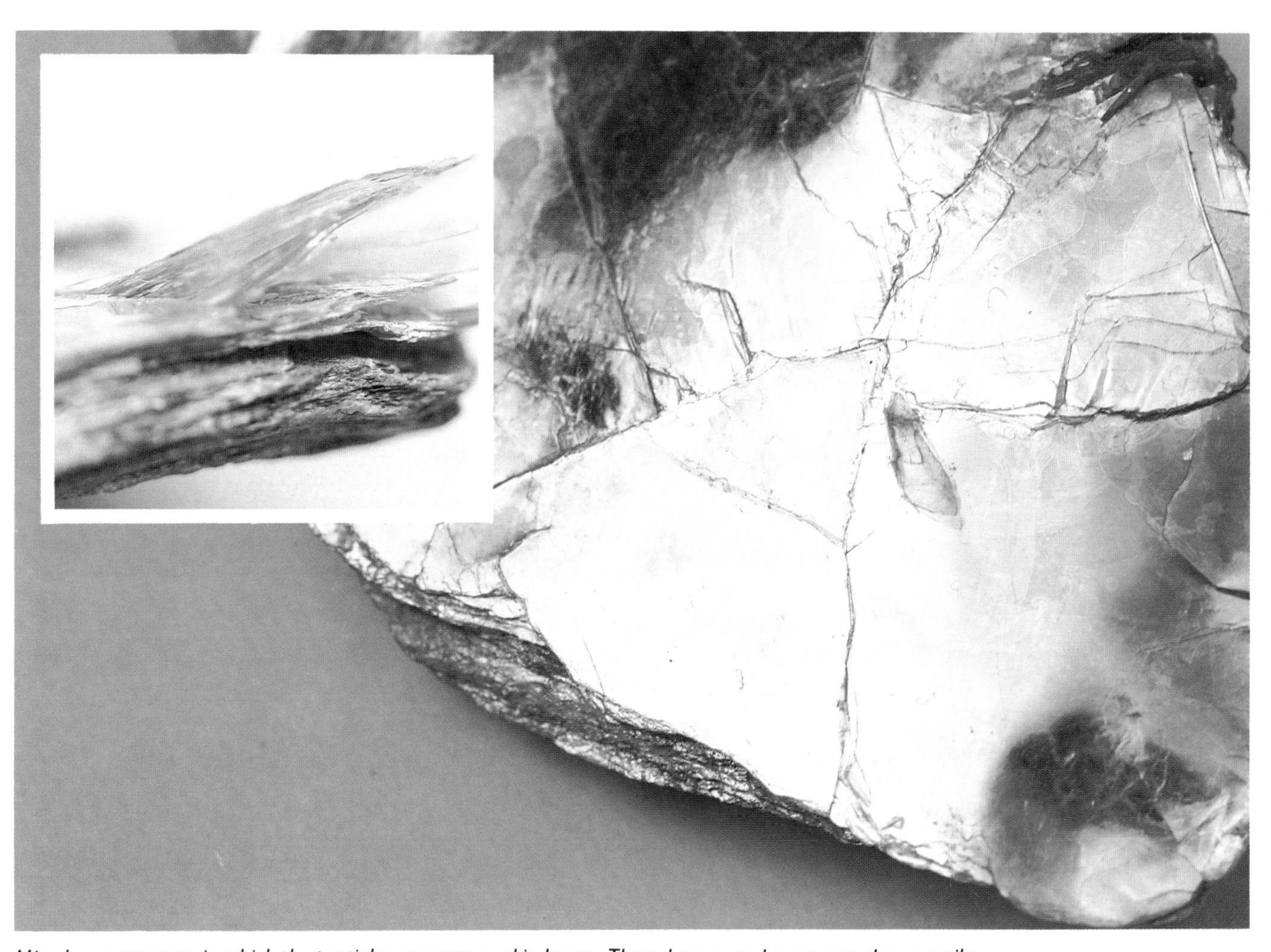

Mica has a structure in which the particles are arranged in layers. These layers can be separated very easily

1 Studying Structures

Naturally occurring crystals of copper

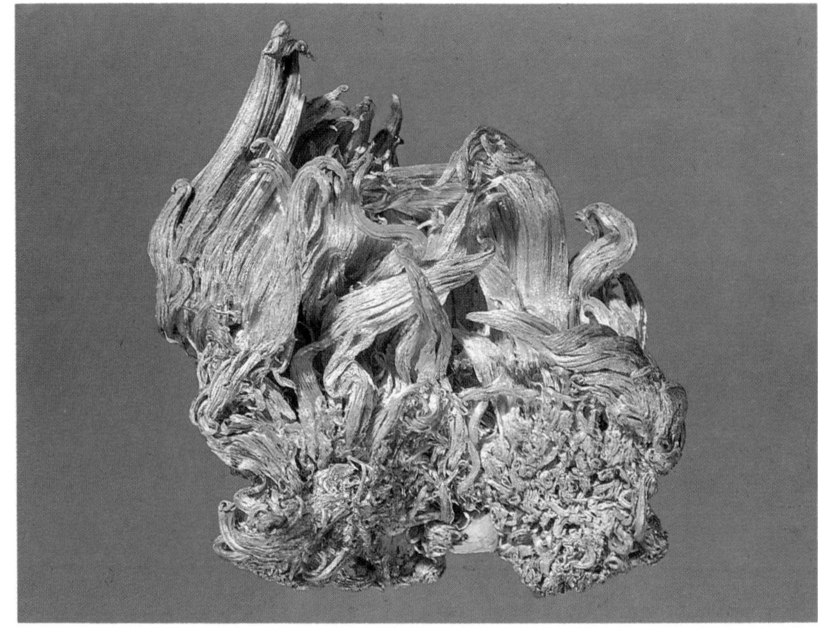

Naturally occurring silver

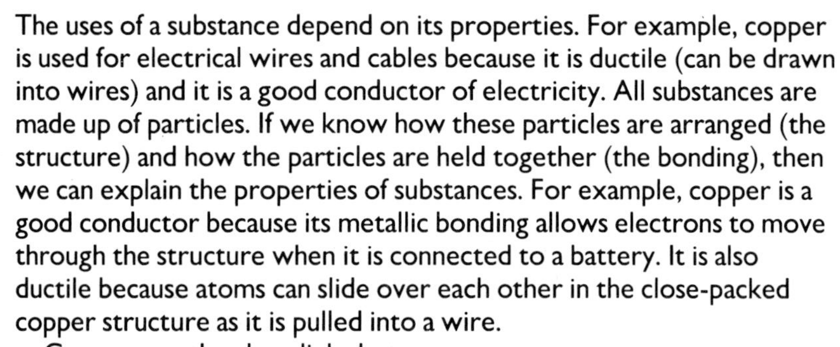

The uses of a substance depend on its properties. For example, copper is used for electrical wires and cables because it is ductile (can be drawn into wires) and it is a good conductor of electricity. All substances are made up of particles. If we know how these particles are arranged (the structure) and how the particles are held together (the bonding), then we can explain the properties of substances. For example, copper is a good conductor because its metallic bonding allows electrons to move through the structure when it is connected to a battery. It is also ductile because atoms can slide over each other in the close-packed copper structure as it is pulled into a wire.

Can you see the close links between:

$$\text{structure} \rightarrow \text{bonding} \rightarrow \text{properties} \rightarrow \text{uses} \quad ?$$

Once we understand the structure and bonding of substances, we can explain why metals are good conductors, why graphite is soft and flaky and why sulphur melts so easily. We can also match the properties of a substance with the job that needs to be done.

We can get some idea about the structure of a substance from the shape of its crystals, but the best way of studying the structure of solids is by **X-ray diffraction**. This was discussed in section F, unit 3. In section F, we also studied the structures of metals in some detail. In this section we will consider the structures of non-metals and compounds. In particular, we shall look at the non-metals sulphur and carbon and the compounds water, tetrachloromethane (carbon tetrachloride) and sodium chloride.

Using X-ray analysis, we can get accurate evidence for the arrangement of particles in a substance (its structure), but it is more difficult to study the forces between particles in the substance (its bonding). Our ideas about bonding are usually worked out from the properties of a substance such as its melting point and boiling point, its conductivity and its solubility in different solvents.

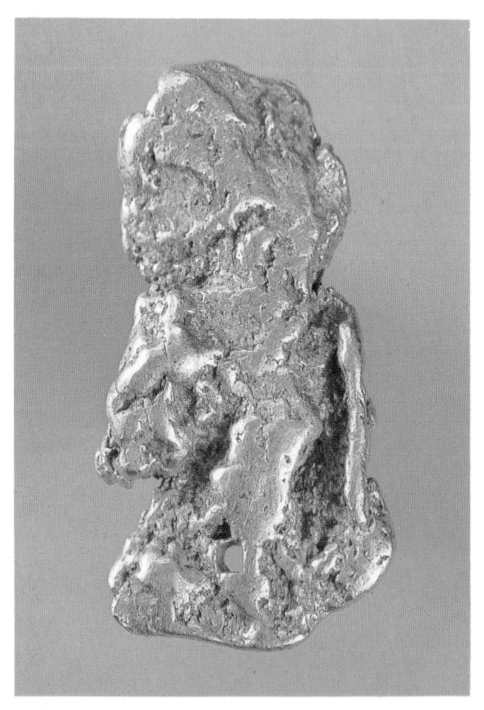

A gold nugget from Australia

From previous sections, you will know that

> *chemical substances are made up of three different types of particle—atoms, ions and molecules.*

These three particles give rise to four different solid structures (see the table).

1 Metallic structures containing metal atoms.

2 Ionic structures in metal/non-metal compounds made up of ions.

3 Simple molecular structures in non-metals like iodine (I_2) and in non-metal compounds like ice (H_2O). These structures have small molecules containing a few atoms.

4 Giant molecular structures in non-metals like diamond and in non-metal compounds like polythene. These structures have very large molecules containing hundreds or even thousands of atoms.

Lead sulphide (galena) crystals on a sample of siderite (iron (II) carbonate)

Type of structure	Particles in the structure	Type of substance	Examples
Metallic	Atoms	Metal	Na, Fe, Cu
Ionic	Ions	Metal/non-metal compound	Na^+Cl^-, $Ca^{2+}O^{2-}$ (salt) (lime)
Simple molecular	Small molecules	Non-metals or non-metal compounds	I_2 (iodine), S_8 (sulphur), H_2O (water), CO_2 (carbon dioxide)
Giant molecular	Very large molecules	Non-metals or non-metal compounds	Diamond, graphite, polythene, sand

The four types of solid structure and the particles they contain

A large natural diamond embedded in an ore, side by side with a cut diamond

Questions

1 Look at section F, unit 3. How are X-rays used to study crystal structures?

2 Look at section F, unit 4.
(a) What are the particles in metal structures?
(b) Why are most metal structures described as close packed?
(c) Explain how the particles are arranged in many metal structures to give a coordination number of 12.

3 What type of structure will the following substances have?
chlorine; calcium carbonate; silver; polyvinylchloride (PVC).

4 How do conductivity tests give evidence for the particles, bonding and structure of substances? (See section D, units 3 and 4.)

2 The Structure of Sulphur

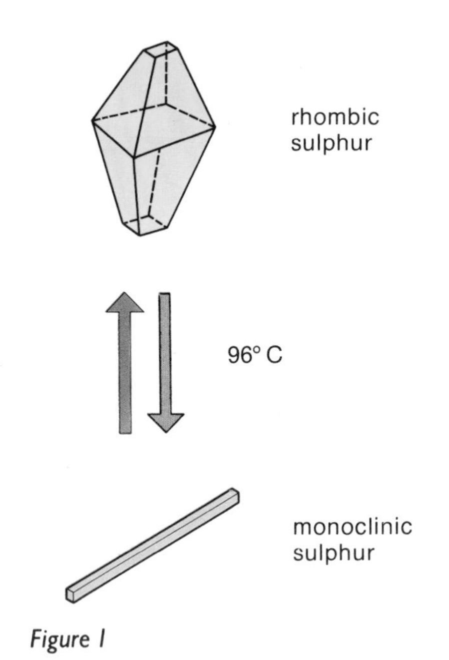

Figure 1

An aboriginal rock painting at Obiri Rock in Australia. The yellow parts of the drawing have been painted using sulphur

Sulphur is one of the most useful and important elements. In prehistoric times, our ancestors used sulphur to colour their cave drawings. About 1500 BC, the Egyptians discovered that the sulphur dioxide made by burning sulphur could be used to bleach cotton and linen. By 500 BC, Chinese warriors were using sulphur in their gunpowder. Around 1800 AD, chemists found a method of manufacturing sulphuric acid. Since then, the increasing uses of sulphuric acid in industry and agriculture have created an increasing demand for sulphur.

Different forms of sulphur

We can make three different forms of sulphur, each with different physical properties.

- **Rhombic sulphur.** Make a solution of sulphur in carbon disulphide. Allow this solution to stand at *room temperature*. Carbon disulphide evaporates leaving pale yellow diamond-shaped crystals of rhombic sulphur (figure 1). **This experiment must be done in a fume cupboard by your teacher** because carbon disulphide is poisonous, flammable and has a very unpleasant smell.

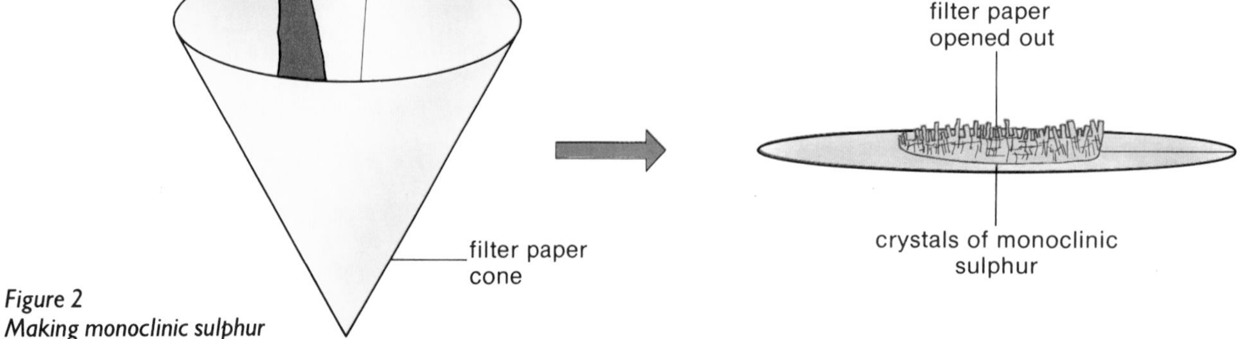

Figure 2
Making monoclinic sulphur

- **Monoclinic sulphur.** Warm some powdered sulphur until it just melts at *about 120°C*. Now pour it into a filter paper cone (figure 2). When a crust has formed on the sulphur, open out the filter paper. Thin needle-shaped crystals of monoclinic sulphur will have formed (figure 2).

- **Plastic sulphur.** Heat some powdered sulphur until it begins to boil. Pour this *hot boiling liquid sulphur* into cold water. Brown, rubbery plastic sulphur forms.

Notice how the different forms of sulphur are produced as sulphur crystallizes at different temperatures. These three forms of solid sulphur are called **allotropes**. Rhombic sulphur forms when solutions of sulphur evaporate at room temperature, monoclinic sulphur forms when molten sulphur cools slowly and plastic sulphur forms when boiling sulphur cools rapidly.

Rhombic sulphur is the most stable allotrope below 96°C and monoclinic sulphur is most stable above 96°C. So, samples of monoclinic sulphur and plastic sulphur slowly change into rhombic sulphur if they are kept at room temperature.

Other elements also have allotropes. Carbon has two allotropes, diamond and graphite. Oxygen has two allotropes, oxygen (O_2) and ozone (O_3). **Allotropes are different forms of the same element in the same state.**

Refining sulphur in the 16th century. Ore containing sulphur was heated in the earthenware pot (A). Sulphur vapour passed down the spout to a second pot (B) where it condensed. The liquid sulphur ran out of B and solidified in a wooden tub

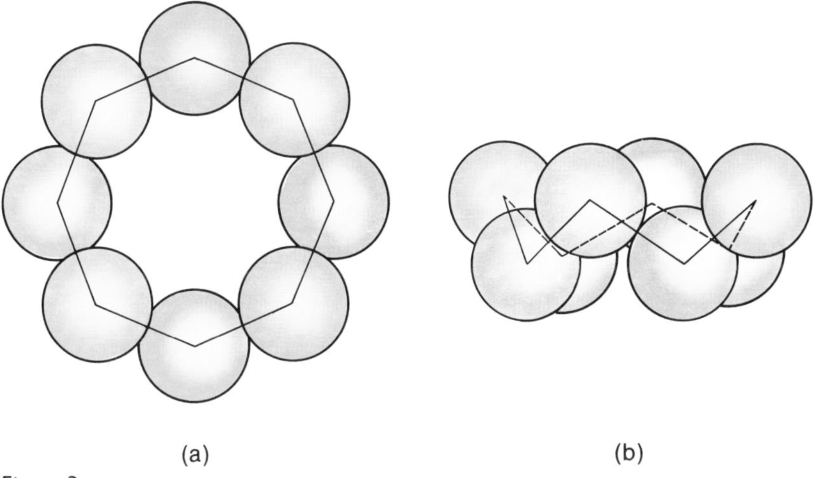

Figure 3
A molecule of sulphur—a ring of eight atoms bonded together. (a) bird's eye view; (b) side view

X-ray studies show that both rhombic sulphur and monoclinic sulphur contain sulphur molecules with eight atoms in a ring (figure 3). These two forms of sulphur come from two different arrangements of these 'rings'. Notice in figure 3(b) that the centres of the eight atoms lie at the points of a crown. The eight atoms are held together in a ring by strong covalent bonds (see section D, unit 8). There are two bonds to each sulphur atom. The eight atoms remain in a ring even when the sulphur is melted. The molecules of sulphur are therefore written as S_8. Substances like sulphur (S_8), chlorine (Cl_2) and water (H_2O) with a few atoms in their molecules are described as **simple molecular**. Their atoms are held together by covalent bonds.

Questions

1 What is meant by the following: *allotrope; allotropy; simple molecular; covalent bond?*

2 (a) Name the two crystalline allotropes of sulphur.
(b) Under what conditions does each allotrope form?
(c) What is the structure of each allotrope?

3 Name three elements, other than sulphur, which exist as simple molecules containing two or more atoms. Write formulas for their molecules.

4 *True or false?*
A Rhombic and monoclinic sulphur have different atoms.
B Rhombic and monoclinic sulphur have different molecules.
C All non-metals have allotropes.
D Steam and ice are allotropes of water.
E Oxygen (O_2) and ozone (O_3) are allotropes.
F Allotropes of the same element always have different physical properties.

3 Properties of Sulphur

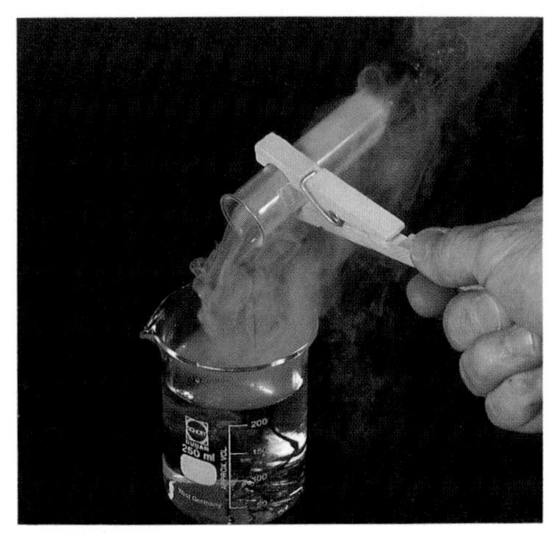

Pouring boiling sulphur into cold water to make plastic sulphur

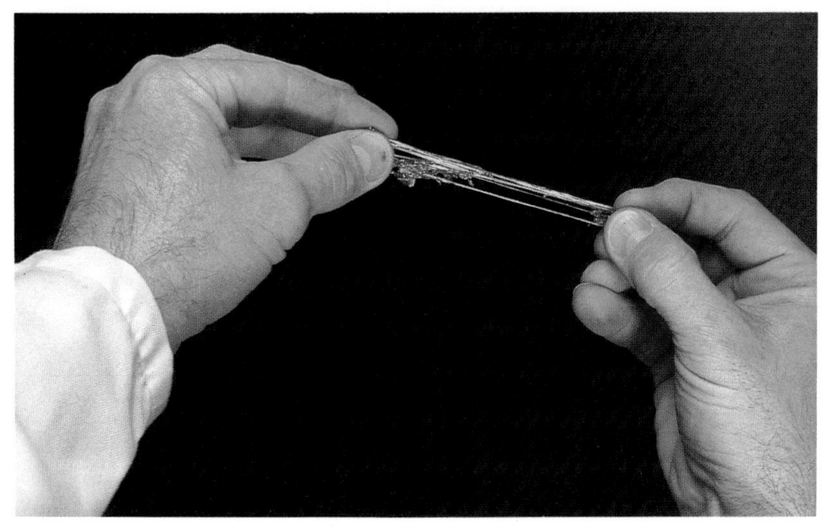

Plastic sulphur

Heating sulphur

When sulphur crystals are heated very gently, we can see four stages clearly.

1 The yellow crystals melt to form an orange runny liquid. This liquid contains S_8 molecules ('rings') that move freely around each other and the liquid is free flowing.

2 Gradually, the orange runny liquid goes dark red and sticky (viscous). The S_8 rings have broken open and then joined up into long chains of sulphur atoms (figure 1). These long chains get tangled up and cannot flow over each other, so the liquid gets very viscous.

3 At higher temperatures, the viscous dark red liquid goes black and runny again. The long chains break up at these higher temperatures forming shorter chains. The shorter chains are less tangled and move over each other easily so the liquid is runny (figure 1).

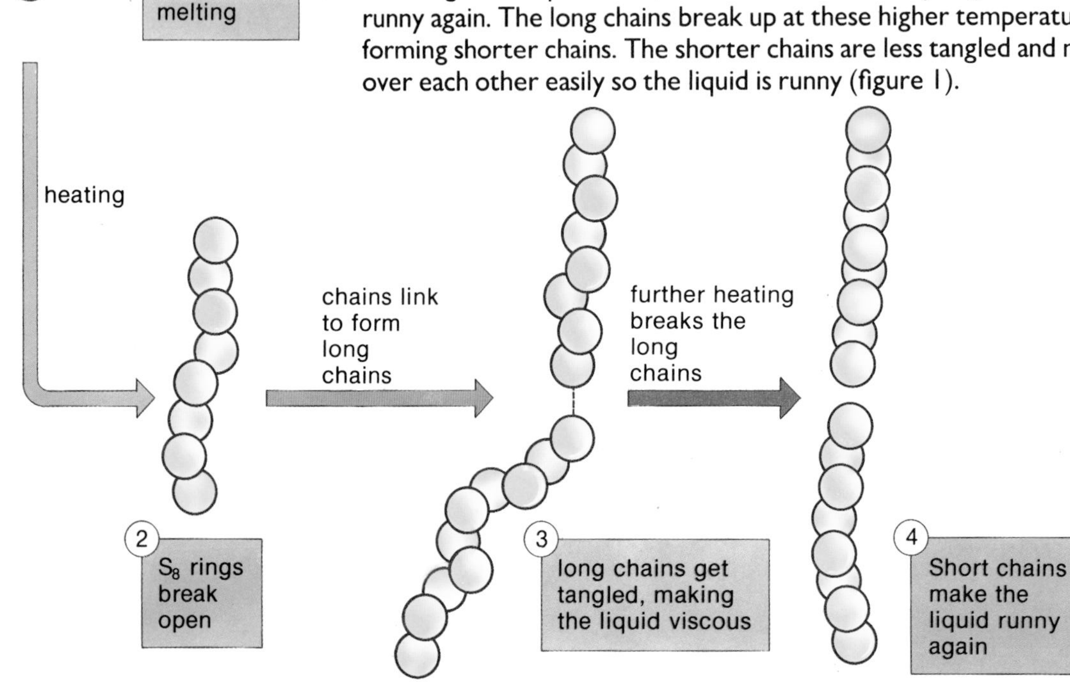

Figure 1

4 The sulphur boils at 445°C, forming an orange vapour. If boiling sulphur is cooled suddenly by pouring it into cold water, the short chains link together forming long tangled chains. The atoms have no time to form neat S_8 rings as in monoclinic or rhombic sulphur. The solid which forms is plastic sulphur. It is elastic like rubber because the long chains can be stretched and then released to form tangled zigzag chains again.

Properties of sulphur

Sulphur is on the right of the periodic table, among the non-metals. It reacts with metals to form sulphides. For example, when iron filings and sulphur are heated, the mixture glows red hot as it reacts to form iron(II) sulphide.

$$Fe(s) + S(s) \rightarrow FeS(s)$$

When sulphur is heated in air or oxygen, it burns with a blue flame forming sulphur dioxide.

$$S(s) + O_2(g) \rightarrow SO_2(g)$$

The sulphur dioxide is a typical non-metal oxide. It is an acidic gas, reacting with water to form sulphurous acid (H_2SO_3).

$$H_2O(l) + SO_2(g) \rightarrow H_2SO_3(aq)$$

Sulphur is less reactive than oxygen, which is directly above it in the periodic table (figure 2). Iron filings spark brightly on heating in oxygen. Compare this with the steady reaction of iron and sulphur described above.

Sulphur (m.pt. 119°C; b.pt. 445°C) is also less volatile than oxygen (m.pt. −219°C; b.pt. −180°C). The big difference in volatility is due to the large difference in the relative molecular masses (M_r) of the two elements (see the table). Sulphur exists as S_8 molecules ($M_r = 8 \times 32 \times 256$). Oxygen exists as O_2 molecules ($M_r = 2 \times 16 \times 32$). So, particles of sulphur are eight times heavier than those of oxygen. As a result, sulphur has a higher melting point and a higher boiling point than oxygen.

	sulphur	**oxygen**
boiling point/°C	445	−180
Volatility	low	high
A_r	32.0	16.0
Molecule	S_8	O_2
Model of molecule	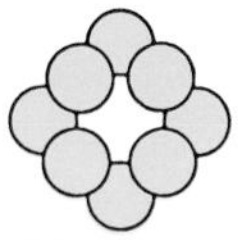	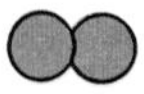
M_r	$8 \times 32 = 256$	$2 \times 16 = 32$

A comparison of sulphur and oxygen

	Group VI	Halogens Group VII	Noble gases Group 0
			He
N	O	F	Ne
P	S	Cl	Ar
As	Se	Br	Kr
Sb	Te	I	Xe

Figure 2

Questions

1 (a) Write the elements of Group VI in order of reactivity (most reactive first).
(b) Do the same for Group VII.

2 (a) Why is sulphur less volatile than oxygen?
(b) Describe and explain the order of volatility from chlorine to bromine to iodine in Group VII.

3 Make a list of the important properties of sulphur.

4 *True or false?*
Sulphur
A dissolves in water.
B dissolves in petrol.
C burns with a blue flame.
D is a good conductor of electricity.
E can be melted by hot water.
F forms an acidic oxide.

4 Sulphur for Industry

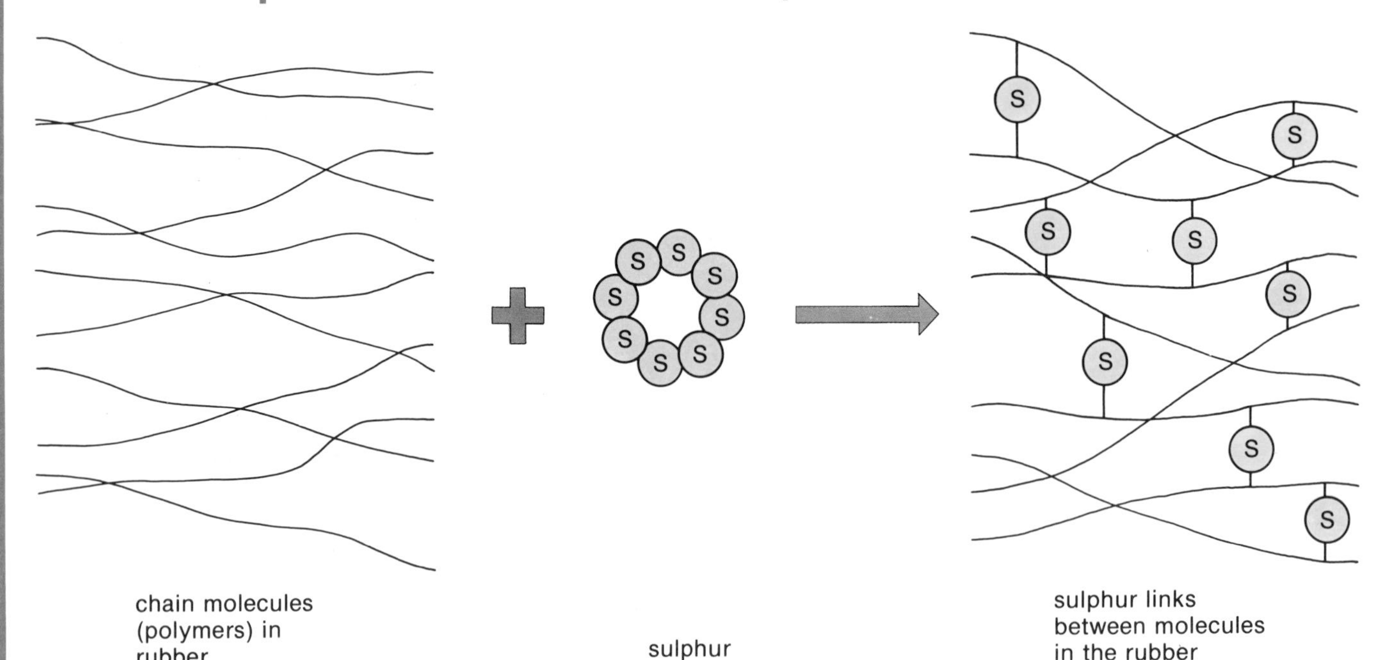

Figure 1
Vulcanizing rubber

The most important use of sulphur is in producing sulphuric acid (section G, unit 1). Other large-scale uses of sulphur are in hardening (vulcanizing) rubber and in making fireworks and matches.

When sulphur is heated with rubber, sulphur atoms form links between long-chain molecules in the rubber (figure 1). These 'crosslinks' stop the chains stretching. This makes the soft elastic rubber hard enough for car tyres. A rubber tyre contains about 2% sulphur. The first chemist to study this way of vulcanizing rubber was Charles Goodyear. Later, he founded the Goodyear Tyre Company.

Smaller quantities of sulphur are used as fungicides to kill fungi in agriculture. These fungicides contain powdered sulphur, suspensions of sulphur in water or sulphur compounds. They are very useful for treating grape vines and hops.

Car tyres contain 2% sulphur

Crops being sprayed with sulphur-based fungicide

Obtaining sulphur

Some sulphur is obtained from hydrogen sulphide in natural gas and from sulphur compounds in crude oil. But most sulphur is obtained from deposits of the element in volcanic areas all over the world. The largest deposits are 150 m underground in Texas, Louisiana (North America) and in Mexico (Central America).

The sulphur cannot be mined by the usual methods because:

1 the layers of limestone and sand above it would 'cave in', and

2 poisonous gases, such as sulphur dioxide, would be produced by drilling.

An American engineer named Frasch devised a clever method of getting sulphur from the underground deposits. He decided to melt the sulphur below the ground and then force the liquid sulphur to the surface using air under pressure. Figure 2 shows a diagram of the process he used. Three pipes (one inside another) are sunk into the sulphur bed. A stream of hot water at 170°C and under ten times atmospheric pressure passes down the outer pipe. This melts the sulphur (m.pt. 119°C). Hot compressed air is blown down the innermost pipe. This air expands and pushes molten sulphur up the middle pipe. At the surface, the molten sulphur runs into large tanks where it solidifies. The sulphur obtained in this way is about 99% pure and can be used for most purposes without further purification.

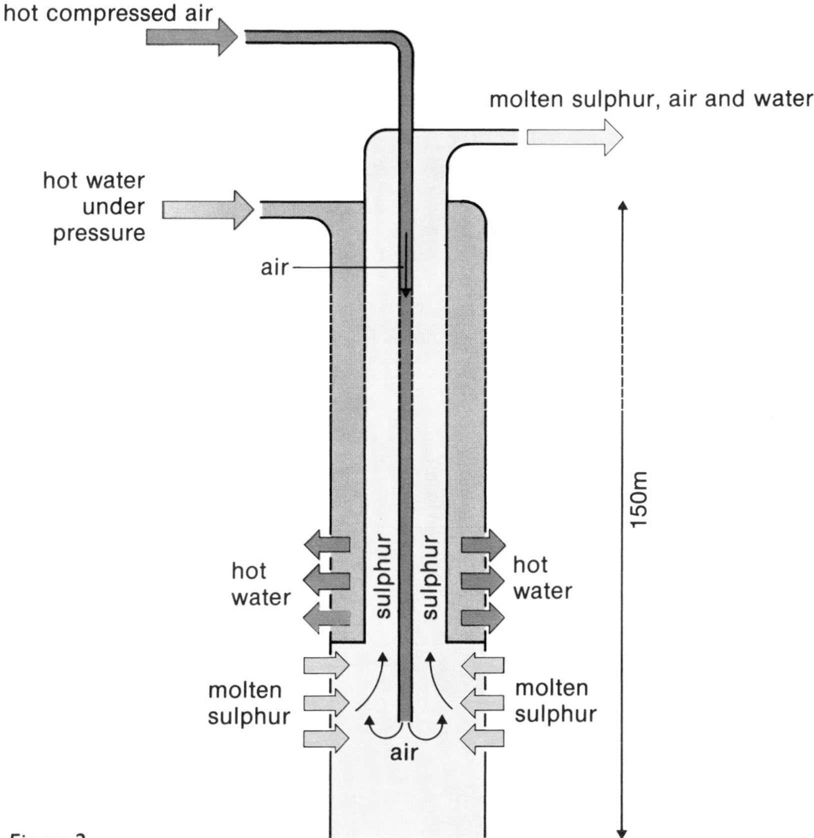

Figure 2
Extracting sulphur by the Frasch process

Sulphur being sprayed from the top of a Frasch plant into tanks

Questions

1 List the main uses of sulphur.
2 (a) Where are the main deposits of sulphur?
(b) Why is the sulphur *not* mined in the same way as coal?
(c) Why is hot *pressurised* water at 170°C used in the Frasch process, rather than boiling water at 100°C?
(d) How does the hot compressed air force the molten sulphur to the surface?
(e) Why is the molten sulphur pumped to the surface through the pipe between the central pipe and the outer pipe?
(f) Suggest possible dangers for workers operating a Frasch process.
3 Why are grape vines sprayed with sulphur or sulphur compounds?
4 Why is natural rubber vulcanized before it is used in car tyres?

5 Sulphuric Acid

The manufacture of sulphuric acid was described in section G, unit 1. About 2½ million tonnes of sulphuric acid are manufactured each year in the UK. Figure 1 shows its main uses.

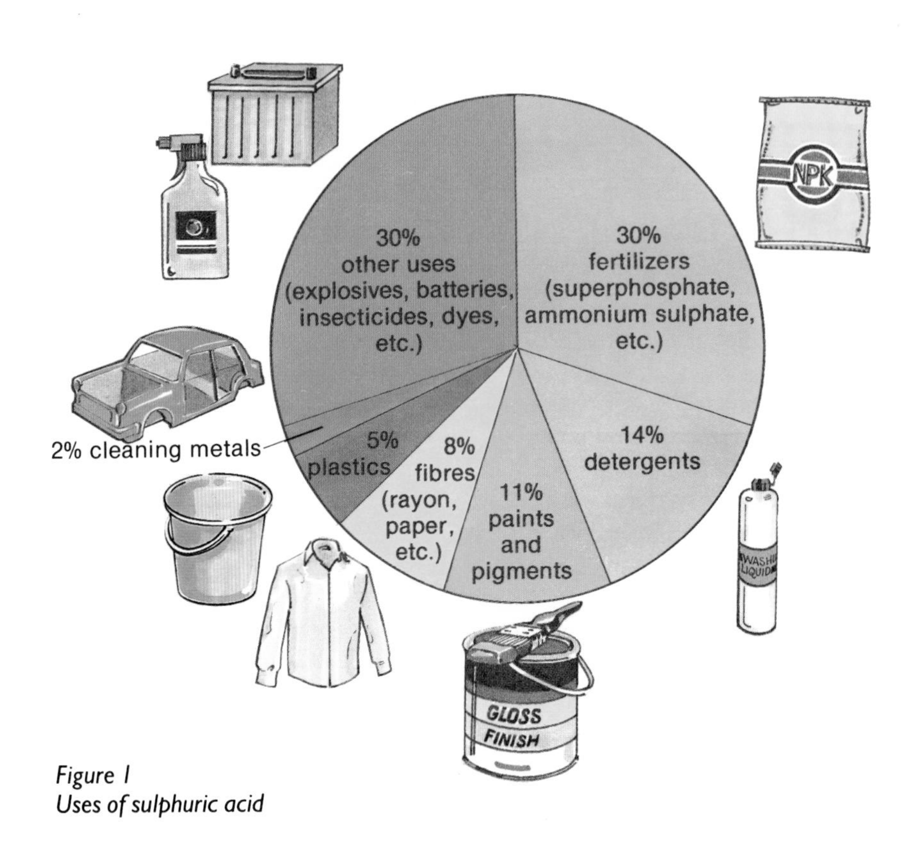

Figure 1
Uses of sulphuric acid

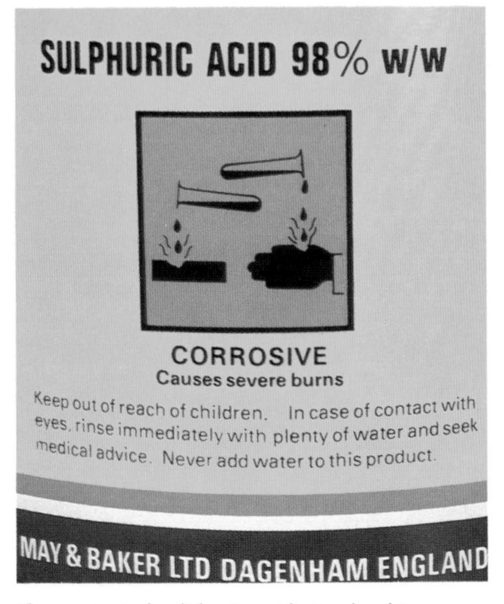

Concentrated sulphuric acid attacks skin, clothing, metals and other materials. Because of these hazards containers for its storage and transport should carry the 'corrosive' warning sign

Reaction with water

Concentrated sulphuric acid contains about 98% H_2SO_4 and only 2% water. **It must be handled carefully wearing eye protection.** It is an oily liquid which reacts with water producing a lot of heat. Because of this,

always add concentrated H_2SO_4 to a large volume of water when mixing the two. Never add water to acid.

If water is added to concentrated H_2SO_4, the heat produced can boil the water and spit out drops of acid.

Pure H_2SO_4 and pure water are both poor conductors of electricity. They contain simple molecules *not* ions. But, a solution of sulphuric acid in water is a good conductor of electricity, so it *must* contain ions. The ions are formed when the H_2SO_4 reacts with water.

$$H_2SO_4(l) \xrightarrow{water} 2H^+(aq) + SO_4^{2-}(aq)$$

Notice that each molecule of H_2SO_4 can produce two H^+ ions when it dissociates (splits up) into ions in water.

Reactions as an acid

Dilute sulphuric acid contains H^+ ions. So, it reacts like a typical acid (section G, unit 3) with

1 indicators

2 metals above copper in the reactivity series, forming a metal sulphate and hydrogen:

$$Mg + H_2SO_4 \rightarrow MgSO_4 + H_2$$

3 bases (metal oxides and hydroxides), forming a metal sulphate and water:

$$MgO + H_2SO_4 \rightarrow MgSO_4 + H_2O$$

4 metal carbonates, forming a metal sulphate, carbon dioxide and water:

$$MgCO_3 + H_2SO_4 \rightarrow MgSO_4 + CO_2 + H_2O$$

Warm the sugar and concentrated sulphuric acid

Concentrated sulphuric acid removes water from sugar and leaves a black mass of carbon

Reactions as a dehydrating agent

Concentrated sulphuric acid reacts violently with water. It absorbs water very rapidly and can be used to dry gases. The concentrated acid also removes water from hydrated salts such as blue copper(II) sulphate ($CuSO_4.5H_2O$), from carbohydrates such as sugar ($C_{12}H_{22}O_{11}$) and from compounds containing hydrogen and oxygen in clothes and skin. This is why it burns and chars clothing and skin. Dilute sulphuric acid does *not* react as a dehydrating agent in this way. When concentrated H_2SO_4 is added to sugar and warmed gently, the reaction gets very hot. The mixture froths up into a steaming black mass of carbon.

$$\underset{\text{sugar}}{C_{12}H_{22}O_{11}} \xrightarrow[H_2SO_4]{\text{concentrated}} \underset{\text{carbon}}{12C} + \underset{\text{water removed by concentrated } H_2SO_4}{11H_2O}$$

Sugar, like other carbohydrates, contains carbon plus hydrogen and oxygen atoms in the ratio 2:1 as in water. Hence the name *carbohydrate*. Our flesh also contains carbohydrates.

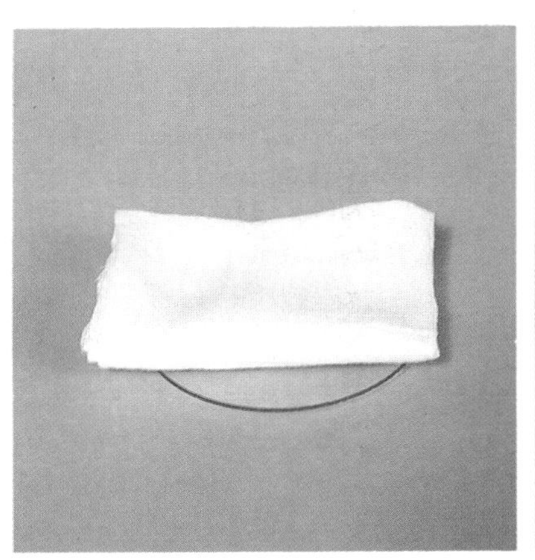

Concentrated sulphuric acid will attack cotton cloth

Questions

1 (a) What is a dehydrating agent?
(b) What uses do dehydrating agents have?
(c) Accurate clocks sometimes have silica crystals (a dehydrating agent) placed near their working parts. Why is this?

2 (a) Write the formula for blue copper(II) sulphate crystals.
(b) What colour will these crystals become when concentrated sulphuric acid is added?
(c) Write an equation for the reaction.

3 Why does dilute sulphuric acid *not* react as a dehydrating agent?

4 (a) Why is concentrated sulphuric acid a poorer conductor of electricity than dilute sulphuric acid?
(b) Why does concentrated sulphuric acid react with magnesium less vigorously than dilute sulphuric acid?
(c) Why does concentrated sulphuric acid burn the skin but dilute sulphuric acid does not?

5 Which of the following are carbohydrates?
Ethanol, C_2H_6O; *Glucose*, $C_6H_{12}O_6$; *Ethene*, C_2H_4; *Glycerine*, $C_3H_8O_3$.

6 Simple Molecular Substances

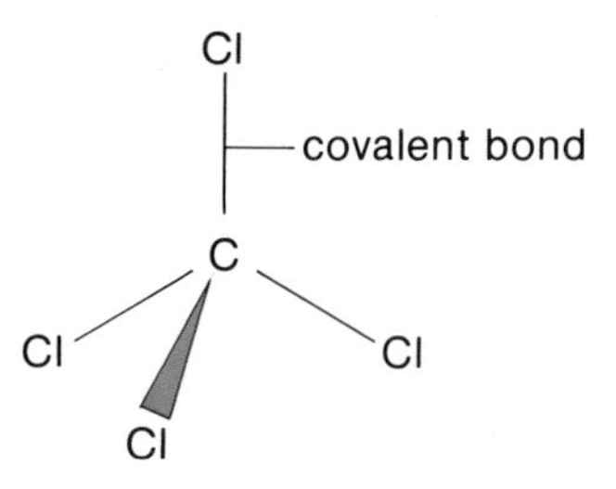

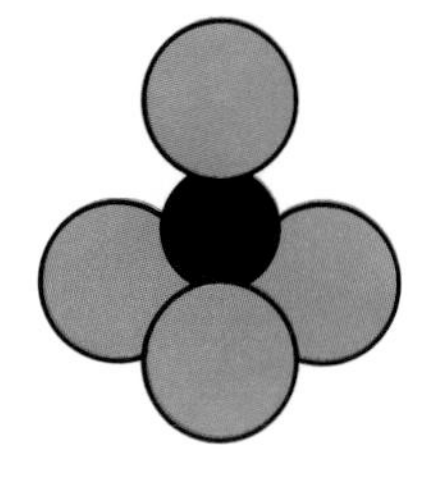

Figure 1
The structure of tetrachloromethane (CCl_4)

Sulphur is a good example of a simple molecular substance. It is made of simple S_8 molecules (each containing eight sulphur atoms). Most other non-metals are also made of simple molecules. For example, hydrogen is H_2, oxygen is O_2 and iodine is I_2. Many non-metal compounds such as water, carbon dioxide, sugar ($C_{12}H_{22}O_{11}$) and tetrachloromethane (CCl_4) are also simple molecular substances.

In these simple molecular substances, the atoms are held together in each molecule by strong **covalent bonds**. (See section D, unit 8.) For example, in tetrachloromethane the carbon atom and the four chlorine atoms are held together by strong covalent bonds (figure 1). The structures of a few simple molecular substances are shown in table 1. (Remember that the combining power (valency) of the element tells you the number of bonds that each atom will have.)

Name	**Molecular formula**	**Structure**	**Model of structure**
Hydrogen	H_2	H – H	
Oxygen	O_2	O = O	
Water	H_2O	H–O–H (bent)	
Sulphur	S_8	ring of eight S atoms	
Methane	CH_4	C bonded to four H	
Hydrogen chloride	HCl	H – Cl	
Iodine	I_2	I – I	
Carbon dioxide	CO_2	O = C = O	

A model of tetrachloromethane

Table 1: the structures of some simple molecular substances

Properties	Iodine	Sugar
Appearance	Soft dark grey crystals	Soft white crystals
Smell	Sharp disinfectant smell	Faint sweet smell
What happens on gentle heating?	Vaporises easily forming a purple vapour	Melts easily to a clear liquid
Does the solid conduct electricity?	No	No
Does the liquid conduct electricity?	No	No

Table 2: properties of iodine and sugar

This butcher is using 'dry ice' (solid carbon dioxide) to keep meat cool and bacteria-free during mincing. After mincing, the 'dry ice', which is a simple molecular substance, will evaporate rapidly without spoiling the meat

Properties of simple molecular substances

Simple molecular substances have some similar properties. These properties are shown by iodine (a non-metal element) and sugar (a non-metal compound) in table 2. Look carefully at table 2. What properties do iodine and sugar have in common? The properties of simple molecular substances can be explained in terms of their structure. The molecules in these substances have no electrical charge (like ions in ionic compounds). So there are no electrical forces holding them together. But as molecular substances do form liquids and solids, there must be some forces holding their molecules together. These weak forces between the separate molecules are called **intermolecular bonds** or **van der Waals' bonds**.

- **Simple molecular substances are soft.** The separate molecules in simple molecular substances, such as I_2, are usually further apart than atoms in metal structures and further apart than ions in ionic structures. The forces between the molecules are only weak and the molecules are easy to separate. Because of this, crystals of these substances are usually soft.

- **Simple molecular substances have low melting points and boiling points.** It takes less energy to separate the molecules in simple molecular substances than to separate ions in ionic compounds, or atoms in metals. So, simple molecular compounds have lower melting points and lower boiling points than ionic compounds and metals.

- **Simple molecular substances do not conduct electricity.** Simple molecular substances have no mobile electrons like metals. They do not have any ions either. This means that they cannot conduct electricity as solids, as liquids or in aqueous solution.

Questions

1 Explain the following: *covalent bond; intermolecular bond; combining power; simple molecule.*
2 List the main properties of simple molecular substances.
3 Simple molecular substances often have a smell, but metals do not. Why is this?
4 Draw structures such as those in table 1 for hydrogen sulphide (H_2S), hydrogen bromide (HBr), ammonia (NH_3) and ethane (C_2H_6). (Remember to use the right number of bonds to each atom.)
5 What properties does butter have that show that it contains simple molecular substances?

7 Carbon: Diamond

An artist using a stick of charcoal

Carbon is another element that has different allotropes. One of the allotropes is diamond, which is hard and clear. The other allotropes, graphite and charcoal, are soft and black. The differences between these allotropes are very great. Diamonds are used to cut and engrave glass, but graphite and charcoal are used by artists to get a soft, shaded effect.

This glass engraving wheel has been toughened with diamond

Diamond, graphite and charcoal are all pure carbon. When they burn in excess oxygen, the only product is carbon dioxide. But these allotropes have different densities. This means that their atoms are not packed in the same way.

The arrangement of carbon atoms in diamond, graphite and charcoal has been studied by X-ray analysis.

Diamond

In diamond, each carbon atom is joined to four other atoms (figure 1). Each atom is at the centre of a tetrahedron surrounded by four others at the corners of the tetrahedron (figure 2). Every carbon atom shares electrons with each of its four neighbours forming strong covalent bonds. The covalent bonds extend through the whole diamond, forming a three-dimensional structure. Thus, a diamond is a single **giant molecule** or **macromolecule**.

Only a small number of atoms are shown in the model in figure 1. In a real diamond, this arrangement of carbon atoms is repeated millions and millions of times.

Figure 1
An 'open' model of the diamond structure

Properties and uses of diamond

- **Diamonds are very hard.** Carbon atoms in diamond are linked by very strong covalent bonds. This makes diamond hard. Another reason for its hardness is that the atoms are not arranged in layers so they cannot slide over one another like the atoms in metals. Diamond is one of the hardest known substances. Most of its industrial uses depend on this hardness. Diamonds which are not good enough for gems are used in glass cutters and in diamond studded saws. Powdered diamonds are also used as abrasives for smoothing very hard materials.

- **Diamond has a very high melting point.** Carbon atoms in diamond are held in the crystal structure by very strong covalent bonds. This means that the atoms cannot vibrate fast enough to break away from their neighbours until very high temperatures are reached. So, the melting point of diamond is very high.

- **Diamond does not conduct electricity.** Diamond does *not* conduct electricity, unlike metals and graphite. In metals and graphite, some of the *outer* electrons are not strongly attached to any nucleus. They move towards the positive terminal when metals and graphite are connected to a battery. In diamond, however, the *outer* electrons of each carbon atom are held firmly in covalent bonds. So, diamond does not conduct electricity.

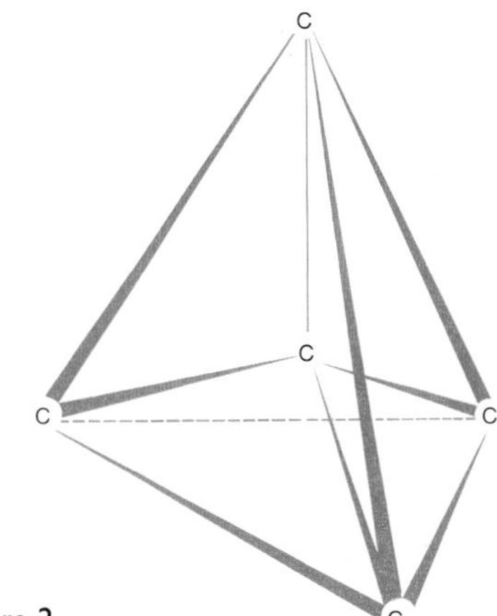

Figure 2

This diamond is being polished

Questions

1 Why is diamond called a giant molecule?

2 (a) Make a list of uses of diamond.
(b) How do these uses depend on the properties and structure of diamond?

3 The largest natural diamond is the Cullinan diamond. This weighs about 600 g.
(a) How many moles of carbon does it contain? (C = 12)
(b) How many atoms of carbon does it contain?

4 Why are diamond cutters used to cut glass?

5 'Diamonds are a girl's best friend'. Is this true? What do you think?

8 Carbon: Graphite

A can of lubricating oil containing graphite

Graphite is the second important allotrope of carbon. Figure 1 shows a model of part of the structure of graphite. Notice that the carbon atoms are arranged in parallel layers. Each layer contains millions and millions of carbon atoms arranged in hexagons. Each carbon atom is held strongly in its layer by covalent bonds, so that every layer is a **giant molecule**. The distance between neighbouring carbon atoms in the same layer is only 0.14 nm, but the distance between the layers is 0.34 nm.

Figure 1
A model of the structure of graphite

Properties and uses of graphite

- **Graphite is a lubricant.** In graphite, each carbon atom is linked by strong covalent bonds to three other atoms in its layer. But, the layers are 2½ times further apart than carbon atoms in the same layer. This means that the forces between the layers are weak. If you rub graphite, the layers slide over each other and onto your fingers. This property has led to the use of graphite as the 'lead' in pencils and as a lubricant. The layers of graphite slide over each other like a pile of wet microscope slides (figure 2). The wet slides stick together and it is difficult to pull them apart, but a force parallel to the slides pushes them over each other easily and smoothly.

- **Graphite has a very high melting point.** Although the layers of graphite move over each other easily, it is difficult to break the bonds between carbon atoms within one layer. Because of this, graphite does not melt until 3730°C and it does not boil until 4830°C. So, it is used to make crucibles for molten metals. The bonds between carbon atoms in the layers of graphite are so strong that graphite fibres with the layers arranged along the fibre are stronger than steel. These fibres are used to reinforce metals and broken bones.

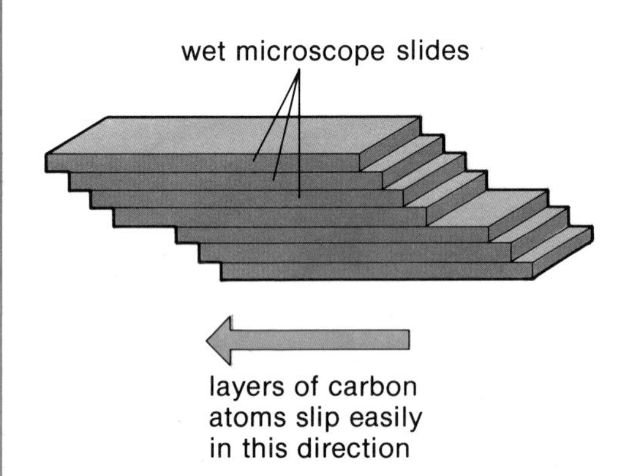

Figure 2
The layers in graphite slide over each other like wet microscope slides

• **Graphite conducts electricity.** The bonds *between* the layers of graphite are fairly weak. The electrons in these bonds move along the layers from one atom to the next when graphite is connected to a battery. So graphite will conduct electricity, unlike diamond and other non-metals. Because of this unusual property, graphite is used as the positive terminal in dry cells (section D, unit 12) and as electrodes in industry.

This large graphite anode is being prepared for use in the electrolytic extraction of aluminium

Graphite fibres have been used to reinforce the shaft of this badminton racket.

Charcoal

X-ray studies show that charcoal contains tiny crystals with a similar structure to graphite. The spaces between the layers of carbon atoms in finely powdered charcoal can trap other atoms and molecules. The biggest single use of powdered charcoal is in the sugar industry where it is used to absorb coloured impurities from brown sugar and syrup. Powdered charcoal can also absorb large volumes of many gases. One gram of charcoal will absorb 380 cm^3 of sulphur dioxide or 235 cm^3 of chlorine at room temperature, but oxygen is not readily absorbed. Because of this, charcoal is used in gas masks to protect the wearer from poisonous gases.

Gas masks contain powdered charcoal which absorbs poisonous gases

Questions

1 Make a list of the important uses of graphite.

2 Make a table to show the similarities and differences between diamond and graphite.

3 (a) Why is a zip-fastener rubbed with a soft pencil to make it move more freely?
(b) Why is pencil used rather than oil?
(c) Why is graphite better than oil for lubricating the moving parts of hot machinery?

4 Graphite powder is mixed with clay to make pencil 'lead'. How does the hardness of a pencil depend on the amounts of graphite and clay in it?

5 (a) Graphite curtains are used to prevent kitchen smells spreading. Why is this?
(b) How does the graphite prevent the smell from spreading?

9 The Structure of Elements

Element	Melting point/°C	Boiling point/°C
Graphite (carbon)	3730	4830
Copper	1083	2600
Sulphur	119	445
Iodine	114	183

Table 1: melting points and boiling points of four elements

Evidence for structure from properties

What happens when iodine, sulphur, graphite and copper are heated? Iodine forms a purple vapour on gentle warming. Sulphur melts at a slightly higher temperature, but graphite and copper do not change until very high temperatures. Iodine is the most volatile. Graphite is the least volatile.

When a solid melts, its particles must have enough energy to break away from their orderly arrangement and move around each other freely. If the melting point is low, then the particles need less energy to break away from their neighbours in the solid. The melting points and boiling points of graphite, copper, sulphur and iodine are listed in table 1.

Look at table 1.

1 Which element is the most difficult to melt?
2 Which element is the least difficult to melt?
3 Which element needs the most energy to separate its particles?
4 Which element has the strongest bonds between its particles?

A model of an iodine molecule

Why is iodine more volatile than graphite?

Iodine contains molecules with two atoms, I_2. Sulphur consists of molecules with eight atoms, S_8. These two elements are like almost all other non-metals. They contain separate small molecules and are **simple molecular structures**. In these structures, the atoms *in* each molecule are held together by strong covalent bonds, but the forces *between* molecules are very weak. It is fairly easy to separate their molecules from each other. Elements with simple molecular structures tend to have melting points below 150°C and boiling points below 500°C. Four more simple molecules are shown in figure 1.

In graphite (unit 8 of this section), the carbon atoms are strongly bonded in layers. Each layer is one big molecule and the layers contain so many carbon atoms that it is difficult to separate them. So, graphite melts at very high temperatures. Graphite is an example of a giant molecule or macromolecule. Diamond and silicon also form giant molecules. The structure of silicon is very similar to that of diamond (unit 7 of this section).

A model of the structure of graphite. The dashed lines show the positions of the atoms in one layer and the next. The dashed lines also show the relatively large distance between the layers compared with the distance between atoms in the same layer.

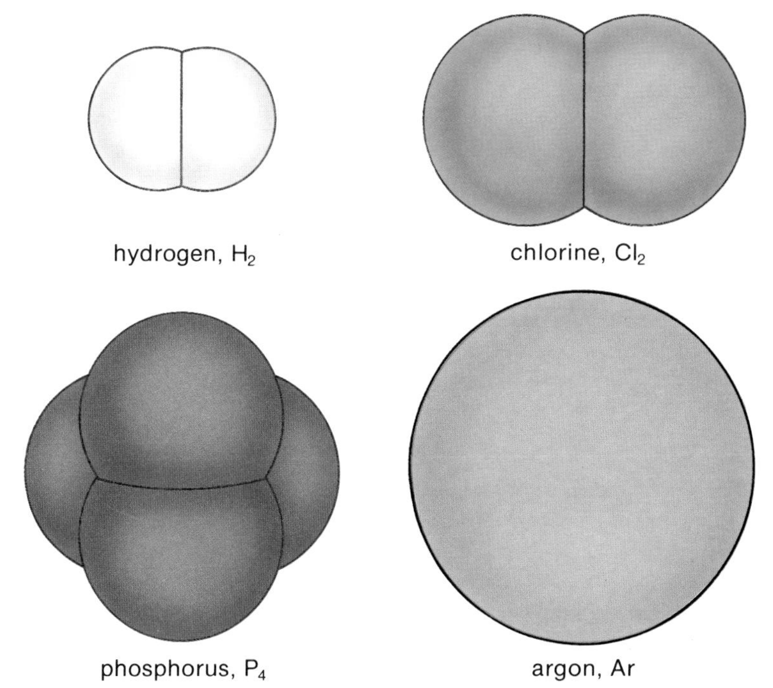

Figure 1
Models of four simple molecules

So far we have looked at non-metals. Now we must look at metals.

1 Look at the melting points and boiling points in table 1. How does copper compare with the three non-metals?
2 What type of structure is likely for copper—a simple molecular structure or a giant structure?
3 How are the atoms held together in a metal?

In metals, the atoms are held together in a giant structure by the attraction of positive ions for mobile electrons. (See section F, unit 5). The structure is described as **giant metallic** or metallic for short. Elements with giant structures, like metals and carbon, have high melting points (above 300°C) and high boiling points (above 1000°C).

Notice how physical properties (such as melting points, boiling points and conductivity) give some evidence for the bonding and structure of elements. The important points from this unit are summarised in table 2.

Element	Structure	Examples	Properties
METAL	Giant metallic	Na, Fe, Cu	High melting points and boiling points. Conduct electricity
NON-METAL	Simple molecular	I_2, S_8, Cl_2 O_2, Ar	Low melting points and boiling points. Do not conduct electricity
	Giant molecular	Diamond, graphite	Very high melting points and boiling points. Do not conduct electricity (except graphite)

Table 2: a summary of the structure of metals and non-metals

Questions

1 Atoms in elements can be arranged in three different structures.
(a) Name the three types of structure.
(b) Give two examples for each structure.
(c) What sort of bonds are present between particles in each structure?
2 Make a list of the important properties of the elements with the three different structures in this unit.
3 Consider sodium, chlorine and sodium chloride.
(a) What type of structure do these three substances have?
(b) Make a table showing the following properties of the three substances: colour; solubility in (or reaction with) water; state at room temperature; volatility; conductivity; reaction with oxygen.
(c) Choose *one* of the three substances and show how its structure can be used to explain its properties.

10 The Structure of Compounds

Evidence for structure

We get the best evidence for the structure of a substance from X-ray studies. As we saw in the last unit, physical properties such as melting points, boiling points and conductivity can also provide clues to the structure of a substance.

Look at the boiling points of the compounds in the table.
1 Which compounds have a high boiling point above 1000°C?
2 Which compounds have a low boiling point below 500°C?
3 What type of compound has a high boiling point?
4 What type of compound has a low boiling point?

The top five compounds in the table have low boiling points. They are non-metal compounds with simple molecular structures (unit 6). The bottom three compounds in the table are metal/non-metal compounds. Their high boiling points show that they have giant structures. In section D, we discovered that metal/non-metal compounds, like sodium chloride and calcium oxide, are composed of positive metal ions and negative non-metal ions. Because of this they are described as **giant ionic structures**. The ions are held together by strong forces of attraction between the positive and negative charges. This electrical force holding the ions together is called an **ionic bond** or electrovalent bond.

Compound	Formula	Boiling point/°C
Water	H_2O	100
Tetrachloromethane (carbon tetrachloride)	CCl_4	77
Ethanol	C_2H_5OH	79
Hydrogen chloride	HCl	−85
Methane	CH_4	−160
Sodium chloride	Na^+Cl^-	1465
Magnesium chloride	$Mg^{2+}(Cl^-)_2$	1418
Calcium oxide	$Ca^{2+}O^{2-}$	2850

The boiling points and formulas of some common compounds

Figure 1
A perfect sodium chloride crystal

The structure of ionic compounds

X-ray studies show that the ions in different ionic compounds are arranged in different patterns. Many ionic compounds, including sodium chloride and calcium oxide, have a cubic structure. The photograph of a sodium chloride crystal in figure 1 shows the cubic shape very well. Figure 2 shows a diagram of the structure of sodium chloride. Each dot shows the centre of an ion. Notice that the ions form a cubic pattern. Although figure 2 shows only twenty-seven ions, there are millions and millions of ions in even the tiniest crystal of sodium chloride.

Ionic compounds, like sodium chloride, are giant structures of ions in the same way that metals are giant structures of atoms. Each Na^+ ion in the structure 'belongs' to all the Cl^- ions around it and vice versa.

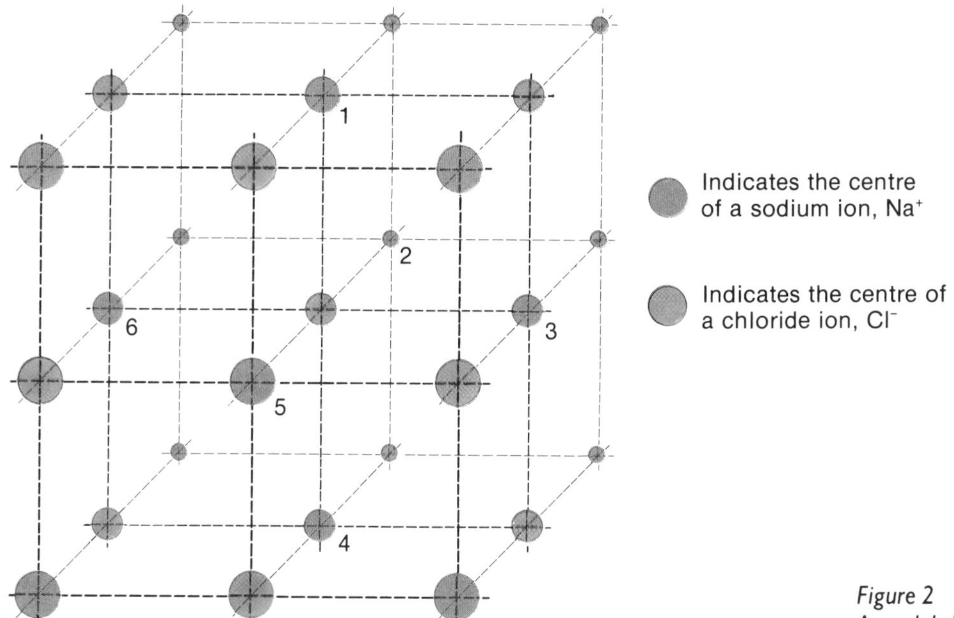

Figure 2
A model showing the positions of Na^+ and Cl^- ions in sodium chloride

Notice that each Na^+ ion in the structure is surrounded by six Cl^- ions, and each Cl^- ion is surrounded by six Na^+ ions. The six Na^+ ions round the central Cl^- ion in figure 2 are numbered. Four of the Na^+ ions (numbered 2, 3, 5 and 6) are in the same horizontal layer as the central Cl^- ion. One Na^+ ion (number 1) is in the layer above and another Na^+ ion (number 4) is in the layer below. Figure 3 shows a model of a crystal of sodium chloride in which the ions are represented by polystyrene balls. The large green balls represent Cl^- ions (A_r = 35.5, the smaller silver balls represent Na^+ ions (A_r = 23.0).

Figure 3
A model of sodium chloride

Questions

1 Explain the following:
giant structure; ionic bond.
2 Make a table for compounds, similar to table 2 for elements in unit 9.
3 Sodium fluoride and magnesium oxide have the same crystal structure and similar distances between ions. The melting point of NaF is 992°C, but that of MgO is 2640°C. Why is there such a big difference in their melting points?
4 A substance is a poor conductor of electricity in the solid state. It melts at 217°C and boils at 685°C. Could this substance be (i) a metal; (ii) a non-metal; (iii) a giant molecule; (iv) an ionic solid; (v) a simple molecular solid?

11 Salt: An Important Ionic Compound

Tennis players usually take salt tablets to replace the salt they lose in their sweat

Ionic compounds are present in the sea and in the Earth's crust. Many rocks contain ionic compounds. These include rock salt (Na^+Cl^-), limestone ($Ca^{2+}CO_3{}^{2-}$) and iron ore ($(Fe^{3+})_2(O^{2-})_3$). Clay, sandstone and granite also contain ionic compounds.

One of the most important ionic compounds is salt (sodium chloride). Salt is an essential mineral in our diet. Most foods contain salt but some foods are saltier than others. Our diet must contain the right amount of salt. Too much salt may cause high blood-pressure. Too little salt causes sharp pains ('cramp') in our muscles. Sweat is mainly salt solution. People who work in hot places eat salt tablets to make up for the salt that they lose by sweating.

Cattle licking a salt block

The table shows the important properties of ionic compounds. These properties can be explained in terms of their giant structure of ions.

- **Hardness.** In order to cut a lump of sodium chloride, the ions must be separated. This is very difficult because each ion is held in the crystal lattice by strong attractions from the six ions of opposite charge around it. So, sodium chloride and other ionic solids are very hard.

- **Melting and boiling.** In ionic solids, the ions vibrate about fixed positions. As the temperature rises, the ions vibrate more and more. Eventually, the ions vibrate so much that they 'escape' from their places in the crystal and slide freely around each other. When this happens, the solid is melting. The forces between ions of opposite charge are so strong that ionic compounds have high melting points and even higher boiling points.

Ionic compounds:
• are hard;
• have high m.pt;
• have high b.pt;
• do not conduct when solid;
• conduct when liquid and in aqueous solution

Properties of ionic compounds

A salt gritting vehicle being loaded, ready for work on icy roads

• **Conductivity.** Solid ionic compounds cannot conduct electricity because the ions are held in the crystal and cannot move towards the electrodes. When the solid melts or when it is dissolved in water, the ions are free to move. So, molten and aqueous ionic compounds conduct electricity. The electrolysis of molten sodium chloride is used to make sodium and chlorine (section D, unit 4).

The electrolysis of saturated aqueous sodium chloride (brine) is used to manufacture chlorine and hydrogen (section E, unit 9). Large amounts of impure sodium chloride (crushed rock salt) are used in de-icing roads. The salt mixes with the ice and lowers its melting point. Mixtures of ice and salt will melt at temperatures down to −22°C. This means that the temperature will have to go well below 0°C before the roads ice up. Unfortunately, this use of salt does have a disadvantage for motorists. The salty water gets splashed onto the steel parts of vehicles and this causes the iron to rust faster than normal.

This picture shows the badly corroded structure of Brighton Pier before extensive repairs were carried out in 1985. Salt in sea water speeds up rusting

Questions

1 Explain the following words: *melting*; *lattice*.

2 Solid sodium chloride does not conduct electricity, but liquid sodium chloride conducts well.

(a) Explain this statement.

(b) Write equations for the processes at the electrodes when liquid NaCl conducts. (Section D may help you here.)

3 Give three important uses of sodium chloride as pure salt, rock salt or brine. Explain why sodium chloride has these uses.

4 Substance *X* melts at a high temperature. Liquid *X* conducts electricity.

(a) Which of the following could be *X*?

calcium chloride; *starch*; *copper*; *sulphur*; *polythene*; *bronze*; *carbon disulphide*; *zinc oxide*.

(b) Explain your answers to part (a).

5 Athletes sometimes need to take salt tablets. Why is this?

Section H: Study Questions

1 Read the passages below and answer the questions that follow.

Carbon. Many of the things that we use contain carbon, for instance glue, wool, rubber, leather and paper. Chemists call these substances organic. The element carbon, in the form of charcoal, was one of the first known elements. The Bible refers to charcoal or coal as a fuel. Prehistoric people used charcoal for drawing as we use graphite today.

Sulphur. This yellow, powdery element is found in Italy. People soon discovered that it would catch fire and burn quickly with a blue flame, giving off a choking gas. Sulphur was often called 'brimstone'. In modern times the chief use for sulphur has been for making sulphuric acid. This chemical is of major importance in making ammonium sulphate. In England, sulphur is found combined with calcium and oxygen as gypsum.

(a) What is an element? (1)
(b) Name another element *not* mentioned in the passages. (1)
(c) What does 'organic' mean? (2)
(d) What does the word 'fuel' mean? (2)
(e) Name *one* substance formed when charcoal is oxidized. (1)
(f) What other elements besides sulphur are present in sulphuric acid? (2)
(g) What is the choking gas given off when sulphur burns? (1)
(h) What can ammonium sulphate be used for? (1)
(i) Give the chemical name for gypsum. (1)

Total [12] **LEAG**

2

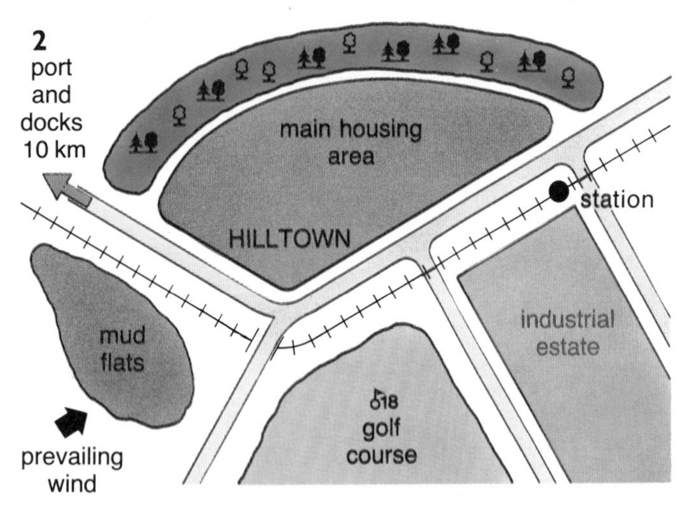

A chemical company (*Sulchem*) wants to build a new factory near Hilltown in order to manufacture sulphuric acid.

(a) Where would you site the new factory?
(b) What are your reasons for your choice of site?
(c) The Council at Hilltown want 'Sulchem' to build their factory on the mud flats. They have offered 'Sulchem' financial help if the factory is sited on the mud flats. What disadvantages will this site have for
(i) Sulchem (ii) Hilltown?

3 Titanium is the seventh most abundant element in the Earth's crust. One form in which it occurs is *rutile*, TiO_2. In extracting titanium from its ore, rutile is first converted to titanium(IV) chloride, $TiCl_4$, and this is then reduced to the metal by heating it with sodium or magnesium in an atmosphere of argon. Titanium(IV) chloride is a simple molecular covalent substance.

(a) Given that the titanium atom has four electrons used for bonding, draw a diagram to show the bonding in titanium(IV) chloride. (Only the outer electrons of the chlorine atoms should be shown.) (2)
(b) Write a balanced equation for the reaction of titanium(IV) chloride with sodium. (1)
(c) (i) In which physical state would you expect to find titanium(IV) chloride at room temperature?
(ii) Explain why the physical state of titanium(IV) chloride differs from that of sodium chloride at room temperature. (3)
(d) Suggest a reason why it is necessary to carry out the reaction of titanium(IV) chloride with sodium in an atmosphere of argon. (1)
(e) Titanium is expensive in spite of the fact that it is relatively abundant in the Earth's crust. Suggest a reason for this. (1)
(f) Titanium is used in the structures of supersonic aircraft and space vehicles. Suggest *two* properties it might have that make it more suitable than other metals for this purpose. (2)
(g) Titanium is a transition metal. State *two* properties, different from those in (f) which you would expect it to have. (2)

Total [12] **LEAG**

4 (a) Give two uses of the element chlorine. (2)
(b) The element astatine (At) is at the bottom of Group VII. Use your knowledge of the rest of the Group to predict some of the properties of this element.
(i) At room temperature would astatine be a solid, liquid or gas? (1)
(ii) Predict the formula of the compound which contains only sodium and astatine. (1)
(iii) Would there be a chemical reaction between bromine and potassium astatide? Give a reason for your answer. (1)
(c) When sulphur is heated in a test tube the yellow crystals melt to form a golden-yellow mobile liquid which changes at 180° C into a dark brown very viscous liquid. More heating to about 400° C produces a brown, less viscous liquid.
(i) What will happen if sulphur is heated in air? (1)
(ii) What is the molecular structure of sulphur in the yellow crystals? (1)
(iii) If the brown liquid at 400° C is cooled rapidly to room temperature, which form of sulphur is produced? (1)
(iv) Explain why the molten sulphur becomes viscous. (2)

Total [10] **NEA**

SECTION 1

Energy, Fuels and Carbon Compounds

A summary of our energy sources

1 Energy in Everyday Life

Bushmen around a fire. The energy given off by the fire keeps the bushmen warm

Everyday we use energy in our homes, schools and industries in thousands of different ways. Energy is often needed to make changes happen. We need it to warm a room, cook a meal or light a torch. Although we use a lot of energy in our homes, industry uses far more. Energy is needed to turn raw materials like clay into useful things like bricks. It is also needed to mine coal and other minerals and to generate electricity.

Transferring energy to or from chemicals is the basis of chemical changes in industry and in living things. Some chemical changes, such as the decomposition of limestone to lime, need heat to make them happen.

$$\underset{\text{limestone}}{CaCO_3} \xrightarrow{heat} \underset{\text{lime}}{CaO} + CO_2$$

Such chemical reactions, which *take in* heat, are described as **endothermic**. Other chemical reactions *give out* heat. In these changes, chemical energy is converted into heat. This happens when fuels such as coal, oil and natural gas are burnt.

$$\underset{\text{natural gas}}{CH_4} + 2O_2 \rightarrow CO_2 + 2H_2O + \text{heat}$$

Chemical reactions like this, which *give out* heat, are described as **exothermic**.

Every year in the UK, we need fuel equivalent to about 300 million tonnes of coal. About one third of this is provided by coal itself, one third by oil, one quarter by natural gas and the rest by nuclear and hydroelectric power (figure 1).

coal 35%
oil 34%
natural gas 24%
nuclear 6%
hydroelectric 1%

Figure 1
Sources of energy in the UK (The figures are for 1983)

Most coal in Britain is used to generate electricity. This is why most power stations have huge coal heaps nearby. This photograph shows the coal-fired power station at Didcot in Oxfordshire

Foods such as fats and carbohydrates are important biological fuels. When they are broken down (metabolized) in our bodies, chemical energy is changed into heat to keep us warm, into mechanical energy to help us move around and into electrical energy to carry messages along our nerves.

The energy we need to heat materials and move objects can come from a chemical reaction in which energy is released (an exothermic reaction). The chemical energy in explosives, petrol and rocket fuels can be converted into mechanical energy for moving objects. The chemical energy stored in the raw materials of cells and batteries can be changed into electrical energy. Figure 2 summarises the most important ways in which chemical energy can be converted into other forms of energy. Notice that energy is converted from one form to another in all these changes. The original source of energy is often a food or a fuel. Chemical energy in these substances gets changed into heat, mechanical energy or electricity. None of the energy is lost.

- *Energy cannot be destroyed. It can only be changed from one form to another. This is sometimes called the* **Law of Conservation of Energy.**

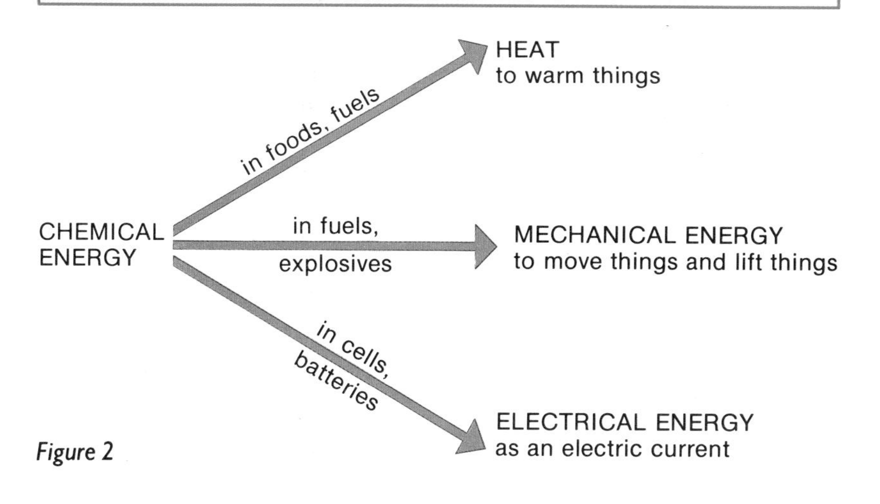

Figure 2

Questions

1 Explain the words: *endothermic; exothermic; fuel.*

2 (a) What is the Law of Conservation of Energy?
(b) Give two examples to illustrate the law.

3 (a) Give one example in each case of a food which contains a high proportion of (i) fat; (ii) carbohydrate.
(b) What elements do fats and carbohydrates contain?
(c) Name two carbohydrates.
(d) What substances are produced when carbohydrates are broken down (metabolized)? (Hint: the carbohydrates act like fuels.)

2 Energy Changes

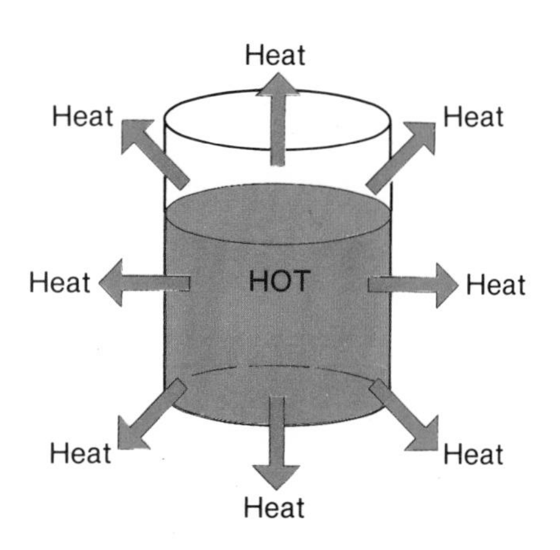

Figure 1

When hydrochloric acid reacts with sodium hydroxide, the reaction is exothermic. Chemical energy is given out by the reactants in forming the products, sodium chloride and water.

$$HCl(aq) + NaOH(aq) \rightarrow NaCl(aq) + H_2O(l)$$

The energy is given out as heat and the temperature of the mixture rises. After a while, the temperature of the products falls to room temperature as the heat from the reaction is lost to the surroundings (figure 1). The surroundings have gained energy but the reaction mixture has lost energy. *Since the total energy of the products is less than that of the reactants*, we say that the *heat change* for the reaction is *negative*. The Greek letter Δ (delta) is often used to mean 'change of'. So, the heat change of a reaction is given the symbol ΔH. Notice that ΔH for every exothermic reaction is negative.

When hydrochloric acid and sodium hydroxide react, the heat lost is 57.9 kJ per mole of each reactant.

$$HCl(aq) + NaOH(aq) \rightarrow NaCl(aq) + H_2O(l) \ \Delta H = -57.9 \text{ kJ}$$

We can show the heat change for the reaction as an *energy level diagram* (figure 2). Chemical energy in the hydrochloric acid and in the sodium hydroxide is partly changed to chemical energy in the sodium chloride and water and partly lost as heat. This heat which is lost from the reactants warms up the products. In time, the products (sodium chloride and water) cool down to room temperature. Overall, energy

This space shuttle needs huge amounts of energy to take off. What do you think provides this energy?

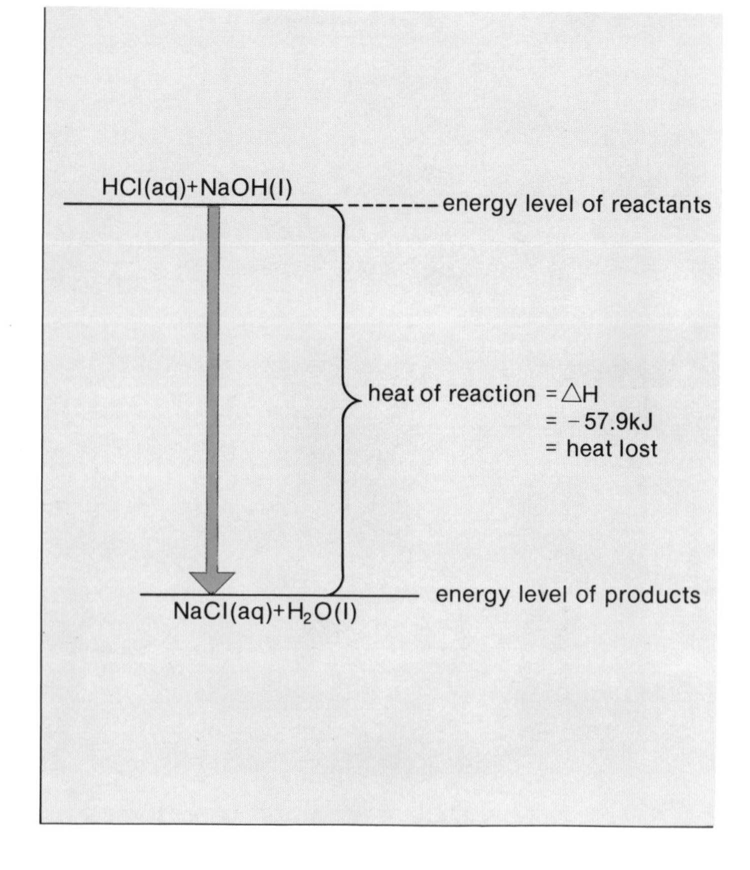

Figure 2
An energy level diagram for the reaction of HCl and NaOH

is transferred from the reacting substances to the surroundings. Thus, the sodium chloride and water have *less* energy and they are more stable than the hydrochloric acid and sodium hydroxide. Because of this, sodium chloride (NaCl) and water (H_2O) are written at a lower level on the energy level diagram. The heat change, ΔH, for the number of moles shown in the equation and on the energy level diagram is usually called the **heat of the reaction**.

When an endothermic reaction occurs, the reactants must take in energy as they form the products. The energy which is needed may be provided by heating the reactants or by electricity. In some endothermic reactions, the temperature falls below room temperature as the heat needed for the reaction is taken from the materials themselves. This happens when ammonium chloride (NH_4Cl) or potassium nitrate (KNO_3) dissolves in water. Eventually the temperature of the products rises to room temperature again as heat is absorbed from the surroundings (figure 3). In this case, the total energy content of the products is greater than that of the reactants. The products are less stable than the reactants and they are at a higher level in the energy level diagram (figure 4). Since the products have *gained* energy, the heat of reaction, ΔH, is *positive* for endothermic reactions. For example,

$$2H_2O(l) \rightarrow 2H_2(g) + O_2(g) \quad \Delta H = +575 \text{ kJ}$$

$$KNO_3(s) \xrightarrow{water} K^+(aq) + NO_3^-(aq) \quad \Delta H = +34.6 \text{ kJ}$$

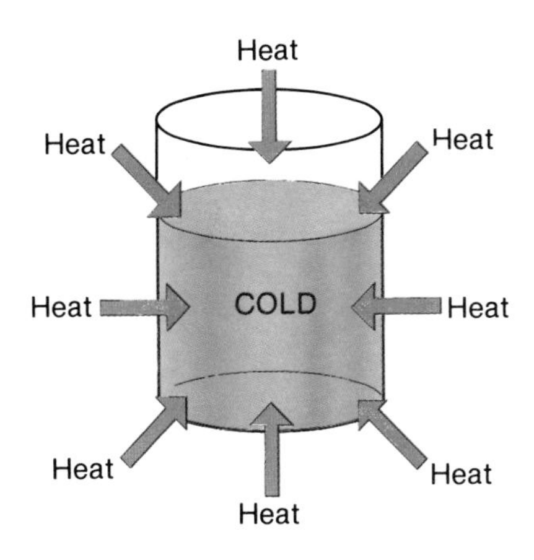

Figure 3

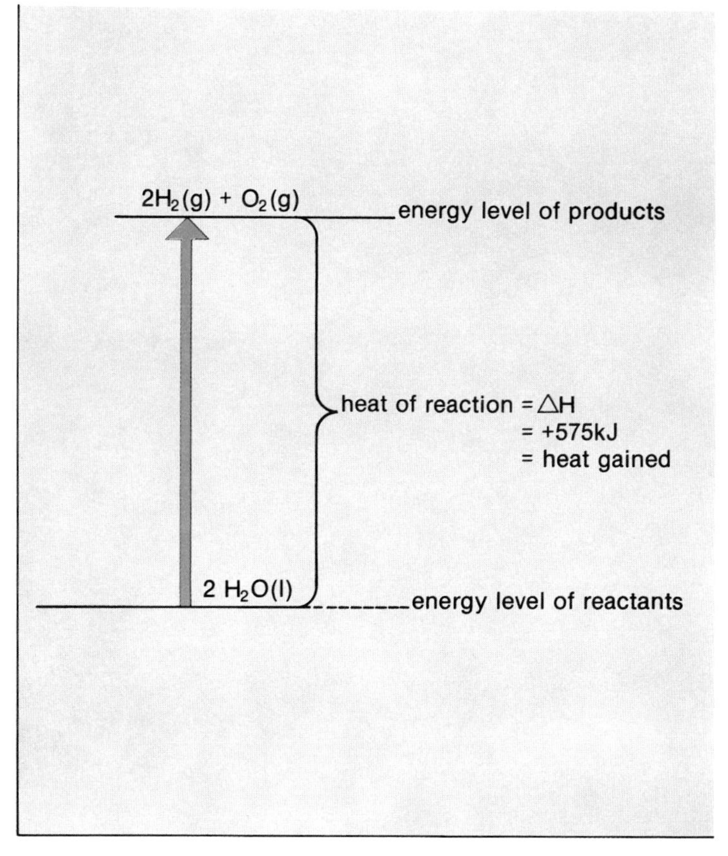

Figure 4
An energy level diagram for the decomposition of water

Questions

1 Why is the symbol ΔH chosen for the heat of a reaction?

2 Compare exothermic and endothermic reactions in terms of (i) heat change with the surroundings; (ii) the sign of ΔH; (iii) energy levels of reactants and products.

3 Energy is *lost* in an exothermic reaction. Does this break the law of conservation of energy? Explain.

4 Draw fully-labelled energy level diagrams for the reactions represented by the following equations:

(i) $CH_4(g) + 2O_2(g) \rightarrow CO_2(g) + 2H_2O(l)$
$\Delta H = -894$ kJ

(ii) $KNO_3(s) \xrightarrow{water} K^+(aq) + NO_3^-(aq)$
$\Delta H = +34.6$ kJ

3 Energy and Change of State

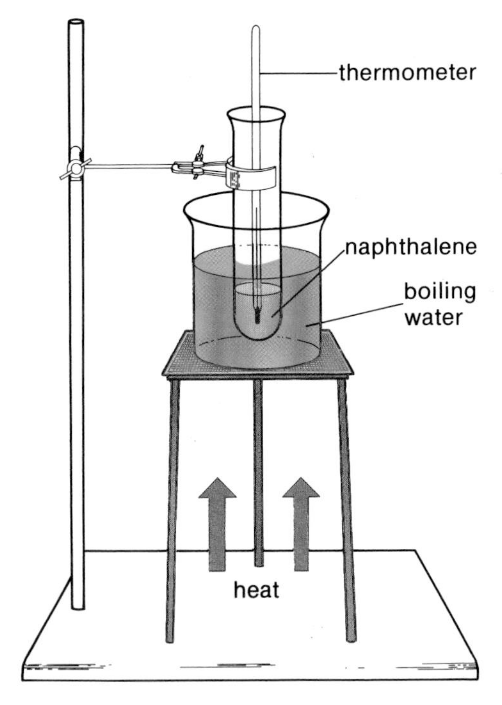

Figure 1

Melting a solid

When a solid is heated, it changes state and forms a liquid. This is an endothermic process. Heat is needed to melt the solid. Figure 1 shows an experiment to investigate how the temperature of naphthalene changes when it is surrounded by boiling water. The temperature of the naphthalene is plotted against time in figure 2.

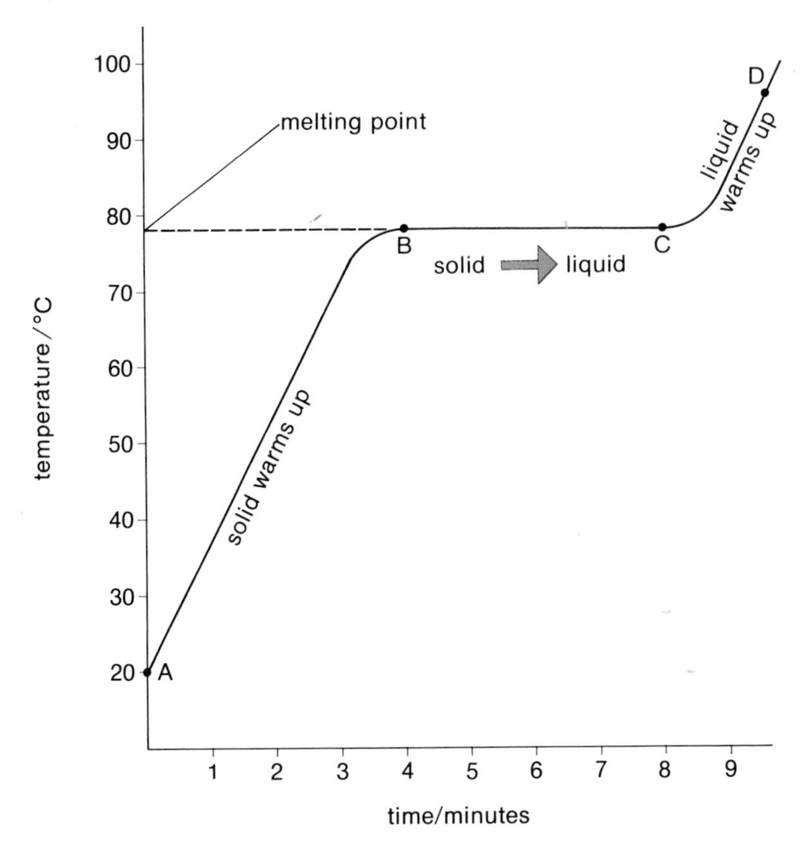

Figure 2

When the solid is heated, its temperature rises along the curve AB in figure 2. As heating continues, the temperature stays constant along BC. There is no rise in temperature even though energy is being supplied to the naphthalene. *Along BC the heat supplied does not raise the temperature. It is needed to break the particles away from their fixed positions in the solid crystals until they can move around each other in the liquid.* The naphthalene is melting along BC and the graph shows that the melting point is 78°C. At C, all the naphthalene has melted. On further heating, the temperature rises again along CD as the heat supplied warms up the liquid naphthalene. Figure 3 shows what happens to the naphthalene particles as the solid melts between B and C.

Vaporising a liquid

When a liquid is heated, the temperature rises until its boiling point is reached. Once the liquid is boiling, the temperature stays constant because the energy being supplied is needed to separate liquid particles so they can move around at high speed in the gaseous state (figure 3).

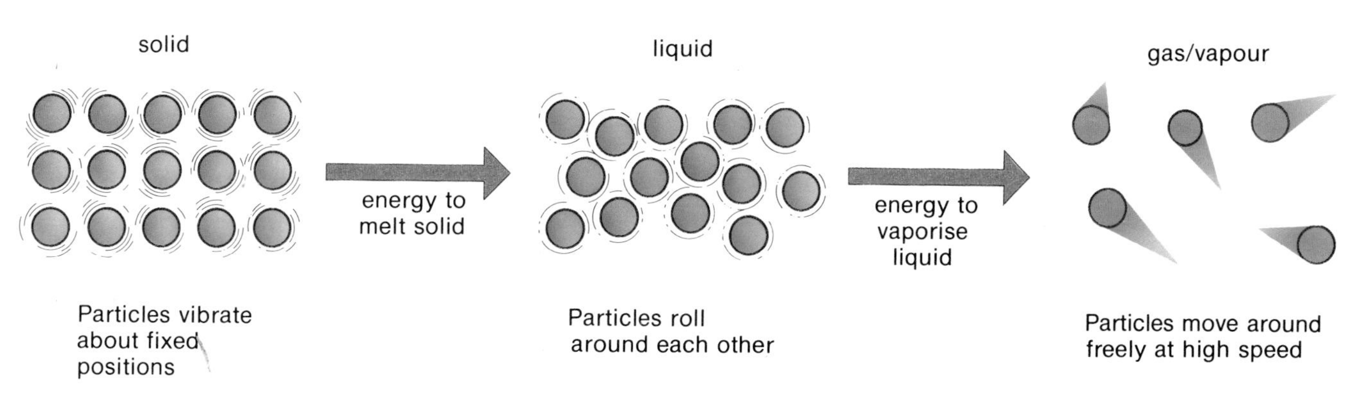

Figure 3

When a substance warms up or melts or vaporises, energy is absorbed. But, when a substance cools down, freezes or condenses, energy is given out. The energy changes between a liquid and a vapour are used in refrigerators (figure 4). The liquids used in refrigerators are called *freons*, which are non-poisonous compounds of carbon, chlorine and fluorine with a low boiling point. Liquid freon vaporises in the coiled pipes around the ice box. As it does so, heat is taken from the refrigerator and the food inside it, which cools them down. The freon vapour is then pumped from the coils inside the refrigerator into the pipes and fins at the back of the machine. Here the pressure rises and the freon vapour turns to liquid, giving out heat to the pipes and the air around them. Finally the liquid freon returns to the pipes around the ice box and the process is repeated.

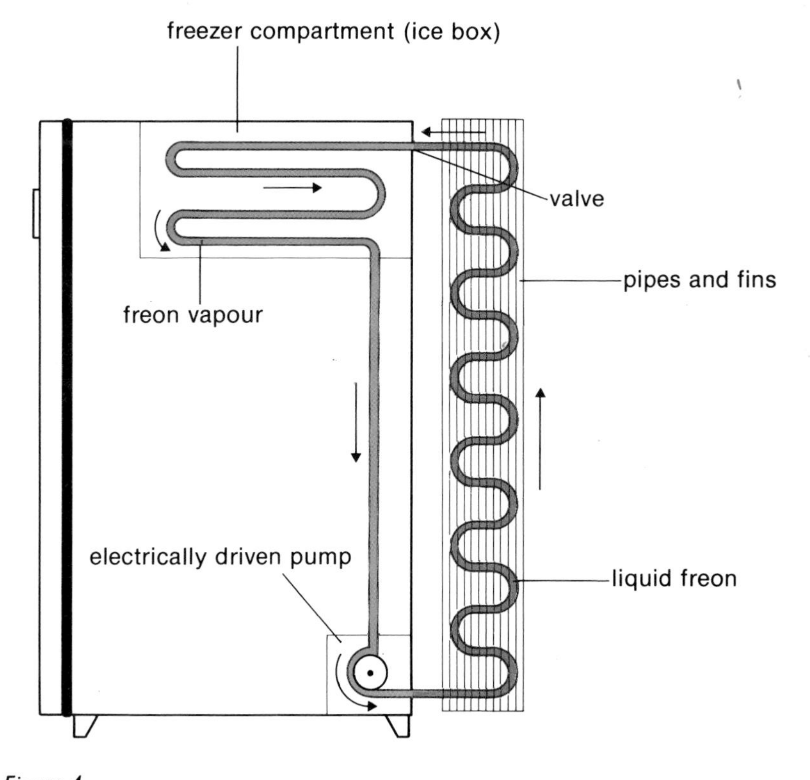

Figure 4
A household refrigerator

Questions

1 How do particles move in the solid, liquid and gaseous states?
2 Why does a solid have a fixed shape whereas a liquid takes the shape of its container?
3 Why does a gas always fill the whole of its container whereas a liquid often fills only part of its container?
4 How does sweating help us to keep cool in hot weather?
5 Why should the liquid used in a refrigerator be (i) volatile; (ii) non-poisonous; (iii) non-flammable?
6 Draw a sketch graph of temperature against time to show what happens when ice is heated from −20°C until the water boils.
7 Why does freon vapour turn to liquid when it is put under increased pressure?

4 Energy From Fuels

It is useful to know how much energy is produced when a fuel burns. This can help us to decide the best fuel to use for a particular job. Figure 1 shows the apparatus we can use to measure the heat given out when a liquid fuel like methylated spirits (meths) burns. The heat produced is used to heat water in the metal can. If we measure the temperature rise of the water, we can work out the heat produced. We can find the mass of meths which is burnt from the loss in weight of the liquid burner. Then we can calculate the heat produced when one gram of the fuel burns.

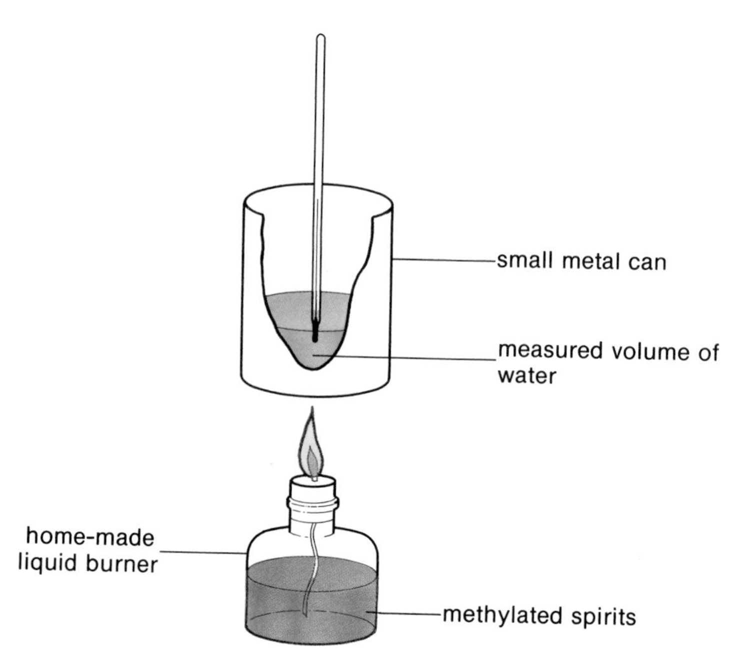

Figure 1

Volume of water in metal can	$= 250\ cm^3$
∴ Mass of water in can	= 250 g
Rise in temp. of water	= 10°C
Mass of meths burnt	= 0.5 g

The results of an experiment to measure the heat produced when meths is burnt

The results from one experiment are shown in the table on the left. In the experiment, 250 g of water are warmed up by 10°C.

We know that 4.2 joules of heat warm up 1 g of water by 1°C.

So, 250×4.2 J warm up 250 g of water by 1°C,

∴ $250 \times 4.2 \times 10$ J warm up 250 g of water by 10°C.

This amount of heat is produced by 0.5 g of meths

0.5g meths produce $250 \times 4.2 \times 10$ J = 10 500 J
= 10.5 kJ

∴ 1 g of meths produces 21 kJ

Methylated spirits is mainly ethanol (alcohol), C_2H_6O.

Suppose that meths is *pure* ethanol.

1 What is the mass of 1 mole of ethanol? (C = 12, H = 1, O = 16).

2 How much heat is produced when 1 mole of ethanol burns? This is called the heat of combustion of ethanol.

> *The* **heat of combustion of a substance, ΔH_c,** *is the amount of heat given out when 1 mole of the substance is completely burnt in oxygen.*

Unfortunately, the simple apparatus in figure 1 does not give very accurate results. An accurate value for the heat produced when meths burns is 30 000 joules per gram. This is about half as much again as the value from our experiment. There are three serious errors in our experiment.

1 Some of the heat produced by the burning meths heats up the can and *not* the water.

2 Some of the heat produced by the burning meths heats up the surrounding air.

3 The meths does not burn fully to carbon dioxide and water. Some of it forms a layer of black soot (carbon) on the can.

Figure 2 shows an apparatus that is specially designed to avoid heat losses and give a more accurate result. The substance to be burnt is placed inside the central glass chamber. Liquids are burnt in a small spirit lamp. Solids, such as coal, sugar and charcoal, are placed in a crucible and lit by a hot wire. Oxygen passes into the central chamber. Hot gases from the burning substance pass through the copper coils and warm up the water between the central chamber and the outer glass. This type of equipment is used by fuel technologists and food scientists to measure the heat produced when fuels and foods burn.

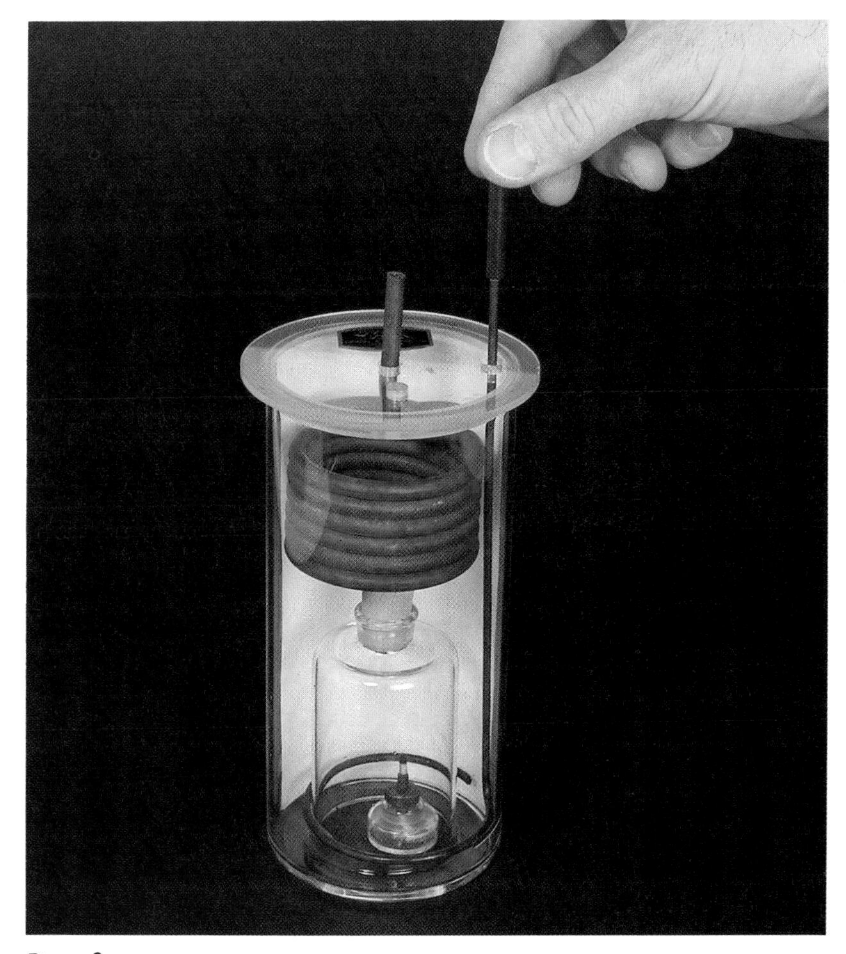

Figure 2

Questions

1 Draw a diagram of the apparatus you would use to find the heat produced when 1 g of a firelighter burns.

2 When 100 cm^3 of 1.0 M hydrochloric acid is added to 100 cm^3 of 1.0 M sodium hydroxide, the temperature of the mixture rises by 6.5°C.

(a) What mass of solution is warmed up? (Assume 1 cm^3 of solution has a mass of 1 g.)

(b) How many joules warm up the mixture? (Assume 4.2 J warms up 1 g of solution by 1°C.)

(c) How many moles of hydrochloric acid react?

(d) How much heat is produced when 1 mole of hydrochloric acid reacts? This is called the heat of neutralization of hydrochloric acid.

(e) Write a balanced equation for the reaction involved in (d).

3 Look at figure 2.

(a) Why are the product gases passed through a coiled tube in the water?

(b) Why is the coiled tube made of copper?

(c) Why is oxygen passed into the space where the fuel or food is burnt?

(d) Why are dietitians interested in the heat produced when foods burn?

5 Fossil Fuels

Fossils in coal show the plants from which the coal was formed

The most commonly used fuels are coal, oil and natural gas. All three contain carbon and carbon compounds. The energy in these fuels can be traced back to the Sun. Using the Sun's energy, plants turn carbon dioxide and water into carbohydrates, like glucose ($C_6H_{12}O_6$) and starch. This process is called **photosynthesis** (unit 7 of this section). After photosynthesis, the carbohydrates provide food for humans and other animals.

When plants and animals die and decay, their carbohydrates usually break down and get oxidized to carbon dioxide and water. Sometimes the plants and animals cannot oxidize when they die. If this happens, the carbohydrates turn into energy-rich substances like coal and oil.

Three hundred million years ago the Earth was covered in forests and the sea was full of tiny organisms. When some of these living things died, they were covered with mud and protected from oxidation. Over millions of years these deposits were changed by bacteria and compressed by the earth and sea above them. They formed coal, oil (petroleum) and natural gas. We now call these deposits **fossil fuels** because they have formed from the remains of dead animals and plants. Coal has formed mainly from plants that grew on land. Petroleum and natural gas have formed mainly from plants and animals that lived in the sea.

When we burn fossil fuels we release the Sun's energy trapped by living things millions of years ago.

fossil fuel + oxygen $\rightarrow$ carbon dioxide + water + energy
(containing C and H)

The importance and uses of petroleum and natural gas are discussed in section J. Coal is mined all over the world. It is possibly the most useful

The effects of coal mining often ruin the landscape. These photographs show the huge spoil heap near Dinnington Mine (South Yorkshire) before and after it was removed

fuel we have, but the effects of coal mining often ruin the landscape. Useless material from the mines is dumped and left as ugly spoil heaps and the land above coal mines is liable to subside.

Most of the coal mined in Britain is burnt at power stations to generate electricity. The rest is used for industrial and domestic heating or for making coke.

If coal is heated in the absence of air it cannot burn. Instead the complex chemicals in the coal are decomposed to simpler substances. The process is called **destructive distillation** and the main products are coke, coal gas and tar. Figure 1 shows you how you can carry out the process on a small scale.

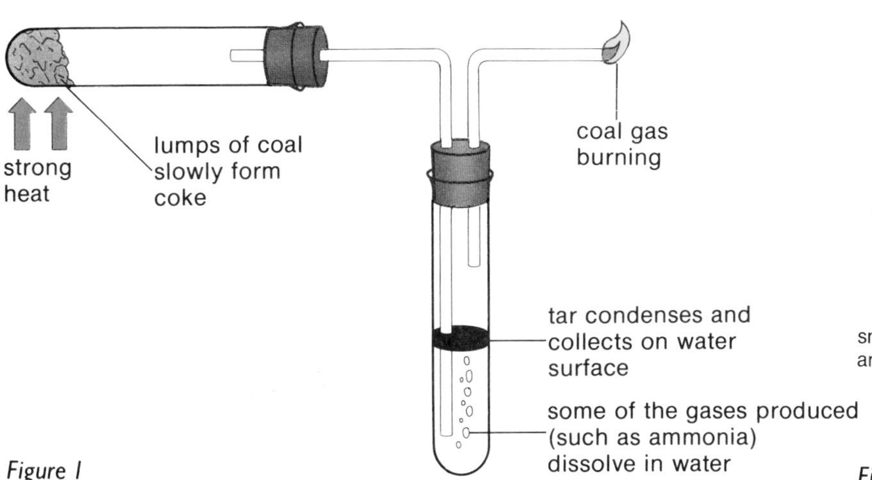

Figure 1
The small scale destructive distillation of coal

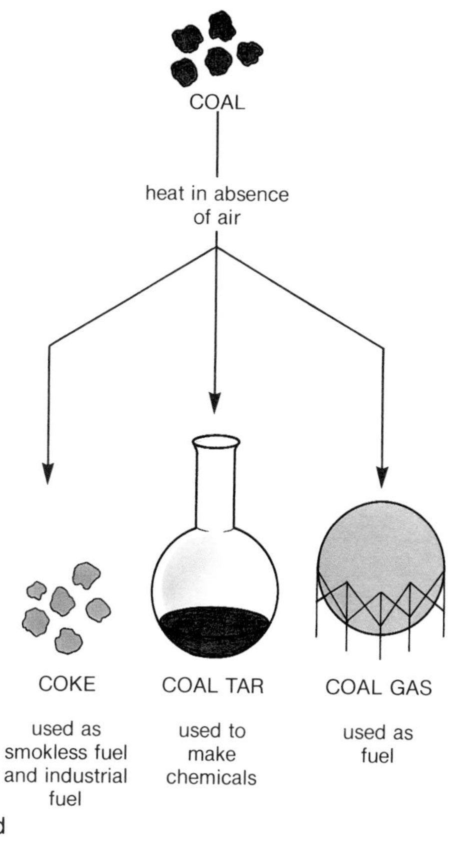

Figure 2
Products from the destructive distillation of coal

In industry, the coal is heated in large coke ovens. It is put in at the top and the ovens are closed and left to heat up. All the products are volatile except coke. Tar and other products are condensed and used to obtain benzene, toluene and creosote. When all the gases have come off, the coke is removed from the ovens and cooled by water sprays to prevent it burning in air. The coke is used as a 'smokeless' fuel and as an important reducing agent in industry. The coal gas is used as a fuel (Figure 2).

Coke being released from coke ovens

Questions

1 Explain the following:
natural gas; carbohydrate; energy-rich compound; fossil fuel; destructive distillation.

2 (a) Draw a flow diagram to show how coal formed.
(b) Why is coal described as a fossil fuel?
(c) Why is good ventilation important when fossil fuels burn?

3 (a) What are the main uses of coal?
(b) Why is coal converted into coke?
(c) What should be done to reduce the unsightly appearance of spoil heaps in coal mining areas?

4 Why is destructive distillation different to normal distillation?

5 What problems does coal mining cause for (i) the safety of miners; (ii) the environment?

6 Other Energy Sources

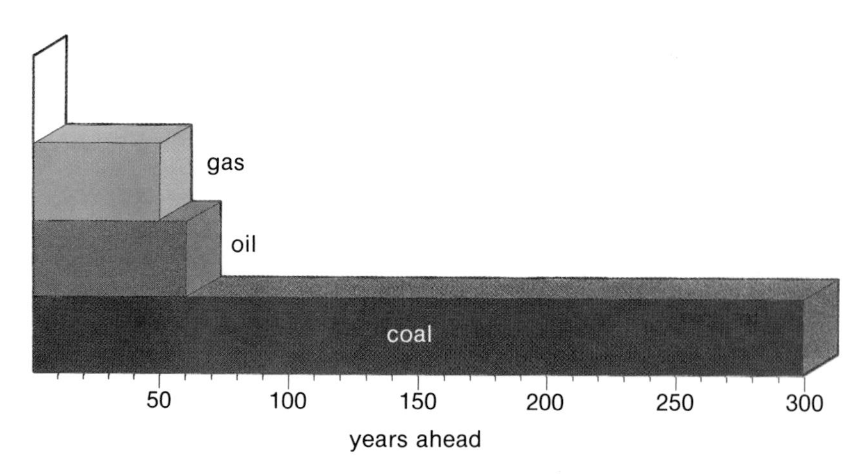

Figure 1
How long will our reserves of fossil fuels last?

For 150 years, industrial countries have relied on fossil fuels for energy. This is still true today, but fossil fuels cannot last forever. Figure 1 gives an estimate of how long the fossil fuels will last if we continue to use them as we do now. Although coal is plentiful, oil and natural gas are likely to run out in 50 or 60 years' time. Unfortunately, oil is the fuel on which we depend most for transport and for chemicals. So, it is important to use fossil fuels carefully and to conserve them for the future. We must also look for alternative energy sources.

- **Nuclear power** is being used more and more to generate electricity. Uranium is used as the fuel. Nuclear energy has many advantages, but some people think that it is unsafe. Nuclear energy is discussed further in section L.

- **Tidal power.** A tidal power station traps the high tides behind a barrage across a river estuary. When the tide falls, the water is made to flow through turbines which generate electricity.

Only a few tidal power stations have been built. This photo shows a tidal barrage in Nova Scotia, Canada

Solar water-heating panels on the roof of a house

The CEGB's first wind turbine which is located at the Carmarthen Bay Power Station in South Wales. The 24 m high machine can supply up to 200 kW of electricity—enough to operate 200 one-bar electric fires

- **Hydro-electric power.** Falling water can be used to drive turbines which generate electricity. Once the power station has been built, the cost of hydro-electricity is fairly cheap. But it needs water falling from a height and this limits its use.
- **Solar power** uses energy straight from the Sun in one of two ways. The Sun's energy can be used to heat water in solar panels or it can be used to make electricity using photo-cells.
- **Wind power.** Windmills, like water mills, have been used as a source of power for hundreds of years. Some scientists believe that giant windmills (or wind turbines as they are called) could be used to generate electricity on a large scale in the future.

Energy resources—savings and income

The Earth's energy resources can be compared to money. Fossil fuels and nuclear fuels are like *savings in the bank*. They represent energy stored up (saved up) over millions of years. They are sometimes called **non-renewable energy sources**. Once used, they are gone forever. We cannot get them back and use them again. Fortunately there are some **renewable energy sources**. Every day the sun shines (somewhere!), the wind blows, the rain falls and the tides come in. So, solar power, wind power, hydro-electric power and tidal power are renewable energy sources. We can use them all the time, but they keep on being replaced. They are like an *income from a job*. In the future, we may have to use these renewable energy sources more and more as our fossil fuel savings run out.

Questions

1 Why is it important to conserve our reserves of fossil fuels?
2 What are the alternative sources of energy to fossil fuels?
3 There are economic and environmental reasons for choosing different energy sources. What are the advantages and disadvantages of the following energy sources?
coal; tidal power; wind power.
4 (a) What is meant by (i) renewable energy sources; (ii) non-renewable energy sources?
(b) Is wood a renewable or a non-renewable energy source? Explain your answer.
(c) Is wave power a renewable or non-renewable energy source? Explain your answer.
5 Why is electricity much cheaper in Norway than in Britain?

7 Foods as Fuels

Daley Thompson, the Olympic decathlon champion. The energy he needs to run and jump is provided by respiration

Respiration

Foods are broken down in our bodies by a chemical process called **respiration**. During respiration, foods react with oxygen forming carbon dioxide and water. This is why we breathe out carbon dioxide and water vapour and why urine is mainly water. *Respiration is also exothermic* and energy is given out. Because of this, foods are sometimes described as 'biological fuels'.

food (containing C and H) + oxygen → carbon dioxide + water + energy

The energy given out during respiration can be used as:

- **heat** to keep us warm;
- **mechanical energy** in our muscles to help us to move around and keep our heart and breathing muscles working;
- **electrical energy** in our nerves to help us to think and respond to stimuli (messages), like heat or pain.

Photosynthesis

Humans and other animals cannot make their own food. To get their food, they must eat other animals or plants. But plants are different. They can make their own food from carbon dioxide in the air and water in the soil. This process involves **photosynthesis** which was mentioned in unit 5 of this section. During photosynthesis, carbon dioxide and water are converted into carbohydrates like glucose ($C_6H_{12}O_6$), sugar ($C_{12}H_{22}O_{11}$) and starch. Oxygen is also produced (figure 1).

$$6CO_2 + 6H_2O \xrightarrow{sunlight} C_6H_{12}O_6 + 6O_2$$

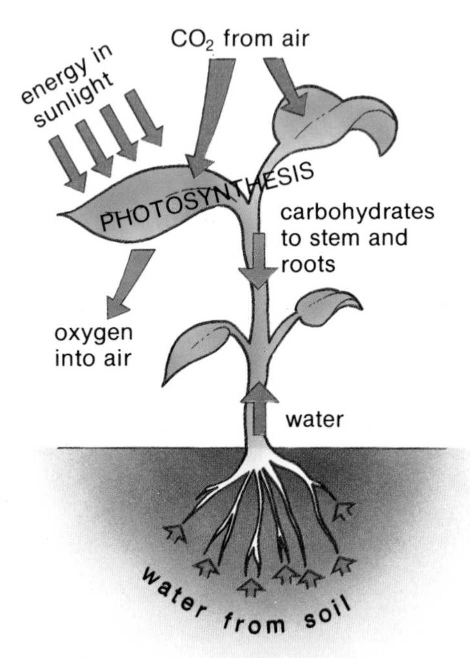

Figure 1 Photosynthesis

Photosynthesis is an endothermic process. The energy needed for the reaction comes from sunlight. Light energy from the sun is absorbed by chlorophyll, a green substance in the leaves of plants.

Carbohydrates like glucose and starch, which are produced during photosynthesis, can be used by plants as foods. They can also be used as foods by animals which eat the plants. Some plants such as potatoes, wheat, rice and sugar beet are grown as a source of food for humans and animals.

Photosynthesis and respiration are very complex processes involving several steps. Photosynthesis is the reverse of respiration. Respiration is exothermic. It uses up food and oxygen and produces carbon dioxide, water and energy. In contrast, photosynthesis is endothermic, using carbon dioxide, water and energy in sunlight to produce food and oxygen.

The two processes can be summarised as:

$$C_6H_{12}O_6 + 6O_2 \underset{\text{photosynthesis}}{\overset{\text{respiration}}{\rightleftharpoons}} 6CO_2 + 6H_2O + \text{energy}$$

Respiration happens in both plants and animals, but photosynthesis can only happen in plants. Respiration and photosynthesis are important in the **carbon cycle** (figure 2). This shows how carbon, in carbon dioxide and in other carbon compounds, is recycled through the various processes mentioned in the last three units.

Chlorophyll is contained in chloroplasts in plants. The chloroplasts can be identified easily in these spirogyra as bright zig-zags across the cells

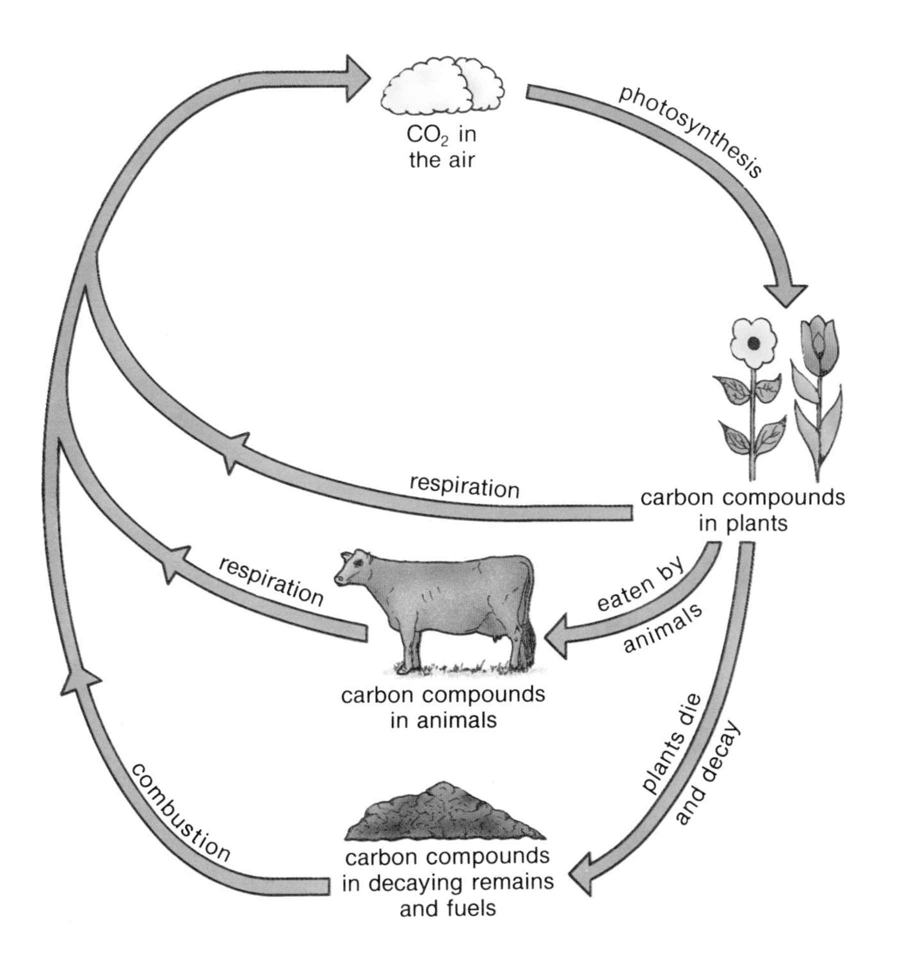

Figure 2
The carbon cycle

Questions

1 Explain the following:
respiration; photosynthesis; carbon cycle; exothermic.

2 How do our bodies use the energy given out during respiration?

3 Why are animals unable to photosynthesize?

4 In what ways is photosynthesis the opposite of respiration?

5 *True* or *false?*
The reaction represented as:

$$C_6H_{12}O_6 + 6O_2 \rightarrow 6CO_2 + 6H_2O$$

A is exothermic.
B is reversible.
C involves a hydrocarbon.
D is called photosynthesis.
E involves several steps.
F involves chlorophyll.

8 Starch

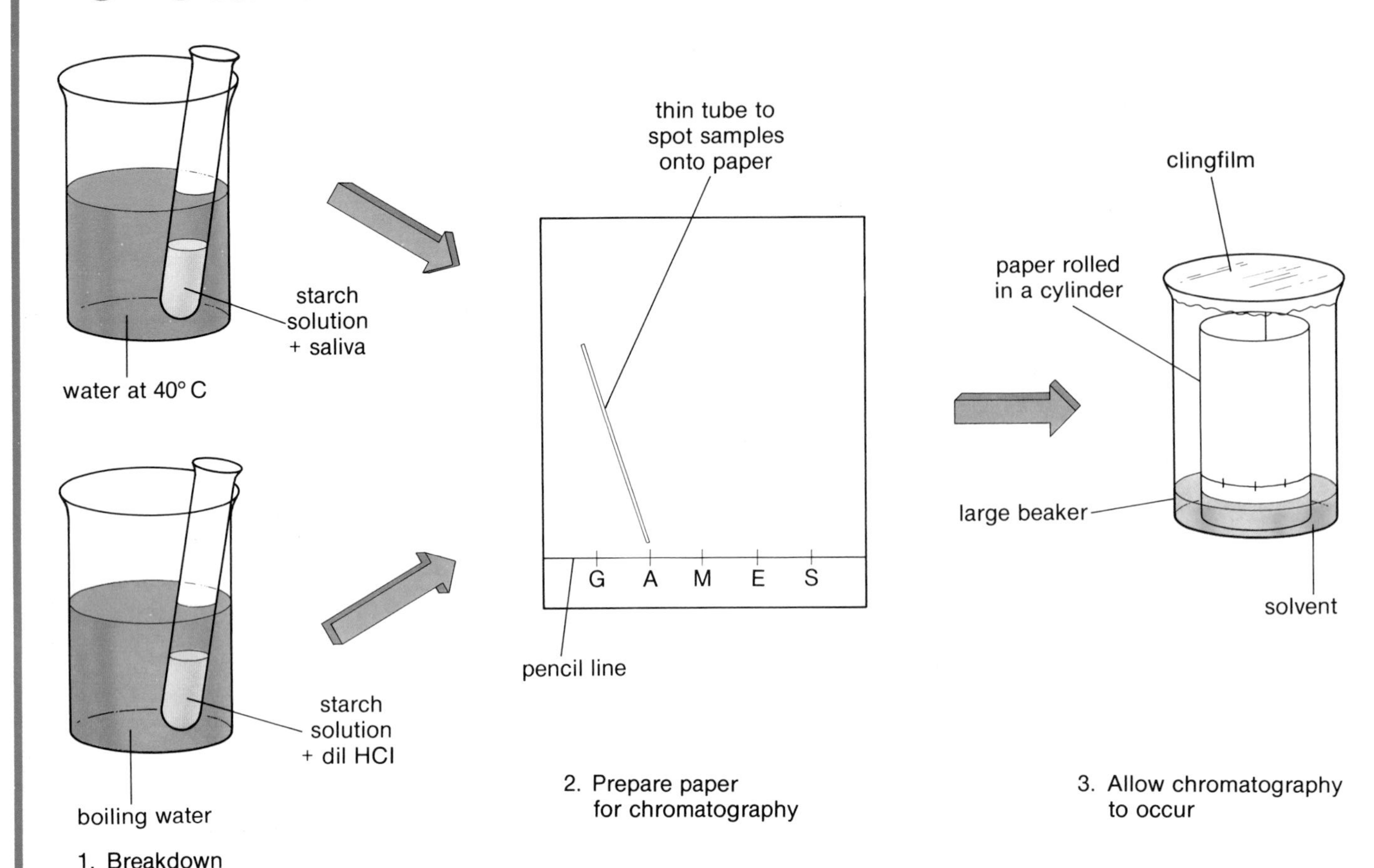

Figure 1
Separating the sugars from the breakdown of starch by chromatography

In the last unit, we saw that starch was produced during photosynthesis and used up during respiration. Starch is a complex carbohydrate. Its relative molecular mass is about 100 000. Plants store most of their carbohydrate as starch and this provides food for humans and other animals. Three important foods that contain starch are bread, potatoes and rice. When we eat them, our bodies break down the starch into smaller molecules. This process of digestion takes place with the help of **enzymes** (section K, unit 5). Enzymes are catalysts that speed up the chemical reactions in living things. If we chew bread for a long time, it begins to taste sweet. This happens because enzymes in our saliva break down starch in the bread to smaller molecules of sugar. Starch is also broken down by acids in our stomach. In the next experiment we can study the breakdown of starch by saliva and by dilute hydrochloric acid.

Which sugars are produced when starch breaks down?

There are several different sugars that might be produced when starch is broken down by saliva or dilute hydrochloric acid. We can identify the sugars that are produced in each case using chromatography. Figure 1 shows the important steps in the first part of the experiment. The spots on the chromatography paper are labelled G, A, M, E and S. 5 drops of glucose solution are spotted on the chromatography paper at G; 5 drops of maltose solution at M; 5 drops of sucrose solution at S; 5 drops of the acid-treated starch at A and 5 drops of the enzyme-treated starch at E.

When the spots have dried, the paper is rolled into a cylinder and then placed in the solvent, as the diagram shows. The solvent is a mixture of propan-2-ol, acetic acid and water (3:1:1 by volume). It is allowed to soak up the paper for 6–12 hours. As it travels up the paper, it carries the substances in the spots with it. Different sugars move up the paper at different speeds. After chromatography, the paper is removed from the solvent and allowed to dry. It is then dipped in the 'locating reagent' and heated in an oven at 100°C. The 'locating reagent' reacts with any sugars on the paper and leaves a brown spot. Figure 2 shows what the chromatography paper finally looks like.

Figure 2 shows that glucose moves up the paper faster than sucrose and sucrose moves faster than maltose. The spot from the acid-treated starch runs up the paper exactly the same distance as glucose. This shows that the acid has broken down the starch to form glucose (figure 3).

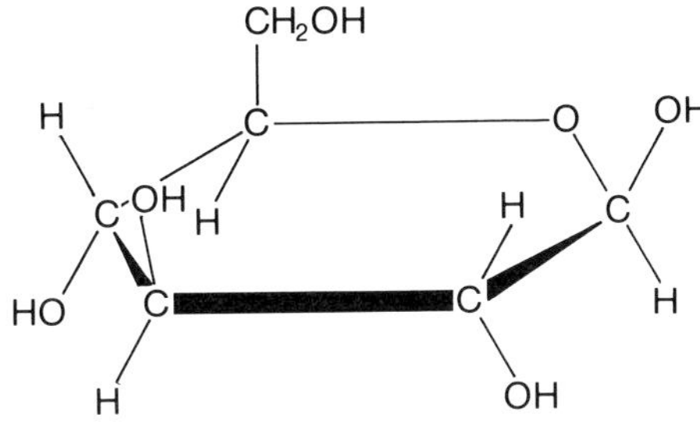

Figure 3
A molecule of glucose ($C_6H_{12}O_6$)

$$\text{starch} \xrightarrow{acid} \text{glucose}$$

The spot from the enzyme-treated starch runs up the paper exactly the same distance as maltose. This shows that the saliva has broken down the starch to form maltose.

$$\text{starch} \xrightarrow{saliva} \text{maltose}$$

Figure 4 summarises the breakdown of starch with acid and saliva. Notice that the symbol for glucose in figure 4 is much simpler than that in figure 3.

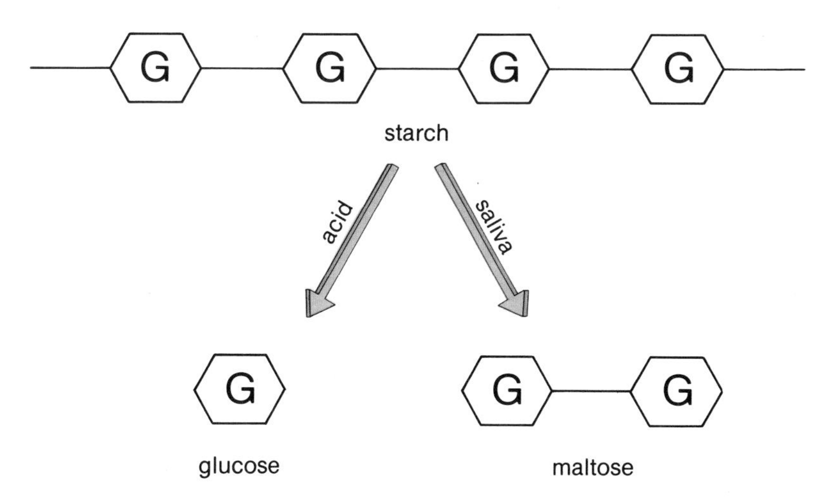

Figure 4
Breaking down starch with acid and saliva

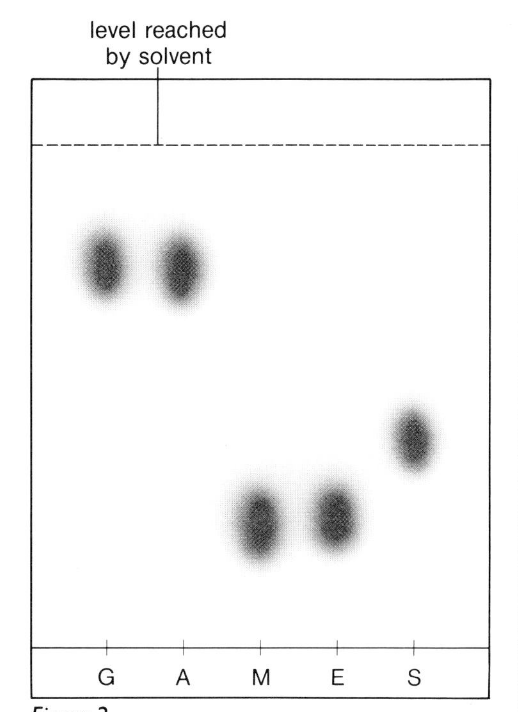

Figure 2
The developed chromatography paper

Questions

1 Starch solution can be used as a test for iodine.
(a) How would you test a solution for starch?
(b) What will happen in your test if the solution contains starch?

2 (a) Name three common foods which contain starch?
(b) Why is starch important to animals and humans?
(c) What is the relative molecular mass of glucose ($C_6H_{12}O_6$)?
(d) The relative molecular mass of starch is about 100 000.
Roughly, how many glucose units will one starch molecule contain?

3 (a) List the main stages in identifying the sugars produced when starch is broken down by saliva.
(b) Why should chromatography papers be marked in pencil and not ink?
(c) How would you keep the spots on the chromatography paper as small as possible?
(d) Why is a 'locating reagent' needed for sugars but not for inks?
(e) Why do you think that different sugars move up the paper at different rates?

9 Carbon Dioxide

When a bottle of soda water is opened, the liquid fizzes because the pressure falls and gas escapes

Carbon dioxide is probably the most important simple carbon compound. It links respiration and photosynthesis and it is produced when carbon compounds burn. Carbon dioxide also has some important uses.

A fireman demonstrates the use of a small carbon dioxide fire extinguisher

- **Soda water and fizzy drinks.** Solutions of carbon dioxide in water have a pleasant taste—the taste of soda water. Soda water and other fizzy drinks are made by dissolving carbon dioxide in them at high pressure. When a bottle of the drink is opened, it fizzes because the pressure falls and carbon dioxide gas can escape from the liquid.

- **Fire extinguishers.** Liquid and gaseous carbon dioxide at high pressure are used in fire extinguishers. When the extinguisher is used, carbon dioxide pours out and smothers the fire. Carbon dioxide is heavier than air so it covers the fire and stops oxygen getting to it. The fire 'goes out' because carbon dioxide does not burn and substances will not burn in it.

- **Refrigeration.** Solid carbon dioxide is used for refrigerating ice-cream, soft fruit and meat. The solid carbon dioxide is called 'dry ice' because it resembles ice. It is colder than ordinary ice and sublimes without going through the messy liquid stage. This is why it is called 'dry ice' or 'Dricold'.

Making carbon dioxide

Strong acids, like hydrochloric acid and nitric acid, react with carbonates to form carbon dioxide and water.

$$\underset{\text{acid}}{2H^+(aq)} + \underset{\text{carbonate}}{CO_3^{2-}(s)} \rightarrow H_2O(l) + CO_2(g)$$

Small amounts of carbon dioxide are usually prepared from marble chips (calcium carbonate) and dilute hydrochloric acid (figure 1).

$$CaCO_3(s) + 2HCl(aq) \rightarrow CaCl_2(aq) + H_2O(l) + CO_2(g)$$

The carbon dioxide may be collected by downward delivery or over water.

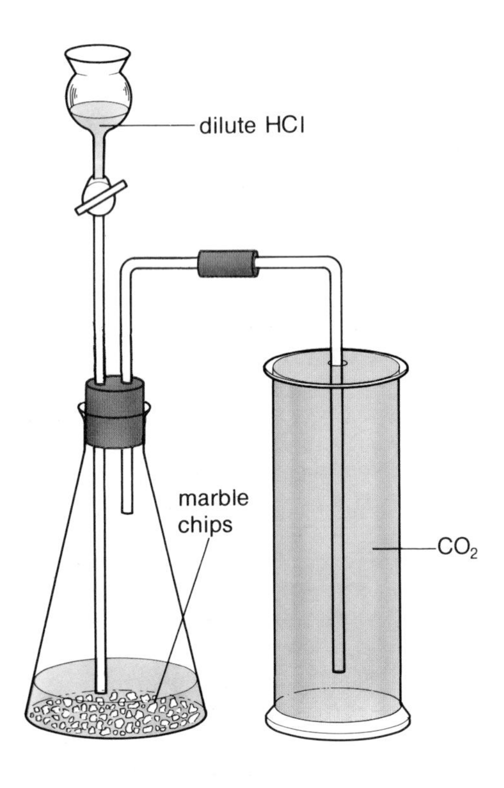

Figure 1
Making CO_2 in the laboratory

Properties of carbon dioxide

Carbon dioxide is a typical non-metal oxide. It is acidic, gaseous and simple molecular. Figure 2 shows some other properties of carbon dioxide. Notice that it is fairly soluble in water. The dissolved gas provides water plants, like seaweed, with the carbon dioxide they need for photosynthesis. About 1% of the gas which dissolves in water reacts to form carbonic acid.

$$H_2O + CO_2 \rightarrow H_2CO_3$$

The solution of carbonic acid is a very weak acid. It turns blue litmus paper only a purplish-red.

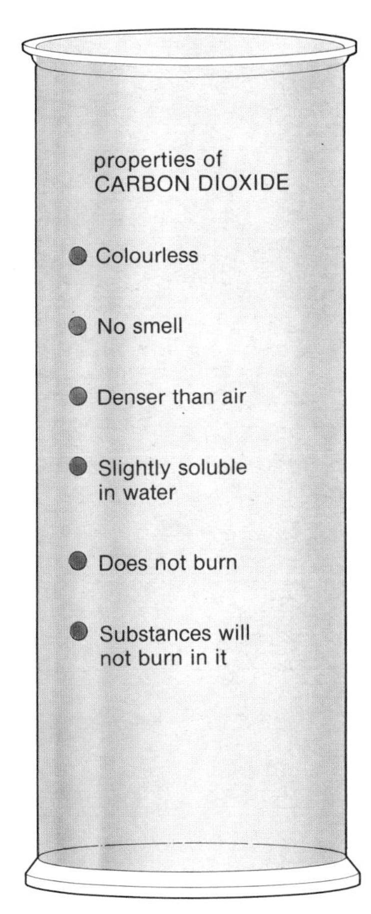

Figure 2

Testing for carbon dioxide with limewater

The test for carbon dioxide uses its acidic property. Lime water is calcium hydroxide solution—a weak alkali. When carbon dioxide is bubbled into lime water, the liquid goes milky with a white precipitate of calcium carbonate. Why does this precipitate form? First, the carbon dioxide reacts with OH^- ions in the alkali to form carbonate.

$$CO_2(g) + 2OH^-(aq) \rightarrow CO_3^{2-}(aq) + H_2O(l)$$

Then, CO_3^{2-} ions react with calcium ions in the lime water to form insoluble calcium carbonate.

$$Ca^{2+}(aq) + CO_3^{2-}(aq) \rightarrow CaCO_3(s)$$

These workmen are using 'dry ice' to shrink-fit an axle into a huge cog wheel. Why does the axle shrink when it is surrounded by solid CO_2?

Questions

1 List the important uses of carbon dioxide. For each use, explain why carbon dioxide is used.

2 How does carbon dioxide link respiration and photosynthesis?

3 Give two reasons why 'dry ice' is better than ordinary ice for refrigeration.

4 Carbon dioxide can be poured from a gas jar onto a lighted candle and the candle goes out. What properties does this simple experiment show for carbon dioxide?

5 (a) How is carbon dioxide obtained from a carbonate?
(b) Write an equation for the reaction in part (a).
(c) How would you show that limestone is a carbonate?

6 Why do you think 'dry ice' is used in drama productions to make mists and fog?

10 Chalk, Limestone and Marble

A limestone quarry

Carbon compounds occur in rocks and mineral ores as well as in fossil fuels and in living things. The most important carbon compounds in rocks are chalk, limestone and marble. These three minerals all contain calcium carbonate, $CaCO_3$. In fact, calcium carbonate is the second most abundant mineral in the Earth's crust after silicates such as clay, sandstone and granite. Chalk is the softest form of calcium carbonate. Deposits of chalk are formed from the shells of dead sea creatures that lived millions of years ago. In some places, the chalk was covered with other rocks and put under great pressure. This changed the soft chalk into a harder rock—limestone. In places where the chalk was under pressure and heat, it was changed into marble.

Limestone: an important industrial compound

Calcium carbonate usually occurs as limestone rather than chalk or marble. As a result, limestone has become an important industrial compound. Figure 1 shows the important uses of limestone and its products. Notice that:

1 the main uses of limestone itself are in making iron and steel (section F, unit 8), in neutralizing acid soil and in fertilizers such as nitro-chalk.

2 the main substances made from limestone are cement, glass, calcium oxide (quicklime) and calcium hydroxide (slaked lime).

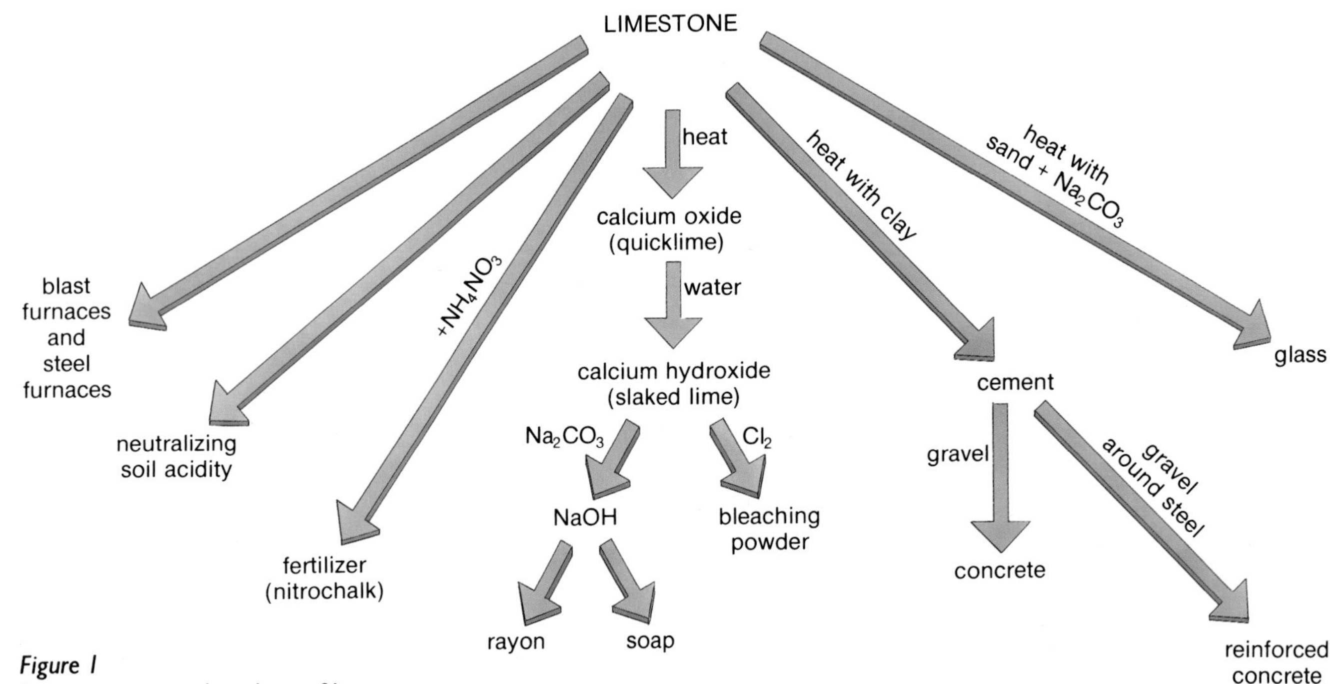

Figure 1
Important uses and products of limestone

• **Calcium oxide (quicklime)** is made by heating limestone in lime kilns (figure 2).

$$CaCO_3(s) \rightarrow CaO(s) + CO_2(g)$$

The quicklime is then reacted with water to make calcium hydroxide (slaked lime).

$$CaO(s) + H_2O(l) \rightarrow Ca(OH)_2(s)$$

• **Calcium hydroxide (slaked lime)** is the cheapest industrial alkali. It is used to make sodium hydroxide, bleaching powder and mortar. Powdered calcium hydroxide is also used to control acidity in the soil. Calcium hydroxide is only slightly soluble in water. The dilute alkaline solution which it forms is called **lime water**.

• **Mortar and cement.** *Mortar* is a mixture of slaked lime and sand with water. The mortar sets by drying out and then hardens by reacting with CO_2 in the air.

$$Ca(OH)_2 + CO_2 \rightarrow CaCO_3 + H_2O$$

Cement is made by heating limestone with clay. It contains a mixture of calcium silicate and aluminium silicate. This mixture reacts with water to form hard interlocking crystals as the cement sets. Cement is mixed with sand and water in a similar way to mortar. But, cement is much harder than mortar and it will set under water because its setting process needs no CO_2.

• **Concrete** is a mixture of cement with gravel, broken stones or bricks. Reinforced concrete is made by allowing concrete to set round a steel framework. It is used in building large structures such as blocks of flats and bridges.

Reinforced concrete being used in building a bridge

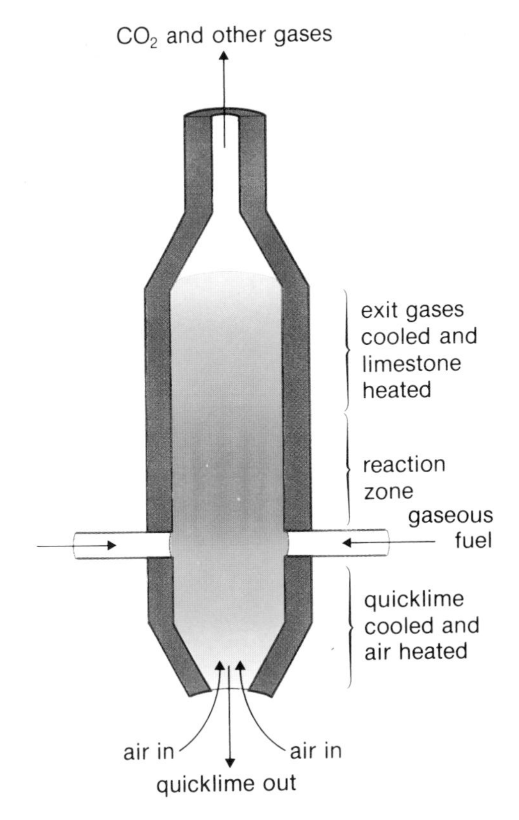

Figure 2
A gas-fuelled lime kiln

Questions

1 Name three forms of calcium carbonate.
2 Why are there three different forms of calcium carbonate which occur in rocks?
3 What are the main uses of limestone?
4 Finely-ground limestone is used to neutralize acids in the soil. How does limestone neutralize the acids?
5 (a) How is quicklime obtained from limestone?
(b) Write an equation to summarise the process.
(c) How is slaked lime obtained from quicklime?
(d) What are the uses of slaked lime?
6 Why is cement better than mortar?
7 (a) What problems are associated with the large-scale quarrying of chalk and limestone?
(b) What steps can be taken to overcome or reduce these problems?

11 Hard Water

At Malham Cove in the Yorkshire Dales, water once poured from the top of the rocks. Over many years carbonic acid in the water has reacted with the limestone and formed caves. The water now flows through these caves and finally emerges at the bottom of the cove

In chalk and limestone areas, soap does not lather easily with the water. The water also leaves a **scum** when it is mixed with soap. This kind of water is known as **hard water**.

Which ions cause hard water?

The table shows what happens when 10 cm^3 of each of the solutions listed are shaken with 0.5 cm^3 of soap solution.

Solution used	Ions present	Reaction with hard water
Sodium chloride	Na^+, Cl^-	No scum, lots of lather
Calcium chloride	Ca^{2+}, Cl^-	Lots of scum, little lather
Potassium nitrate	K^+, NO_3^-	No scum, lots of lather
Magnesium nitrate	Mg^{2+}, NO_3^-	Lots of scum, little lather
Sodium sulphate	Na^+, SO_4^{2-}	No scum, lots of lather
Iron(II) sulphate	Fe^{2+}, SO_4^{2-}	Lots of scum, little lather

The reactions of some solutions with soap

Look closely at the table.

1 Which solutions cause hard water?

2 Which of the following ions cause hard water: Na^+; Cl^-; Ca^{2+}; K^+; NO_3^-; Mg^{2+}; SO_4^{2-}; Fe^{2+}?

3 Which ions are most likely to cause hard water in the UK?

Why does scum form?

The main cause of hard water in the UK are calcium ions, Ca^{2+}. Soaps contain salts such as sodium palmitate and sodium stearate. When hard water is mixed with soap, Ca^{2+} ions in the hard water react with palmitate and stearate ions in the soap forming a precipitate of calcium stearate and calcium palmitate. This precipitate is scum.

$$\underset{\text{in hard water}}{Ca^{2+}(aq)} + \underset{\text{stearate/palmitate ions in soap}}{2X^{-}(aq)} \rightarrow \underset{\text{scum}}{CaX_2(s)}$$

Soaps and detergents are both used for cleaning, but they differ in *one* important way. Detergents, like washing-up liquid, do *not* give a scum with hard water. Unlike soaps they do not contain ions which react with Ca^{2+} ions in hard water to form a precipitate.

In limestone areas, caves have formed as carbonic acid in rain water has reacted with limestone. These caves are at Castleton in Derbyshire

How does hard water form?

When rain falls, it reacts with carbon dioxide in the air to form carbonic acid.

$$H_2O(l) + CO_2(g) \rightarrow H_2CO_3(aq)$$

When this dilute solution of carbonic acid flows over limestone or chalk it reacts with calcium carbonate in the rocks to form calcium hydrogencarbonate.

$$\underset{\text{in limestone}}{CaCO_3(s)} + \underset{\text{in rain water}}{H_2CO_3(aq)} \rightarrow \underset{\text{in hard water}}{Ca(HCO_3)_2(aq)}$$

Unlike calcium carbonate, calcium hydrogencarbonate is soluble in water and the calcium ions make the water hard.

Calcium carbonate in chalk and limestone is the main cause of hard water. In some areas, calcium sulphate which occurs as gypsum ($CaSO_4.2H_2O$) and anhydrite ($CaSO_4$) also caused hardness. Calcium sulphate is only slightly soluble in water but enough will dissolve to make the water hard.

Carbonic acid in rain water has helped to dissolve the limestone in the cracks of this limestone pavement

Questions

1 Explain the following:
hard water; scum; soap; detergent.

2 (a) Why does scum form?
(b) Write an equation for the reaction involved.
(c) Why is scum a nuisance?

3 What is the main advantage of detergents over soaps?

4 How does hard water form?

5 Describe some of the geological features in limestone areas that have resulted from the effects of dissolved carbon dioxide on limestone.

12 Softening Hard Water

'Fur' inside a kettle is a deposit of calcium carbonate from hard water

Hard water usually tastes better than soft water. The dissolved substances in hard water also help to produce strong teeth and bones which contain calcium carbonate and calcium phosphate. But hard water has several disadvantages compared to soft water.

1 It uses more soap than soft water.

2 It produces scum, which looks unsightly.

3 It is necessary to remove the hardness from the water in some areas and this causes extra expense.

4 It results in the formation of 'scale' in water pipes and 'fur' in kettles. The 'scale' may block pipes and 'fur' will reduce the efficiency of a kettle.

When hard water warms up or boils, the calcium hydrogencarbonate in it is decomposed to calcium carbonate, water and carbon dioxide.

$$Ca(HCO_3)_2(aq) \rightarrow CaCO_3(s) + H_2O(l) + CO_2(g)$$

The calcium carbonate is insoluble and forms a deposit inside the pipe or kettle. This reaction is the reverse of that which forms hard water. The reaction also explains the formation of stalagmites and stalactites in limestone caves. The temperature inside the cave is just warm enough for some of the hard water to decompose and leave a tiny deposit of calcium carbonate. More water drips down and the deposit gets larger. A deposit also forms where the drops hit the floor. After hundreds of years, the deposits will grow into large stalagmites and stalactites.

Stalagmites and stalactites in Cox's Cave, Cheddar, Somerset

How is hard water softened?

In some areas, substances that cause hardness have to be removed from the water. This is called **water softening**. In order to soften hard water we must take out the Ca^{2+} ions. We can do this in various ways.

- **By boiling.** Boiling decomposes calcium hydrogencarbonate forming insoluble calcium carbonate.

$$Ca(HCO_3)_2(aq) \rightarrow CaCO_3(s) + H_2O(l) + CO_2(g)$$

This removes the hardness caused by calcium hydrogencarbonate, but boiling does not remove the hardness caused by calcium sulphate. Therefore, the hardness from calcium sulphate is called **permanent hardness**. The hardness caused by calcium hydrogencarbonate (which is removed by boiling) is called **temporary hardness**.

- **By adding washing soda.** Washing soda and bath salts contain sodium carbonate (Na_2CO_3). Adding these to hard water removes all the Ca^{2+} ions as a precipitate of calcium carbonate.

$$\underset{\text{in hard water}}{Ca^{2+}(aq)} + \underset{\text{in washing soda}}{CO_3^{2-}(aq)} \rightarrow CaCO_3(s)$$

- **By ion-exchange.** The most convenient way of softening water is to use an ion-exchange column. The water passes through a column containing a special substance called a **resin** (figure 1). The resin contains sodium ions which are exchanged for calcium ions as hard water passes through the column. Na^+ ions do not cause hardness, so the water is now 'soft'.

$$\underset{\text{in hard water}}{Ca^{2+}(aq)} + \underset{\text{on resin}}{2Na^+(s)} \rightarrow \underset{\text{on resin}}{Ca^{2+}(s)} + \underset{\text{in the water}}{2Na^+(aq)}$$

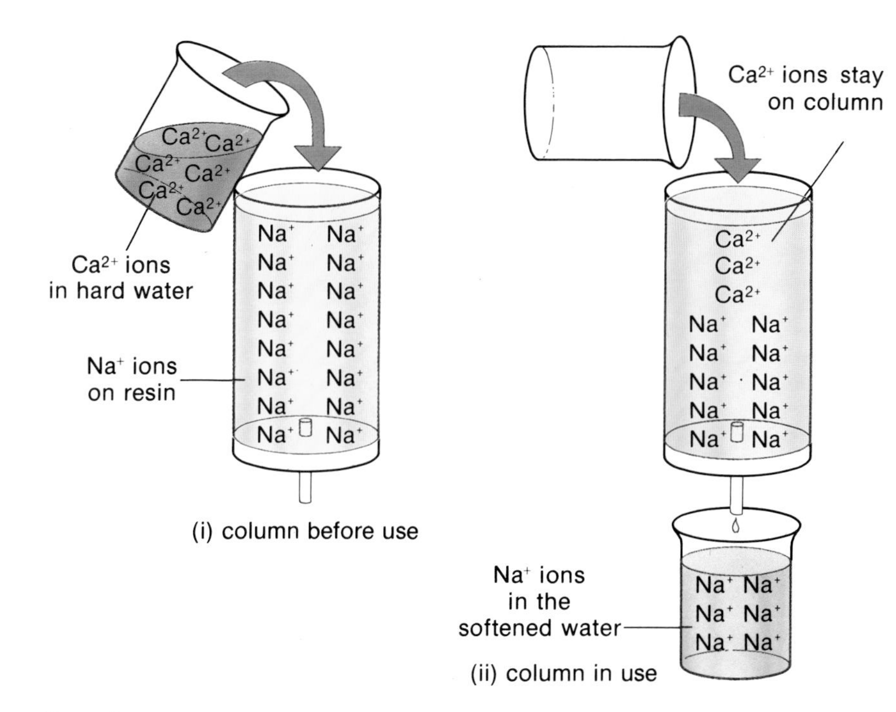

Figure 1
Using an ion-exchange column

Questions

1 Explain the following:
water softening; permanent hardness; temporary hardness; ion-exchange resin.

2 (a) Give *two* advantages of hard water.
(b) Give *three* disadvantages of hard water.

3 (a) How do stalagmites and stalactites form?
(b) Write an equation for the reaction involved in (a).

4 (a) List the main methods of softening water.
(b) For each method write an equation to summarise the reaction involved.

5 Water in London is much harder than water in Manchester. Why is this?

6 Limescale often forms on taps in hard water areas. Usually, there is more scale on the hot tap than on the cold tap.
(a) How does limescale form?
(b) Why is there usually more on the hot tap?

Section 1: Study Questions

1 The graph shows the rise in temperature of a piece of ice as it is gradually heated in a container until steam is formed.

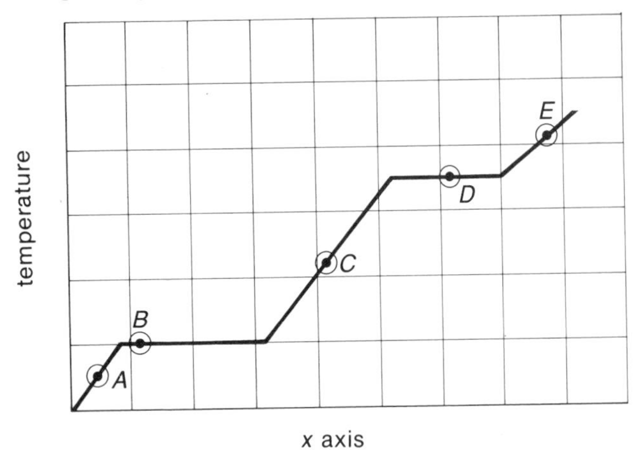

Use the graph to answer the questions.

(a) What label should be placed on the *x* (horizontal) axis of the graph? (1)
(b) What will be the temperature at point *B*? (1)
(c) What will be the temperature at point *D*? (1)
(d) At which of the points *A* to *E* is there a mixture of ice and water in the container? (1)
(e) At which of the points *A* to *E* is there only steam present in the container? (1)

LEAG

2 Carbon dioxide can be prepared by adding hydrochloric acid to calcium carbonate.

(a) Name the salt produced by the reaction. (1)
(b) What else is produced by the reaction, besides the salt and the carbon dioxide? (1)
(c) When a gas jar containing carbon dioxide is held over a burning splint, the flame goes out.

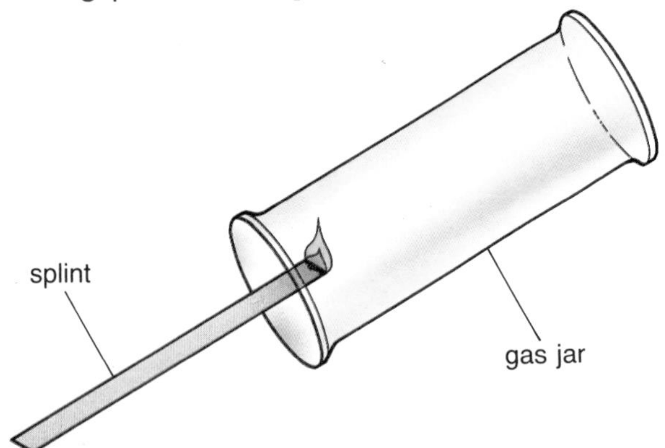

What two properties of carbon dioxide does this illustrate? (2)
(d) If a piece of magnesium is burned in a gas jar of carbon dioxide, a white powder and particles of a black solid are formed.
(i) What is the black solid? (1)
(ii) What is the white powder? (1)
(e) Write a word equation for the reaction in (d). (2)

Total [8] **LEAG**

3 This question is about a threatened strike at *Lichem*, a large chemical company which produces lime and cement. Suppose you are:

1. The Chairperson of Lichem
2. The chief chemist at Lichem
3. The Trade Union Representative at Lichem.

(a) What are your main worries?
(b) What plans would you make in case the strike happens?

4 The major source of the world's energy supply is fossil fuels. The chemical energy that they contain is converted to other more useful forms, e.g. heat and electrical energy. During the last thirty years there has been a considerable increase in demand for electrical energy and now other sources of energy are being investigated as a matter of urgency.

(a) Explain what is meant by the term 'fossil fuel'. (1)
(b) Name *three* fossil fuels in use at the present time, each one existing in a different state of matter at room temperature and pressure. (2)
(c) (i) Name the two elements that are present in the highest proportion in fossil fuels. (1)
(ii) Write the symbol equations for the complete combustion of each of these elements in oxygen. (2)
(d) Both of the reactions represented in part (c) are exothermic. Draw an energy-level diagram to show the energy changes that take place when one of the elements is burned in oxygen. Indicate clearly on the diagram the value for the heat of combustion of the element by labelling it $\triangle H$. (2)
(e) Why are urgent efforts being made to find sources of energy as an alternative to the use of fossil fuels? (1)
(f) State one source of energy, other than direct solar energy, which may be used as an alternative to fossil fuels. (1)

Total [10] **NEA**

5 (a) Three unlabelled bottles are known to contain soft water, permanently hard water and temporarily hard water. Describe how you would distinguish between these different samples of water using the following tests.
Test 1: a small amount of soap was added to each sample of water which was then shaken.
Test 2: each sample of water was boiled, allowed to cool, and a small amount of soap was added. (6)
(b) The water supply to a house passes through a container labelled *'Ion Exchange Resin'*.
(i) What does this resin do to the water?
(ii) Name an ion removed from the water by this resin.
(iii) Give one disadvantage of treating domestic water supplies in this way.
(iv) How do phosphates get into the water system?
(v) What would be the effect of lowering the amount of dissolved oxygen in streams and lakes? (5)

MEG

SECTION J

Chemicals from Crude Oil

1 Crude Oil
2 Alkanes
3 Properties and Uses of Alkanes
4 Petrol from Crude Oil
5 Ethene
6 Products from Ethene
7 Fermentation
8 Ethanol

A crude fractionating column in an oil refinery in Singapore

1 Crude Oil

An aerial view of an oil refinery near Stavanger, Norway. Notice the fractionating column, the storage tanks and the jetty at which tankers can berth

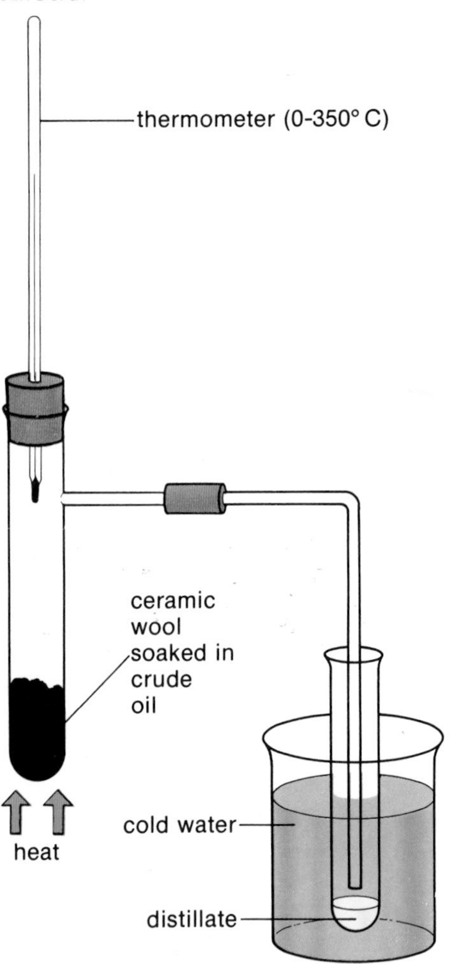

Figure 1
The small scale fractional distillation of crude oil

Crude oil (petroleum) is the main source of fuel and organic chemicals in the UK. The crude oil comes to our refineries, from the North Sea and the Gulf area in the Middle East, as a sticky, smelly, dark-brown liquid. It contains hundreds of different compounds, from simple substances like methane (CH_4) to complicated substances with long chains and rings of carbon atoms. Nearly all the substances in crude oil contain carbon. These carbon compounds are often called **organic compounds**.

A century ago oil was almost unknown. Now we could hardly survive without it. It is almost as important to our lives as air and water. In the UK, 70% of all organic chemicals come from oil. Antifreeze, brake fluid, lipstick, nylon, explosives and paint are all made from it. You may be dressed entirely in oil-based textiles, like Terylene or nylon. Without oil, most transport would come to a standstill and any machine larger than a toy car would seize up from lack of lubricant. We have enough oil to last another 60 years and new reserves are being discovered all the time. Even so, we are using up oil reserves so fast that we need to use crude oil more economically.

Boiling range	20–70°C	70–120°C	120–170°C	170–240°C
Name of fraction	Petrol	Naphtha	Paraffin	Diesel oil
Colour	Pale yellow	Yellow	Dark yellow	Brown
Viscosity	Runny	Fairly runny	Fairly viscous	Viscous
How does it burn?	Easily, with clean yellow flame	Quite easily, yellow flame, some smoke	Harder to burn, quite smoky flame	Hard to burn, smoky flame

Table 1: the properties of fractions obtained by the small-scale fractional distillation of crude oil

At oil refineries, crude oil is separated into fractions by fractional distillation (section A, unit 9). Then it is treated and purified to produce different fuels and chemicals. Figure 1 shows the small scale fractional distillation of crude oil. The ceramic wool, soaked in crude oil, is heated very gently at first and then more strongly so that the distillate slowly drips into the collecting tube. Four fractions are collected, with the boiling ranges and properties shown in table 1. Table 1 also shows the industrial fractions to which our fractions correspond. Notice how the properties of the fractions gradually change.

Bitumen is mixed with stone chippings and used to surface roads

Constituents and uses of the different fractions

Although single pure substances can be obtained from petroleum, it is usually separated and used as fractions containing mixtures of similar substances. These fractions contain compounds with roughly the same number of carbon atoms. Table 2 shows what each fraction contains and what it is used for. The uses of the fractions depend on their properties. Petrol vaporises easily and is very flammable, so it is ideal to use in car engines. Lubricating oil, which is very viscous, is used in lubricants and in central heating. Bitumen, which is solid but easy to melt, is used for waterproofing and asphalting on roads.

Fraction	Boiling range	Number of carbon atoms in the constituents	Uses
Fuel gas	−160 to 20°C	1–4. Mainly methane (CH_4), ethane (C_2H_6), propane (C_3H_8) and butane (C_4H_{10})	Fuels for gas ovens, LPG, GAZ, chemicals
Petrol (gasoline)	20 to 70°C	5–10. E.g. octane (C_8H_{18})	Fuel for vehicles, chemicals
Naphtha	70 to 120°C	8–12.	Chemicals
Paraffin (Kerosine)	120 to 240°C	10–16.	Fuel for central heating and jet engines, chemicals
Diesel oils and lubricating oils	240 to 350°C	15–70.	Fuel for diesel engines, trains and central heating, chemicals, lubricants
Bitumen	above 350°C	More than 70.	Roofing, waterproofing, asphalting on roads

Table 2: the constituents and uses of fractions from crude oil

Questions

1 Explain the following:
organic compounds; fractional distillation; fractionating column.

2 (a) Why is crude oil important?
(b) Why should we try to conserve our reserves of crude oil?
(c) How is crude oil separated into various fractions at a refinery?

3 (a) List the main fractions obtained from crude oil.
(b) Give the main uses of each fraction.

4 (a) What does the word 'organic' mean in everyday use?
(b) Why do you think carbon chemistry is called 'organic chemistry'?
(c) Organic compounds are simple molecular compounds. What physical properties would you expect them to have?

5 What should be done to ensure that crude oil is used more economically?

2 Alkanes

There are millions of different carbon compounds. Why can carbon form so many compounds? The main reason is that carbon atoms can form strong covalent bonds with each other. Atoms of other elements cannot do this. Remember diamond and graphite—giant structures of carbon atoms joined together by strong covalent bonds? Because of these strong C–C bonds, carbon forms molecules containing long chains of carbon atoms. There are millions of compounds containing just hydrogen and carbon. These are called **hydrocarbons**.

The four simplest hydrocarbons are methane, ethane, propane and butane. Figure 1 shows the formulas and molecular models for these four hydrocarbons. The structural formulas show which atoms are

Name	methane	ethane	propane	butane
Formula	CH_4	C_2H_6	C_3H_8	C_4H_{10}
Structural formula	H | H – C – H | H	H H | | H – C – C – H | | H H	H H H | | | H – C – C – C – H | | | H H H	H H H H | | | | H – C – C – C – C – H | | | | H H H H

Figure 1

A model of methane

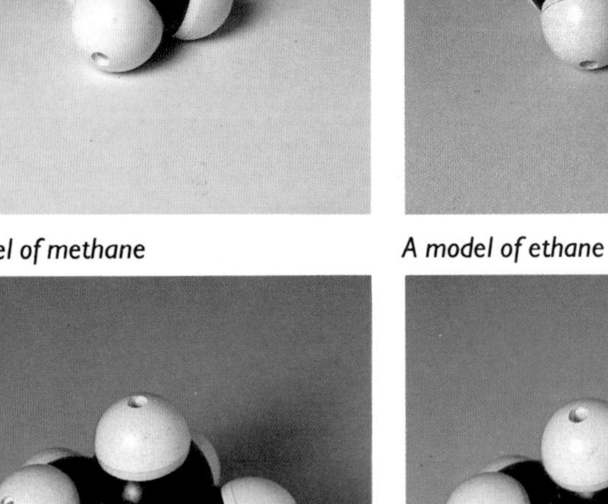

A model of ethane

A model of propane

A model of butane

attached to each other. But they cannot show the correct three-dimensional structure of the molecules. There are four covalent bonds to each carbon atom. Each of these bonds consists of a pair of electrons shared by two atoms. The four pairs of electrons around a carbon atom repel each other as far as possible. So, the bonds around each carbon atom spread out tetrahedrally, as in diamond (figure 2).

Methane, ethane, propane and butane are members of a series of compounds called **alkanes**. All the other members of the series are named from the number of carbon atoms in one molecule. So C_5H_{12} is *pent*ane, C_6H_{14} is *hex*ane, C_7H_{16} is *hept*ane, etc. The names of all alkanes end in *-ane*.

Look at the formulas of methane, CH_4, ethane, C_2H_6, propane, C_3H_8 and butane C_4H_{10}. Notice that the difference in carbon and hydrogen atoms between methane and ethane is CH_2. The difference between ethane and propane is CH_2 and the difference between propane and butane is also CH_2. This is an example of a **homologous series**. A homologous series is a series of compounds with similar properties in which the formulas differ by CH_2.

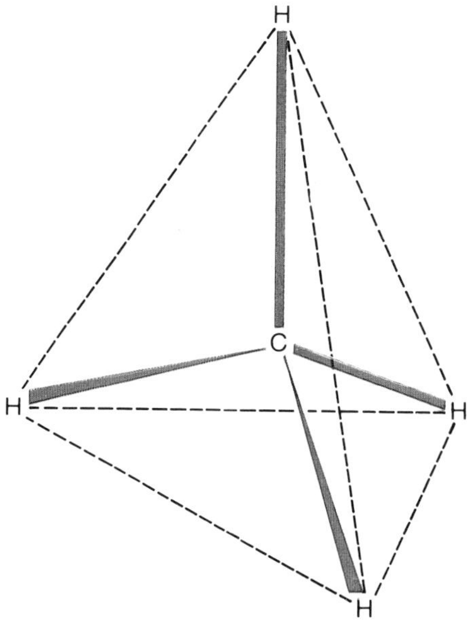

Figure 2
The tetrahedral arrangement of bonds in methane

Isomerism

Earlier in this unit, we noticed that carbon could form a large number of compounds. Even more compounds are possible than we might expect, because it is sometimes possible to join the same set of atoms in different ways. For example, take the formula C_4H_{10}. We know already that butane has a formula C_4H_{10}. But there is another compound which also has this formula. This other compound is called methylpropane (figure 3). Notice that each carbon atom in butane and methylpropane has 4 bonds and each H atom has 1 bond. Butane and methylpropane are two distinct compounds with different properties (see figure 3). Compounds like this with the same molecular formula, but different structures and different properties are called **isomers**.

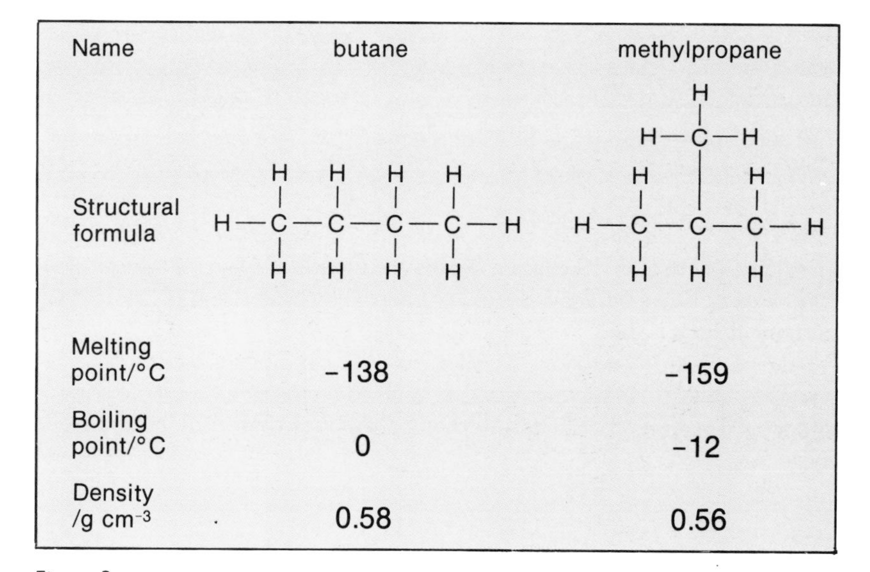

Name	butane	methylpropane
Structural formula	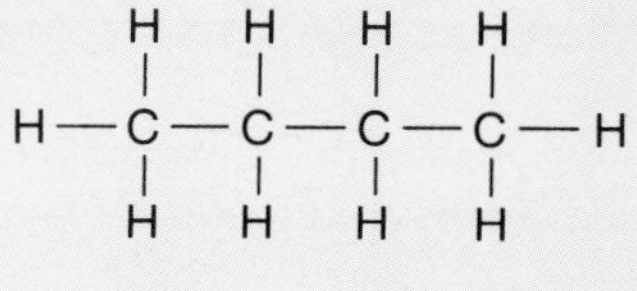	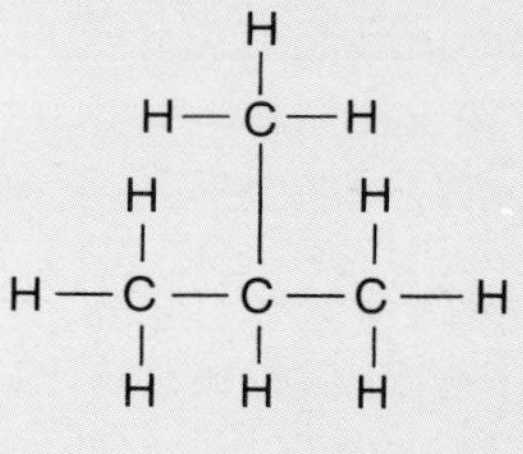
Melting point/°C	-138	-159
Boiling point/°C	0	-12
Density /g cm^{-3}	0.58	0.56

Figure 3
Isomers with the formula C_4H_{10}

Questions

1 Explain the following: *hydrocarbon; alkane; homologous series; isomerism.*
2 (a) What is C_8H_{18} called?
(b) Draw a structural formula for C_8H_{18}.
(c) How many H atoms will the alkane with 10 carbon atoms have?
(d) If an alkane has *n* carbon atoms, how many H atoms will it have?
3 Why can a homologous series of compounds be compared to a group of elements in the periodic table?
4 Draw the structural formulas for (i) the two isomers with the formula C_3H_7Cl; (ii) the two isomers with the formula C_2H_6O; (iii) the three isomers with the formula C_5H_{12}.

3 Properties and Uses of Alkanes

Alkanes are typical molecular (non-metal) compounds. Some of their properties are described below.

Volatile compared with ionic compounds

Tar and bitumen contain alkanes with a relative molecular mass of more than 500. Even so, they begin to melt on very hot days. Compare this with an ionic compound like sodium chloride. NaCl has a relative formula mass of only 58.5, yet its melting point is 808°C.

Petroleum and natural gas are the main sources of alkanes. Alkanes with up to four carbon atoms in each molecule are gases at room temperature. Methane (CH_4) and ethane (C_2H_6) are the main constituents of natural gas. Propane (C_3H_8) and butane (C_4H_{10}) are the main constituents of 'liquefied petroleum gas' (LPG). The best known use of LPG is 'Calor gas' and GAZ for camping, caravans and boats. Butane is also used as lighter fuel and in portable hair tongs.

A small blue butane cylinder used to fuel a gas barbecue

Two red propane cylinders, stored outside, and used to fuel a domestic gas cooker

Alkanes with 5 to 17 carbon atoms are liquids at room temperature. Mixtures of these liquids are used in petrol, in paraffin and in lubricating and engine oils.

Alkanes with 18 or more carbon atoms per molecule are solids at room temperature. Notice that the alkanes become less volatile and change from gases to liquids and then to solids as their molecular size increases.

Insoluble in water

Alkanes are insoluble in water, but they dissolve in organic solvents such as benzene and tetrachloromethane (carbon tetrachloride).

Poor reactivity

Alkanes do not contain ions and their C—C and C—H bonds are very strong. So they have very few reactions. They do not react with metals, aqueous oxidizing agents, acids or alkalis. It may surprise you, but petrol will not react with sodium, potassium permanganate or concentrated sulphuric acid. Ask your teacher to show you.

Concorde being refuelled. Kerosine (paraffin) is used as the fuel in aircraft

Combustion

The most important reaction of alkanes is combustion. They burn in oxygen, producing carbon dioxide and water.

$$\underset{\text{octane in petrol}}{C_8H_{18}} + 12\tfrac{1}{2}O_2 \rightarrow 8CO_2 + 9H_2O$$

The combustion reactions are very exothermic, so alkanes in natural gas and crude oil are used as fuels. When there is too little oxygen for combustion, carbon (soot) and carbon monoxide form as well as carbon dioxide. Carbon monoxide is very poisonous, so it is dangerous to burn carbon compounds in a poor supply of air.

Reaction with halogens

Chlorine and bromine are even more reactive than oxygen, so these halogens will also react with alkanes. Ethane reacts with bromine to form bromoethane and hydrogen bromide.

$$\underset{\text{ethane}}{CH_3-CH_3} + \underset{\text{bromine}}{Br_2} \rightarrow \underset{\text{bromoethane}}{CH_3-CH_2Br} + \underset{\text{hydrogen bromide}}{HBr}$$

One of the H atoms in ethane is substituted by a bromine atom forming bromoethane. This is called a **substitution reaction**. If there is enough bromine, more hydrogen atoms can be replaced by bromine atoms. Chlorine reacts in a similar way to bromine. The reactions of alkanes with chlorine and bromine are very slow at room temperature, but they go much faster when the mixture is heated or exposed to ultraviolet light.

Questions

1 (a) What is meant by a substitution reaction?
(b) Write an equation for the reaction of methane with chlorine.
(c) Write the formulas of all possible substitution products when methane reacts with chlorine.

2 (a) Write an equation for the complete combustion of butane (GAZ) in oxygen.
(b) What are the products when butane burns in a poor supply of oxygen?
(c) Why is it dangerous to allow a car engine to run in a garage with the doors closed?

3 Why do alkanes change from gases to liquids and then to solids as their molecular size increases?

4 Petrol from Crude Oil

Fraction	Approximate % in	
	crude oil	everyday demand
Fuel gas	2	4
Petrol	6	22
Naphtha	10	5
Kerosine	13	8
Diesel oil	19	23
Fuel oil and bitumen	50	38

Relative amounts of different fractions in crude oil and the demand for each fraction

After 1930 the demand for petrol increased much faster than the demand for heavier fractions which make up three quarters of crude oil (see the table). So at first, refineries were left with large surpluses of the heavier fractions. Fortunately, chemists have found ways of converting the heavier fractions into petrol and other useful products.

One method of splitting (cracking) large alkane molecules in these heavier fractions into smaller molecules is **catalytic cracking**.

```
     H   H   H   H   H   H   H   H   H   H
     |   |   |   |   |   |   |   |   |   |
 H — C — C — C — C — C — C — C — C — C — C — H
     |   |   |   |   |   |   |   |   |   |
     H   H   H   H   H   H   H   H   H   H
```

decane

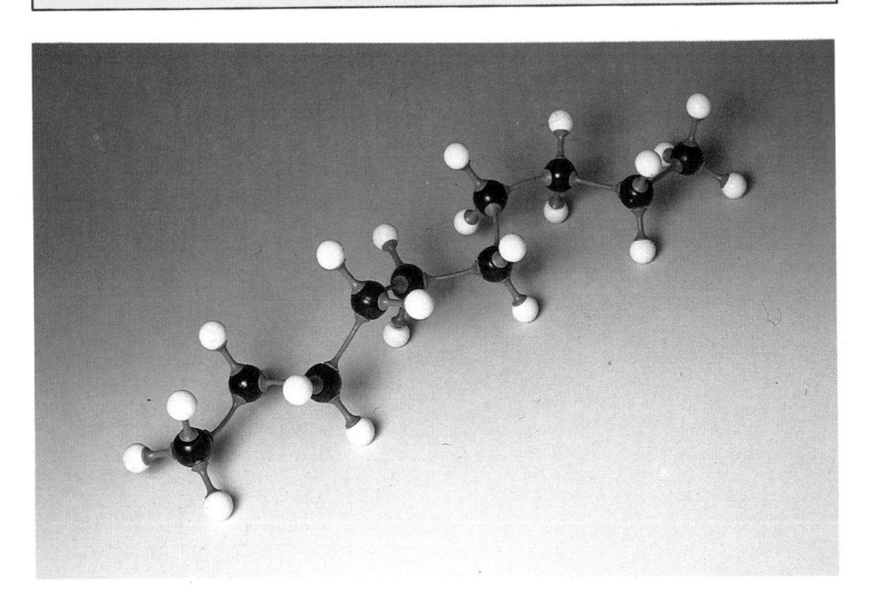

Figure 1 Decane

What are the products of cracking?

Look at the long molecule of decane ($C_{10}H_{22}$) in figure 1. It has a chain of ten carbon atoms with twenty-two hydrogen atoms. Imagine that decane is cracked (split) between two carbon atoms. This cannot produce two smaller alkane molecules because there are not enough hydrogen atoms to go round. But suppose that one product is the alkane, octane (C_8H_{18}). If C_8H_{18} is split off from $C_{10}H_{22}$, the molecular formula of the remaining part is C_2H_4 (figure 2). The chemical name for C_2H_4 is **ethene**. Notice in figure 2 that ethene has a *double* bond between the two carbon atoms. This double bond allows all the carbon atoms in the products to have four bonds.

Hydrocarbons such as ethene, which contain a double bond (C = C), are known as **alkenes**. Their names come from the alkane with the same number of carbon atoms, using the ending *-ene* rather than *-ane*. Organic compounds, like alkanes, which have four single covalent bonds to all their carbon atoms, are described as **saturated compounds**. Alkenes, which have double bonds between some carbon atoms, are examples of **unsaturated compounds**.

Why is 2-star petrol cheaper than 4-star?

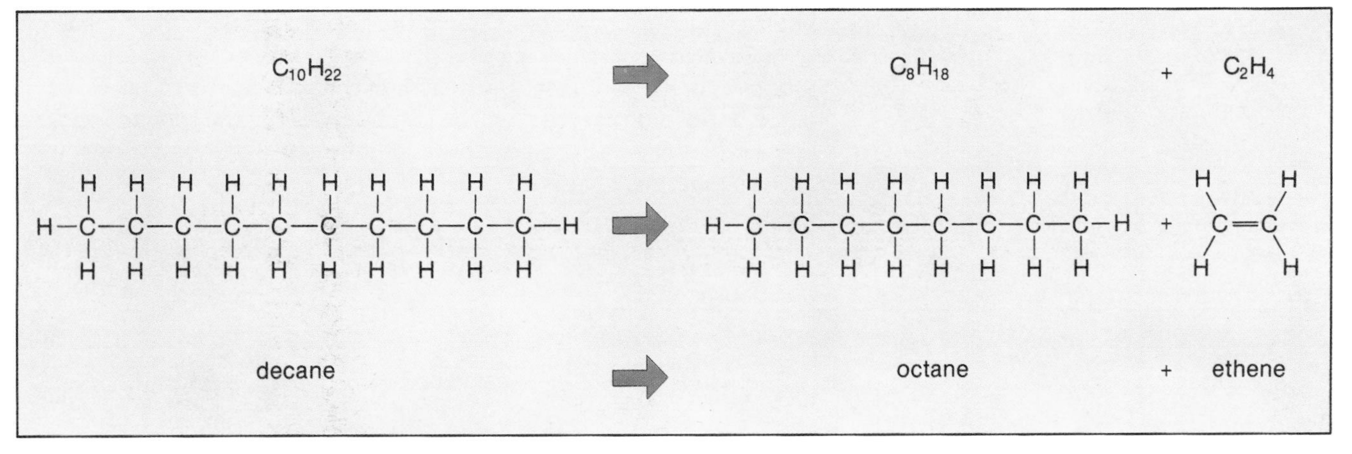

Figure 2

How are molecules cracked?

Unlike distillation, cracking is a chemical process. It involves breaking a strong covalent bond between two carbon atoms. This requires high temperatures and a catalyst. See section K, unit 5. At high temperatures the larger alkane molecules have more energy and they break apart into two or more smaller molecules. The catalyst is finely-powdered aluminium oxide and silicon(IV) oxide. These substances do not react with the crude oil fractions but they do provide a hot surface that speeds up the cracking process.

Cracking helps to produce more petrol. The petrol obtained in this way is of a better quality than that obtained by the distillation of crude oil. Cracked petrol is therefore blended with other petrols to improve their quality.

The catalytic cracking plant at the Wilmington refinery, Connecticut

Questions

1 Explain the words: *cracking; alkene; unsaturated.*

2 (a) Why is cracking important?
(b) What conditions are used for cracking?
(c) Write an equation for the cracking of octane in which one of the products is pentane.
(d) Draw the structural formulas of the products in the equation in (c).

3 (a) What is the name of the alkene of formula C_3H_6?
(b) Draw the structure of this alkene.
(c) Why is there no alkene called methene?

4 Design a simple experiment in which you could attempt to crack some kerosine (paraffin) in order to obtain ethene. (Ethene is a gas at room temperature.)

5 What changes do you think there will be in our lives when crude oil begins to run out?

5 Ethene

Ethene is manufactured by cracking the heavier fractions from crude oil. It can be made on a small scale by cracking paraffin oil using the apparatus in figure 1. Heat the middle of the tube below the aluminium oxide or porous pot. Heat will be conducted along the tube to vaporise the paraffin. The main gaseous product is ethene. Figure 2 shows some of the properties of ethene.

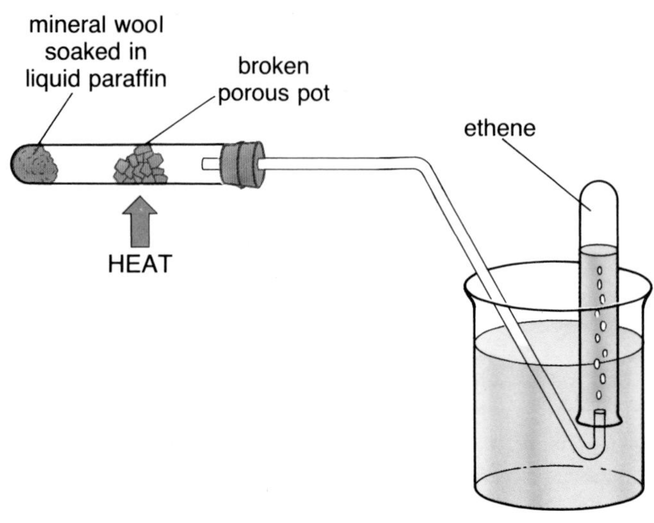

Figure 1
Preparing ethene by cracking paraffin oil

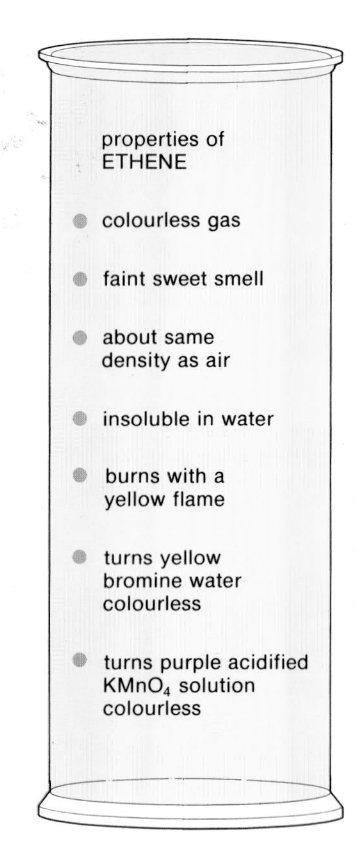

Figure 2

Reactions of alkenes

Alkenes, such as ethene, are much more reactive than alkanes. The most stable arrangement for the four bonds to a carbon atom is a tetrahedral one. This means that a C = C bond is unstable. Other atoms can add across the double bond to make two single bonds. So, alkenes readily undergo **addition reactions**.

This explains why ethene decolorizes bromine water and acidified potassium permanganate solution. When ethene is shaken with bromine water the yellow colour disappears. The bromine molecules add across the double bond in ethene forming 1,2-dibromoethane.

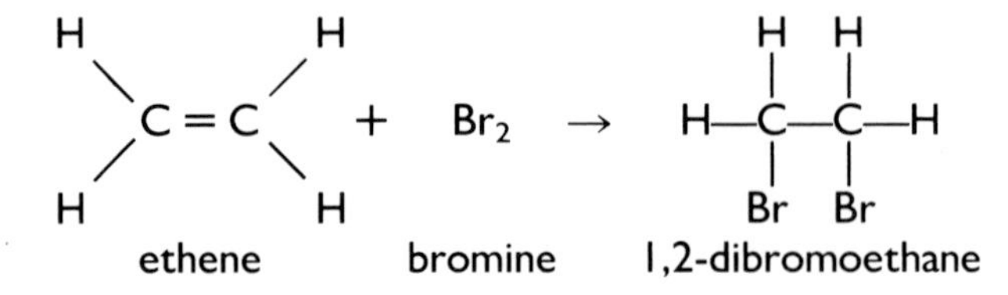

1,2-dibromoethane is an important additive in petrol. Tetraethyllead(IV), $Pb(C_2H_5)_4$, is also added to petrol to help it burn smoothly and prevent 'knocking'. Unfortunately the tetraethyllead(IV) causes lead to be deposited in the cylinder and round the sparking plugs. 1,2-dibromoethane reacts with the deposit of lead to form volatile lead bromide that passes out with the exhaust gases.

Ethene also has an addition reaction with hydrogen at 150°C using a nickel catalyst. The product of this reaction is ethane.

$$\begin{array}{ccc} H & & H \\ & \diagdown \quad \diagup & \\ & C=C & \\ & \diagup \quad \diagdown & \\ H & & H \end{array} \quad + H_2 \quad \rightarrow \quad \begin{array}{ccccc} & H & & H & \\ & | & & | & \\ H- & C & - & C & -H \\ & | & & | & \\ & H & & H & \end{array}$$

This process is known as **catalytic hydrogenation**. It is important in making lard and margarine from vegetable oils, such as palm oil and coconut oil. The vegetable oils are liquids containing alkenes. During hydrogenation these alkenes are converted to alkanes. This change in structure can turn an oily liquid into a harder, fatty solid that can be used to make margarine.

By controlling the amount of hydrogen added to vegetable oils, the margarine can be made as hard or as soft as required. Soft margarine is being made in this factory

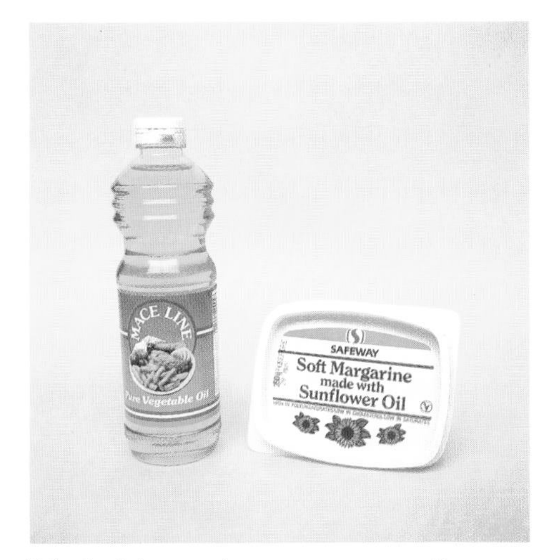

Which of these is the most unsaturated?

Comparing ethene with ethane

Ethane and other alkanes are unreactive. They react only with oxygen (burning) and other reactive non-metals like chlorine and bromine. They will not react with bromine water or dilute potassium permanganate ($KMnO_4$) solution.

In comparison, ethene and alkenes are very reactive because of the addition reactions that take place across their C = C bonds. They readily decolorize bromine water and dilute potassium permanganate and these reactions are used to test for alkenes.

Questions

1 (a) Explain what 'substitution reaction' and 'addition reaction' mean.
(b) Use these words to compare the way in which ethane and ethene react with bromine.

2 (a) Write the structural formula for propene, C_3H_6.
(b) Write equations for the reactions of propene with (i) hydrogen; (ii) bromine.
(c) Draw the structural formulas of the products in part (b).

3 Three cylinders of gas are known to contain ethane, ethene and carbon dioxide but their labels are not clear. What simple tests would you make to show which gas is which?

4 Why are alkenes much more reactive than alkanes?

5 Why do some people object to the addition of tetraethyllead(IV) to petrol?

6 (a) How is margarine made from vegetable oils?
(b) Explain what happens during the reaction in (a).
(c) How can the melting point of the margarine be controlled?

6 Products from Ethene

Ethene is one of the most valuable raw materials for the chemical industry. Because it is so reactive, ethene is used to make a number of useful products including

1 Polythene (polyethene)

2 PVC (polychloroethene)

3 Ethanol (methylated spirits)

4 1,2-dibromoethane (see unit 5).

Clingfilm is made from polythene

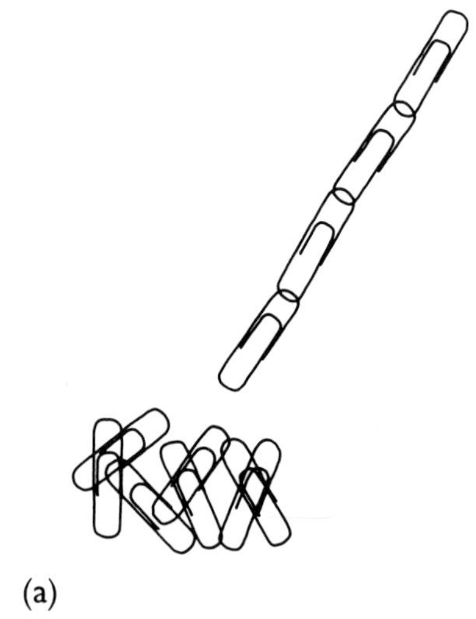

(a)

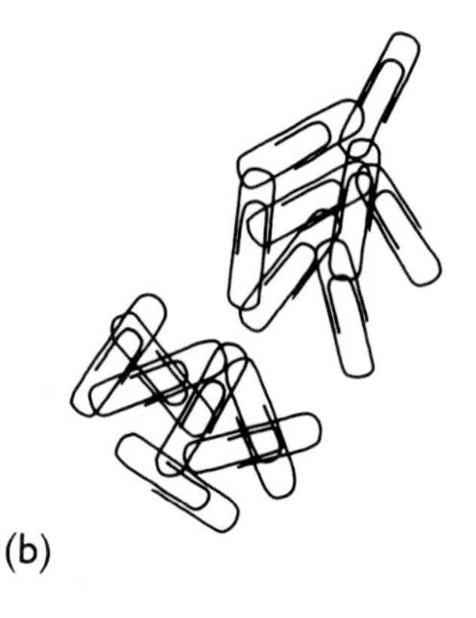

(b)

Figure 1
(a) Correct conditions for polymerization;
(b) Unsuitable conditions for polymerization

• **Polythene (polyethene).** Molecules of ethene have addition reactions with other substances such as hydrogen and bromine. If the conditions are right, molecules of ethene will also add to each other. Hundreds of ethene molecules may join together, forming a giant molecule of polythene.

$$n\left(\begin{array}{c} H \qquad\quad H \\ \diagdown \quad\;\; \diagup \\ C = C \\ \diagup \quad\;\; \diagdown \\ H \qquad\quad H \end{array}\right) \xrightarrow[\text{+ catalyst}]{\text{high temp, high pressure}} \left(\begin{array}{c} H \quad H \\ | \quad\; | \\ -C-C- \\ | \quad\; | \\ H \quad H \end{array}\right)_n$$

In the structure of polythene, n is between 500 and 1500. This process, in which small molecules of the same substance add to each other to form a giant molecule, is called **addition polymerization**. The giant molecule is called a **polymer**. The small molecules which add to each other are called **monomers**. It is essential that the conditions used for polymerization are correct. Figure 1 shows a model (using paper clips) of what happens when unsuitable conditions are used.

The table on page 225 shows the properties of polythene. These have led to many different uses and polythene is the most important plastic at present. It is used as thin sheets for packing and coating materials. It is moulded into beakers, buckets and troughs. It is made into pipes and used to insulate underwater cables.

- **PVC (polychloroethene).** PVC is almost as widely used as polythene. PVC fabric is made into rainwear, wallpaper, curtains, furniture upholstery and protective clothing in industry. Rigid PVC is used for records and for gas and water pipes. Flexible PVC is used for toys and for insulating cables. PVC is made by the polymerization of chloroethene which itself comes from ethene.

$$n\left(\begin{array}{ccc} H & & H \\ & C=C & \\ H & & Cl \end{array}\right) \xrightarrow{\textit{polymerization}} \left(\begin{array}{cc} H & H \\ | & | \\ -C- & C- \\ | & | \\ H & Cl \end{array}\right)_n$$

All plastics (like polythene, PVC, perspex and polystyrene) and manmade fibres (like nylon and Terylene) are polymeric materials. The main raw material for all these products is petroleum. Although these polymers have provided us with many new materials, they have one big disadvantage. They are non-degradable. This means that they are not decomposed by the weather or by bacteria. As litter, they often pollute the environment.

The upper walls and roof of this house have an outer covering of PVC

- **Ethanol.** Ethanol is manufactured by the addition of water to ethene. The reaction is carried out at 300°C and very high pressure using a catalyst of phosphoric acid.

$$\begin{array}{ccc} H & & H \\ & C=C & \\ H & & H \end{array} + H_2O \rightarrow \begin{array}{ccc} & H & H \\ & | & | \\ H- & C- & C-H \\ & | & | \\ & H & OH \end{array}$$

ethene   water   ethanol

Ethanol is the main constituent of methylated spirits, the most widely-used industrial solvent. Methylated spirits is used as a solvent for paints, resins, soaps and dyes. Pure ethanol is used as a solvent for perfumes, cosmetics and after-shave lotions.

POLYTHENE
- tough
- light
- flexible
- easily moulded
- transparent
- easily coloured
- good insulator
- no reaction with water acids or alkalis

Properties of polythene

Questions

1 Explain the following:
polymer; addition polymerization; non-degradable.

2 PVC is made by the polymerization of chloroethene ($CHCl = CH_2$). Draw simple diagrams to show the structure of: (i) a molecule of chloroethene; (ii) a section of the PVC molecule.

3 (a) How is ethene manufactured?
(b) Why is ethene important in industry?

4 *True* or *false?*
Polyethene
A has the same empirical formula as ethene.
B has the same molecular formula as ethene.
C is an alkane.
D is an alkene.
E is a hydrocarbon.
F undergoes addition reactions like ethene.

5 Give one advantage and one disadvantage of plastic rubbish.

7 Fermentation

Cider apples arriving in the silos at the Bulmer cider mill. Fifty thousand tonnes are processed each season

The earliest method of making ethanol was by fermenting carbohydrates. At one time, even industrial alcohol was made by this method. Nowadays, ethanol is manufactured from ethene (unit 6 of this section). But fermentation is still important in winemaking, brewing and bread making. The starting material for fermentation is usually starch, sucrose or glucose. In wine making, a sweet sugary liquid is extracted from grapes. In beer making, starch is extracted from barley by soaking in hot water. The starch is then heated to 55°C with malt. The malt contains enzymes that break down the starch to maltose.

$$\text{starch} \xrightarrow[\text{in malt}]{\text{enzymes}} \text{maltose } (C_{12}H_{22}O_{11})$$

This reaction is very similar to the effect of saliva on starch (see section I, unit 8).

$$\text{starch} \xrightarrow[\text{in saliva}]{\text{enzymes}} \text{maltose } (C_{12}H_{22}O_{11})$$

Finally, yeast is added to the maltose. Yeast contains enzymes which break down the maltose first to glucose and then to ethanol and carbon dioxide.

$$\underset{\text{maltose}}{C_{12}H_{22}O_{11}} + H_2O \rightarrow \underset{\text{glucose}}{2C_6H_{12}O_6}$$

$$\underset{\text{glucose}}{C_6H_{12}O_6} \rightarrow \underset{\text{ethanol}}{2C_2H_5OH} + 2CO_2$$

The fermented liquid contains only 5 to 10% ethanol. Ethanol can be obtained from this liquid by fractional distillation. In alcoholic drinks it is the ethanol from the fermentation that is important (making the drinks intoxicating). In bread making, the carbon dioxide from fermentation is important because this makes the bread rise before it is baked.

Foam in copper brewery tanks

Alcoholic drinks, like beer and wine, that contain ethanol can be pleasant in small amounts. They make people more relaxed and they have different tastes. But alcoholic drinks taken in excess can be very dangerous. They make us slow to react, they damage the liver and they can be addictive like drugs. There is more about ethanol in unit 8 of this section.

Making ethanol by fermentation

See if you can make some ethanol from glucose by fermentation. Set up the apparatus in figure 1 and leave it in a warm place for 3–7 days. Then filter the solution in the conical flask and separate ethanol from the filtrate by fractional distillation (figure 2). Collect the first few drops of distillate on a crucible lid and see if they will burn like ethanol.

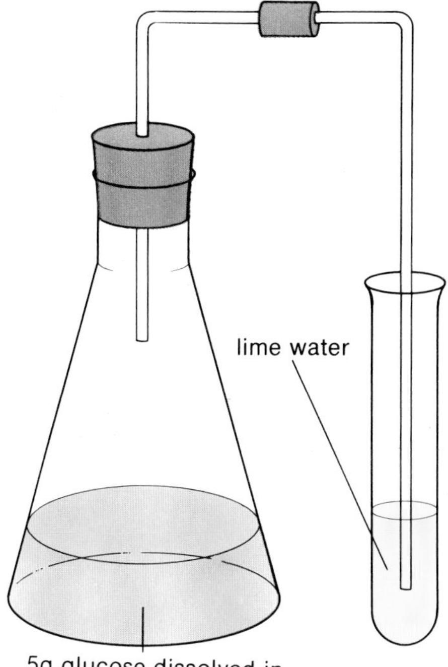

Figure 1
Making ethanol by fermentation. Why does the lime water go milky?

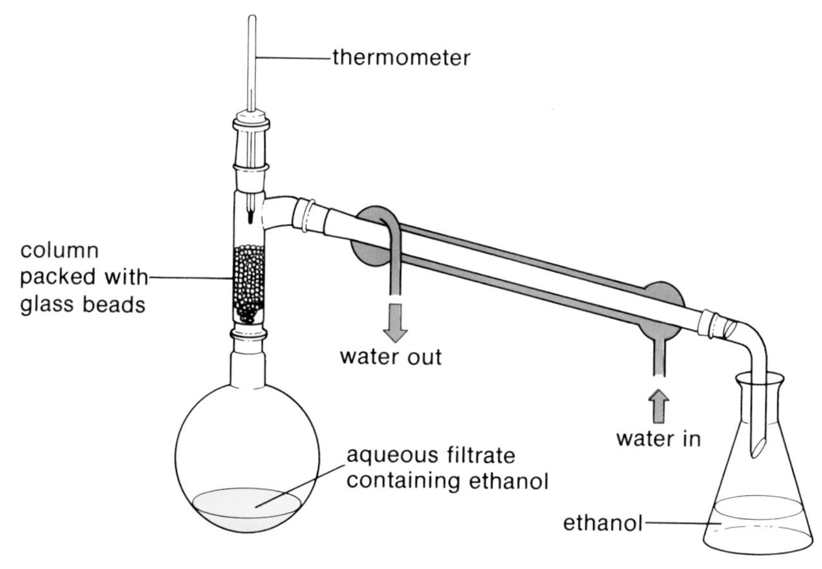

Figure 2
Fractional distillation of ethanol from an aqueous solution

In this unit and in unit 8 in section 1, we have studied the breakdown of starch to maltose and glucose and then to ethanol. These changes are summarised in figure 3.

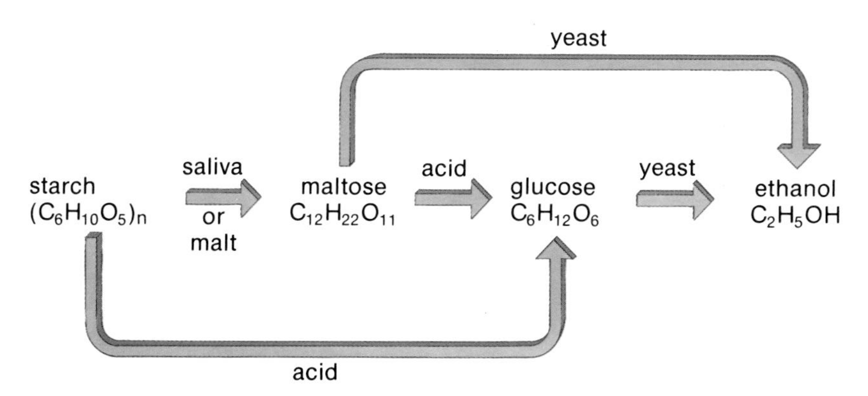

Figure 3
The breakdown of starch to ethanol

Questions

1 Explain the words:
fermentation; enzyme.

2 (a) Why is fermentation important?
(b) What are the main products of the fermentation of starch with yeast in anaerobic (no air) conditions?
(c) What causes the fermentation process?

3 Look at figures 1 and 2 which relate to the fermentation of glucose.
(a) Why does the lime water turn milky?
(b) Why is fermentation carried out in a warm place?
(c) Why is *fractional* distillation needed to separate the ethanol in figure 2?
(d) How would you test the final distillate to show that it is (i) free of water; (ii) pure ethanol?
(e) Write an equation for the fermentation of glucose by yeast under anaerobic conditions.

4 The adverts say 'Don't drink and drive'. Why is this important?

8 Ethanol

CH_3OH methanol
CH_3CH_2OH ethanol
$CH_3CH_2CH_2OH$ propanol
$CH_3CH_2CH_2CH_2OH$ butanol

The homologous series of alcohols

Ethanol is a member of a large class of compounds called alcohols. When people talk about 'alcohol' in drinks, they really mean ethanol. All alcohols contain an —OH group attached to a carbon atom. The simplest alcohol is methanol, CH_3OH.

Alcohols form a homologous series like alkanes (see the table). They have similar properties and their structures increase by units of CH_2.

```
   H   H                H   H
   |   |                |   |
H—C—C—O—H          H—C—C—H          H—O—H
   |   |                |   |
   H   H                H   H

  ethanol              ethane           water
```

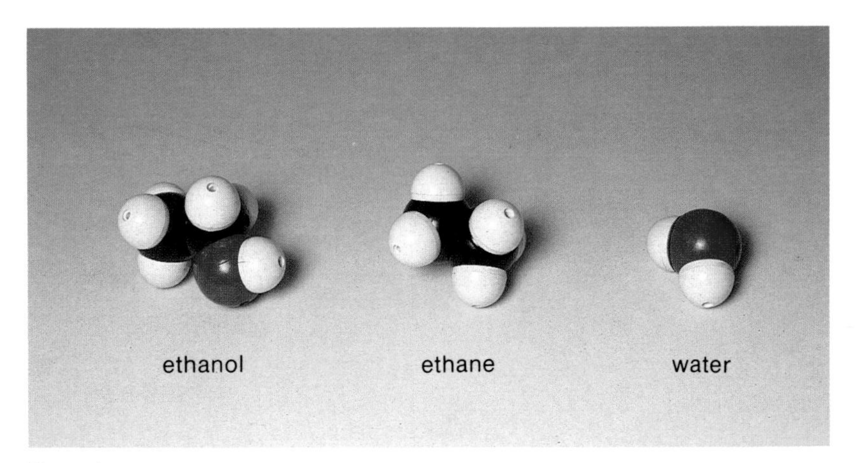

Figure 1

What are the properties of ethanol?

Look at the structures of ethanol, ethane and water in figure 1. The structure of ethanol has similarities to ethane and to water. It contains a C_2H_5 group like ethane

```
   H  H
   |  |
H—C—C—
   |  |
   H  H
```

and an —OH group like water. So we would expect ethanol to have some properties like ethane and some like water.

1 Ethanol is a colourless liquid (boiling point 78°C). It mixes with water in all proportions.

2 It is a very good solvent for molecular substances. Methylated spirits, which is widely used as a solvent in industry, contains 95% ethanol.

3 Ethanol burns very easily (like ethane) with a pale yellow flame forming carbon dioxide and water.

$$C_2H_6O + 3O_2 \rightarrow 2CO_2 + 3H_2O$$

4 Ethanol, unlike ethane or water, can be oxidized. The product is ethanoic acid (acetic acid).

```
     H  H              H
     |  |              |    //O
  H—C—C—OH   →     H—C—C
     |  |              |    \OH
     H  H              H
   ethanol          ethanoic acid
```

Wine goes sour after the bottle has been opened because the ethanol in it is oxidized to ethanoic acid. This is how wine vinegar is produced. In this case, the oxidation is catalysed by enzymes in the wine using oxygen in the air. Malt vinegar is made by a similar process using beer instead of wine. In the laboratory, ethanol is oxidized to ethanoic acid using acidified potassium permanganate or potassium dichromate. The reaction of ethanol with dichromate was once used in the breathalyser test. The driver breathes out through a tube containing yellow crystals of dichromate into a polythene bag. Any ethanol in the breath reacts with the yellow dichromate to produce a green chromium salt. The more ethanol in the breath, the greener the colour.

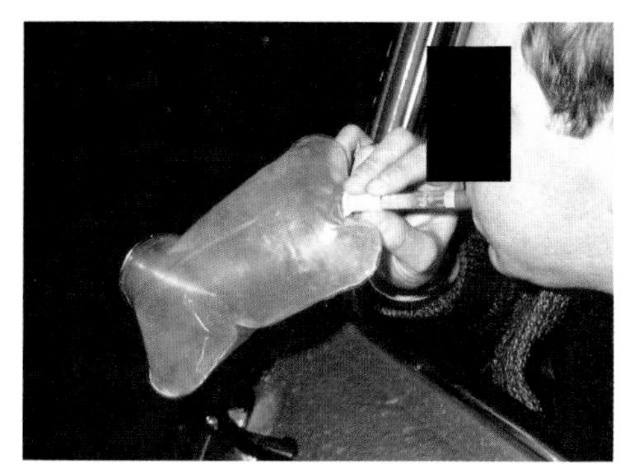

A motorist being breathalysed using the old technique

5 Ethanol can also be dehydrated to form ethene. This involves removing water from ethanol.

```
     H  H         H         H
     |  |          \       /
  H—C—C—OH   →      C = C      + H2O
     |  |          /       \
     H  H         H         H
```

In chemistry, the usual dehydrating agent is concentrated sulphuric acid. This could be used with ethanol, but it is safer and easier to use aluminium oxide, porous pot or porcelain chips (figure 2). Heat the tube below the porous pot. Heat will be conducted along the tube to vaporise the ethanol. After the air has been driven out of the apparatus, you can collect the ethene.

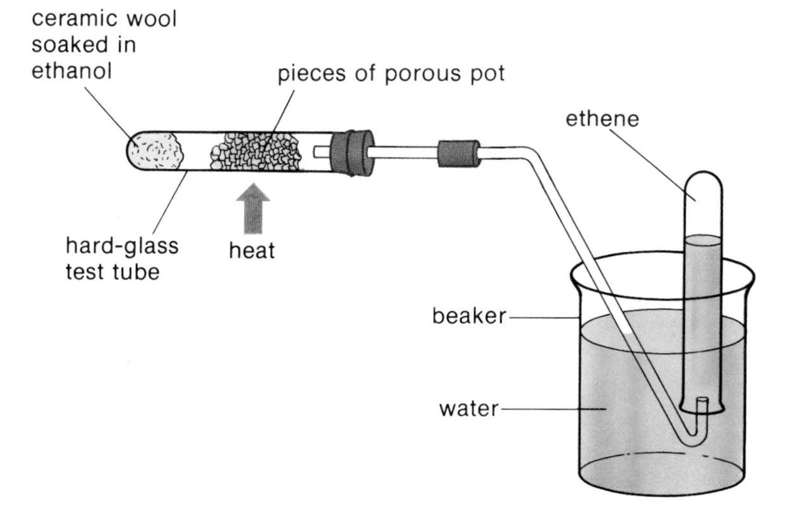

Figure 2
Preparing ethene by dehydration of ethanol

Questions

1 Explain the following:
alcohol; homologous series; dehydration.

2 (a) If an alcohol has *n* carbon atoms, how many hydrogen atoms will it have?
(b) Write a general formula for the family of alcohols.

3 Two substances have the molecular formula, C_2H_6O. Draw structural formulas for these two substances.

4 Write an equation for (i) the combustion of methanol; (ii) the oxidation of propanol to propanoic acid; (iii) the dehydration of propanol.

Section J: Study Questions

1 Give two reasons why, in a fire, plastic furniture is more dangerous than wooden furniture. (2) **SEG**

2 (a) Petroleum is a mixture of an *homologous series* of *alkanes*. When petroleum is subjected to *fractional distillation*, a series of fractions is obtained, some of which are used as *fuels*. Some of the higher fractions are subjected to *catalytic cracking* to make chemical feedstocks including ethene. Explain the terms in italics. (12)
(b) State and explain what is involved in the polymerisation of ethene to polythene. (5)
(c) Explain, with the aid of simple diagrams and named examples, the difference between thermoplastic and thermosetting polymers. (5)
(d) Write brief notes on the environmental problems created by
(i) oil slicks
(ii) the increased use of plastic materials. (5)
Total [27] **WJEC**

3 The following list shows some of the processes which are used in industry:
cracking
fermentation
polymerization
Using *one* example from industry in each case, describe briefly how each process may be used to manufacture a product useful in everyday life. (20) **LEAG**

4 (a) The following diagram shows some of the reactions of ethene.

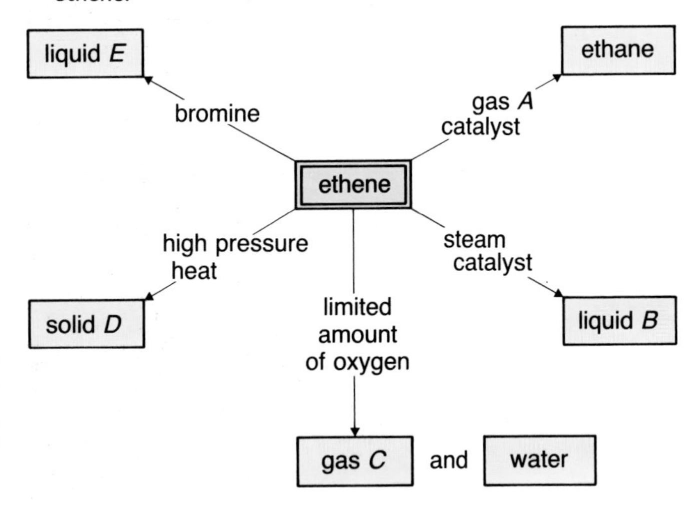

(i) Draw the structural formulas for ethane and ethene. (2)
(ii) Give the names for:
gas *A* gas *C*
liquid *B* solid *D*. (4)
(iii) Write a symbol equation for the reaction between ethene and water to form liquid *B*. (2)
(iv) Give the name and draw the structural formula for liquid *E*. (3)
(v) Ethene is an unsaturated compound. By means of gas *A*, ethene is converted into ethane, a saturated compound. Give one commercial application of this *type* of reaction. (1)

(b) Liquid paraffin, which is a mixture of alkanes, is decomposed into products which include ethene. This process can be carried out in the apparatus below.

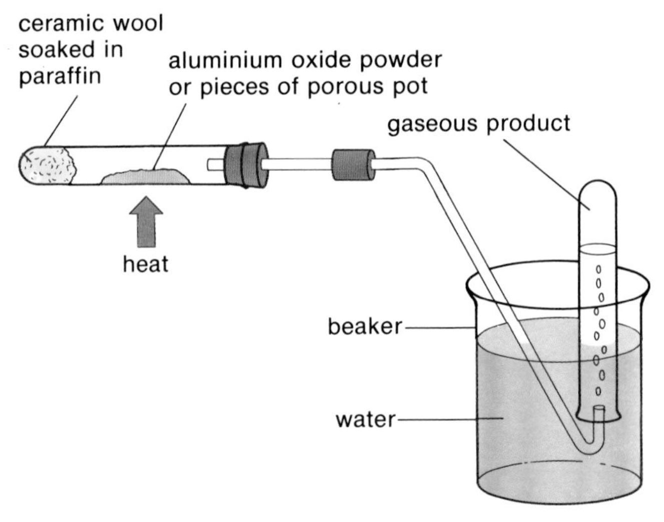

(i) What is the purpose of the pieces of porous pot? (1)
(ii) What is the name given to this example of thermal decomposition? (1)
(iii) Explain why changing alkanes into simpler alkanes and alkenes is important in the petrochemical industry. (2)
(c) Esters may be prepared by a reaction which is represented by the following equation:

organic acid + alcohol → ester + water

This reaction takes place when the mixture is heated in the presence of concentrated sulphuric acid.
(i) What is the purpose of the concentrated sulphuric acid? (1)
(ii) Name both the acid and the alcohol which would produce the ester ethyl ethanoate ($CH_3COOC_2H_5$). (2)
(iii) Give a use of esters. (1)
Total [20] **NEA**

5 Petrol is a mixture of liquids which are all hydrocarbons.
(a) Name the *two* elements that are found in hydrocarbon particles.
(b) When petrol vapour is exploded in a car engine, the elements mentioned in part (a) form exhaust gases. Give the name of *one* of these gases.
(c) Name *one* very poisonous gas found in car exhaust fumes. **LEAG**

6 The substance ethanediol ($C_2H_6O_2$) is used as an antifreeze. It is made from ethene according to this equation:

$$\underset{\text{ethene}}{C_2H_4} \xrightarrow{\text{water, air, catalyst}} \underset{\text{ethanediol}}{C_2H_6O_2}$$

The relative atomic masses are C = 12; H = 1; O = 16.
(a) State *one* use of antifreeze. (1)
(b) What is a catalyst? (2)
(c) Calculate the relative molecular mass of ethene. (1)
(d) Calculate the relative molecular mass of ethanediol. (1)
(e) Calculate the mass of ethanediol that could be made from 5.6 g of ethene. (2)
Total [7] **SEG**

SECTION K

Reaction Rates

Explosives are used in mining to break up rock. Any explosion is an extremely fast reaction

Stalagmites and stalactites in Gough's Cave, Cheddar, Somerset. Stalagmites and stalactites form by a very slow reaction. Limestone dissolves in rain water and then gets precipitated again

1 How Fast?

Limestone is weathered so slowly that it takes centuries before the effect is noticed

Some reactions, like explosions,.happen almost instantaneously

Everyday we are concerned about how fast things happen. We want to know how fast we can get to school, how fast a car is travelling or how fast we can run a hundred metres. Chemists are particularly interested in how fast chemical reactions happen. They want to know how quickly steel rusts, how quickly food cooks, and how fast stone buildings are weathered by 'acid rain'.

Different chemical reactions happen at different rates. Some reactions, like explosions, are so fast that they are almost instantaneous. For example, when a burning splint is put into a mixture of hydrogen and chlorine, there is a loud bang and hydrogen chloride is produced.

$$H_2(g) + Cl_2(g) \rightarrow 2HCl(g)$$

Other reactions, like the rusting of steel and the weathering of limestone on buildings, happen so slowly that it may be years or even centuries before we notice their effects.

Most reactions take place at speeds somewhere between those described in the last two paragraphs. The reactions which take place when coal is burnt and when food is cooked are good examples of reactions which occur at steady rates.

Why bother about reaction rates?

Most of the reactions in our bodies would never happen if the reacting substances were just mixed together. Each reaction is helped along by its own special **catalyst**. Catalysts allow the chemicals to react more easily. The catalysts in living things are called **enzymes**. Without enzymes the reactions in your body would stop and you would die. One of these enzymes is amylase. Amylase is present in saliva. It speeds up the first stage in the breakdown of starch in foods such as bread, potatoes and rice.

Industrial chemists are not usually satisfied with just turning one substance into another. They want to carry out reactions faster and more cheaply. In industry, speeding up slow reactions makes them more economical, as saving time saves money.

The key reaction in the manufacture of sulphuric acid is the contact process (section G, unit 1). This involves converting sulphur dioxide and oxygen to sulphur trioxide.

$$2SO_2 + O_2 \rightarrow 2SO_3$$

At room temperature, this reaction will not happen. But chemical engineers have found that the reaction takes place quickly at 450°C if a catalyst of vanadium(V) oxide or platinum is used. By speeding up the reaction in this way, sulphuric acid can be made faster and more cheaply. This is important because sulphuric acid is a major industrial chemical.

The reactions which take place when food is cooked occur at a steady rate

A workman fitting a special catalyst section to the exhaust system of a car. The catalyst removes nitrogen oxide from the car's exhaust fumes

Questions

1 Name *two* reactions that use a catalyst and give the name of the catalyst in each case.

2 (a) State *three* ways in which food can be preserved and stored without 'going bad'.
(b) What conditions slow down the rate at which foods deteriorate?

3 How do you think a pressure cooker speeds up the rate at which food is cooked?

4 Why are reaction rates important?

5 How do gardeners speed up the growth rate of plants?

2 Studying Reaction Rates

A chemical reaction cannot happen unless particles in the reacting substances collide with each other.

This statement explains why reactions between gases and liquids usually happen faster than reactions involving solids. Particles in gases and liquids can mix and collide much more easily than particles in solids. In a solid, only the particles on the surface can react. During a reaction, reactants are being used up and products are forming. So, the amount and the concentration of the reactants fall as the amount and the concentration of the products rise. The reaction rate tells us how fast the reaction is taking place. We can measure reaction rates by measuring how much of a reactant is used up or how much of a product forms in a given time.

$$\therefore \text{Reaction rate} = \frac{\text{change in amount (or concentration) of a substance}}{\text{time taken}}$$

For example, when 0.1 g of magnesium was added to dilute hydrochloric acid, the magnesium reacted and disappeared in 10 seconds.

$$\therefore \text{Reaction rate} = \frac{\text{change in mass of magnesium}}{\text{time taken}}$$

$$= \frac{0.1}{10} \text{ g magnesium used up per second}$$

$$= 0.01 \text{ g s}^{-1}$$

Strictly speaking, this is the *average* reaction rate over the 10 seconds for all the magnesium to react. Although reaction rates are usually measured as changes in mass (or concentration) with time, we can also use changes in volume, pressure, colour and conductivity with time.

Time /min	Mass of flask and contents /g	Decrease in mass /g	Decrease in mass for each minute interval /g
0	78.00	0	
1	76.50	1.50	1.50
2	75.50	2.50	1.00
3	74.95	3.05	0.55
4	74.60	3.40	0.35
5	74.41	3.59	0.19
6	74.33	3.67	0.08
7	74.30	3.70	0.03
8	74.30	3.70	0

The results of one experiment to measure the rate of reaction between marble chips and dilute hydrochloric acid

Measuring reaction rates

The rate of reaction between small marble chips (calcium carbonate) and dilute hydrochloric acid can be studied using the apparatus in figure 1. As the reaction occurs, carbon dioxide escapes from the flask and so the mass of the flask and its contents decrease.

$$\underset{\text{marble chips}}{CaCO_3(s)} + 2HCl(aq) \rightarrow CaCl_2(aq) + H_2O(l) + CO_2(g)$$

The cotton wool in the mouth of the flask stops liquid escaping from the flask as the mixture fizzes. The results of one experiment are given in the table. These results have been plotted on a graph in figure 2. During the first minute there is a decrease in mass of 1.5 g as carbon dioxide escapes.

∴ Average rate of reaction in the first minute

$$= \frac{\text{change in mass}}{\text{time taken}} = \frac{1.5}{1} = \frac{\text{1.5g of carbon dioxide}}{\text{lost per minute}}$$

(Notice that the units for reaction rate are grams per minute this time.) During the second minute (from time = 1 minute to time = 2 minutes), 1.0 g of carbon dioxide escapes.

∴ Average rate of reaction in the second minute

$$= \frac{1.0}{1} = 1.0 \text{ g of } CO_2 \text{ lost per minute.}$$

Notice that the reaction is fastest at the start of the reaction when the slope of the graph is steepest. During the reaction the rate falls and the slope levels off. Eventually the reaction rate becomes zero and the graph becomes flat with a slope (gradient) of zero.

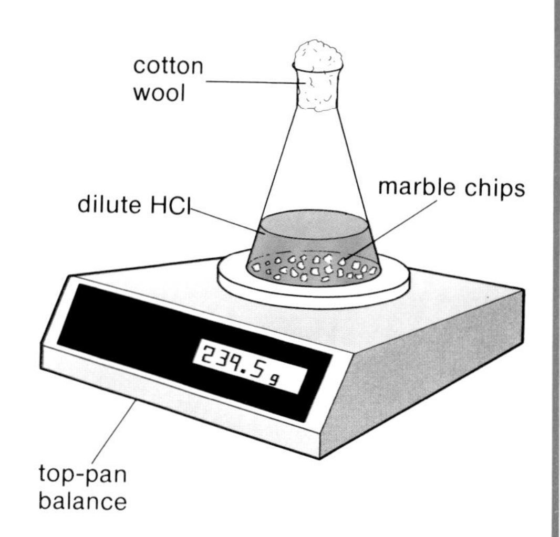

Figure 1

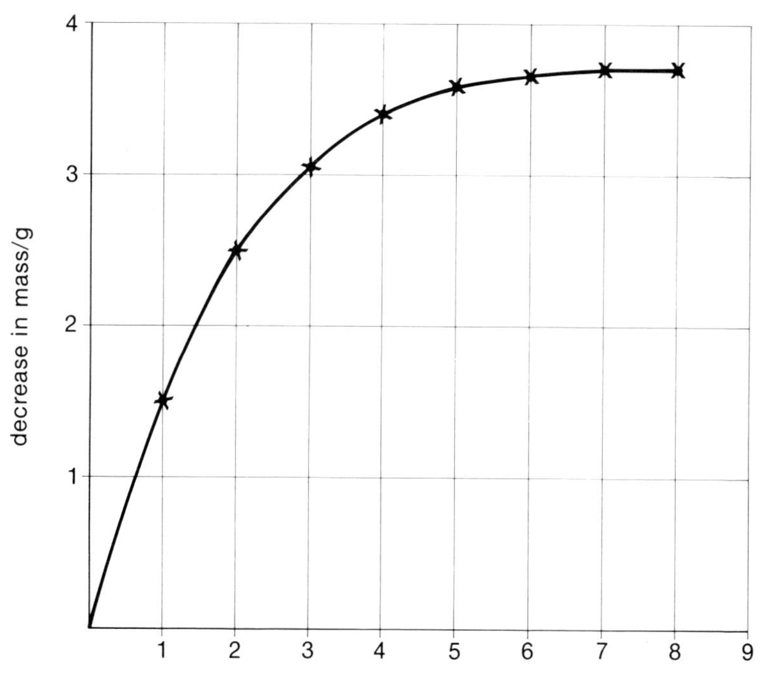

Figure 2

Questions

1 (a) When magnesium reacts with dilute hydrochloric acid, does the magnesium react faster at the start of the reaction or at the finish?
(b) Give two reasons for your answer in part (a).

2 Look at the results in the table and the graph in figure 2.
(a) What mass of carbon dioxide is lost from the flask in (i) the third minute (time 2 to 3 min); (ii) the fourth minute (time 3 to 4 min); (iii) the fifth minute (time 4 to 5 min)?
(b) What is happening to the reaction rate as time passes?
(c) Explain the change in reaction rate with time.
(d) Why does the graph become horizontal after a while?

3 Selective weedkillers can be added to a lawn either as solid pellets or as aqueous solutions. Which method will affect the weeds faster? Explain your answer.

3 Making Reactions Go Faster

It's easier to get a fire started with sticks rather than logs

Anyone who has tried to light a fire knows that it is easier to burn firewood than logs. The main reason for this is that the firewood has a greater surface area.

> *In general, reactions go faster when there is more surface area to react.*

Surface area and reaction rate

The reaction between marble chips (calcium carbonate) and dilute hydrochloric acid can also be used to study the effect of surface area on reaction rate.

$$CaCO_3(s) + 2HCl(aq) \rightarrow CaCl_2(aq) + H_2O(l) + CO_2(g)$$

During the reaction, carbon dioxide escapes from the reacting mixture so there is a decrease in mass. The results are shown in figure 1. In experiment I, thirty *small* marble chips (with a total mass of 10 g) reacted with 100 cm^3 of dilute hydrochloric acid. In experiment II, six *large* marble chips (with a total mass of 10 g) reacted with 100 cm^3 of the same hydrochloric acid. There are more than enough marble chips in both experiments, so the acid will be used up first.

1 Why is the overall decrease in mass the same in both experiments?
2 Why do the graphs become flat?
3 Which graph shows the greater decrease in mass per minute at the start of the experiment?
4 Which experiment begins at the faster rate?
5 Why is the reaction rate different in the two experiments?

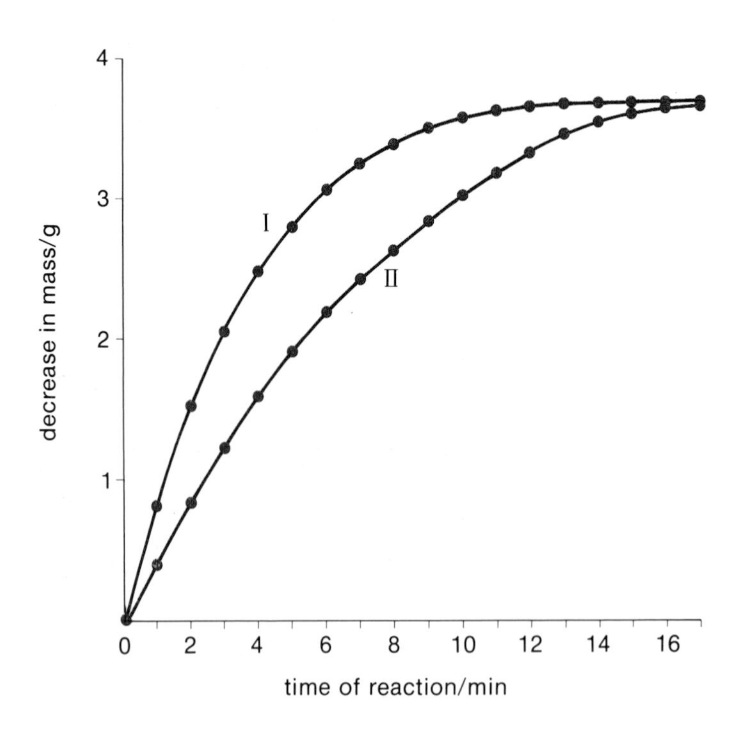

Figure 1

Concentration and reaction rate

Substances that burn in air burn much more rapidly in oxygen. Charcoal in a barbecue normally burns very slowly with a red glow. But, if you blow onto it so that it gets more air and more oxygen, it glows much brighter and may burst into flames. In oxy-acetylene burners, acetylene burns in pure oxygen. These burners produce such high temperatures that the flame will cut through sheets of metal.

$$\underset{\text{acetylene}}{C_2H_2} + 2\tfrac{1}{2}O_2 \rightarrow 2CO_2 + H_2O + \text{heat}$$

Chemical reactions occur when particles of the reacting substances collide with each other. Collisions between acetylene molecules and oxygen molecules occur more often when oxygen is used instead of air. So, the reaction happens faster and gives off more heat when the concentration of oxygen is increased by using the pure gas. Pure oxygen can also be used to speed up chemical changes in the body. This can help the recovery of hospital patients, such as those suffering from extensive burns.

Pure oxygen is used in oxygen tents like this one, to speed up the recovery of hospital patients

> *In general, reactions go faster when the concentration of reactants is increased.*

In reactions between gases, the concentration of each gas can be increased by increasing its pressure. Some industrial processes use very high pressures. For example, in the Haber process (unit 7 of this section), nitrogen and hydrogen are made to react at a reasonable rate by increasing the pressure to 250 times atmospheric pressure.

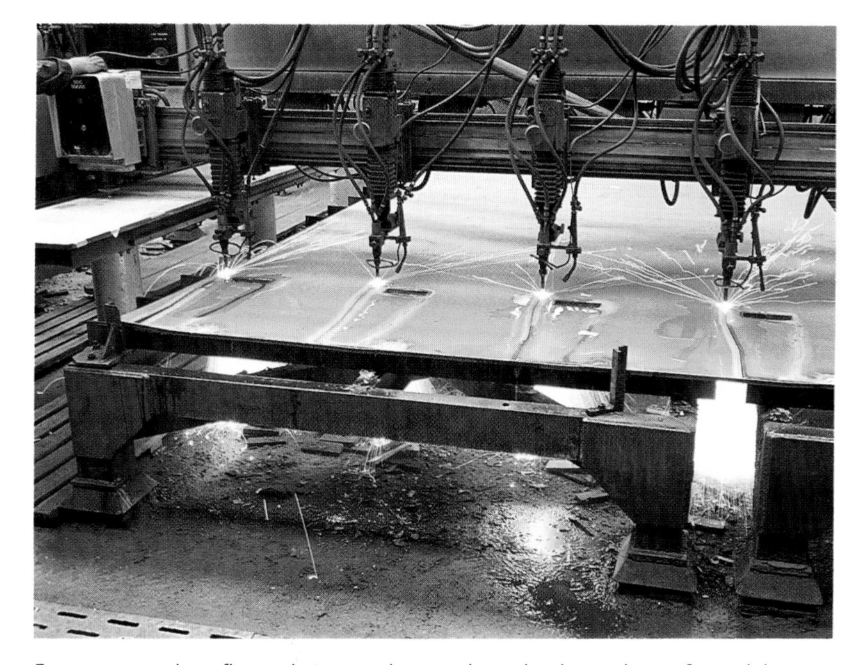

Four oxy-acetylene flames being used to cut through a large sheet of metal. In oxy-acetylene burners, acetylene burns very rapidly in pure oxygen. The flame is hot enough to melt the metal

Questions

1 It takes about 10 minutes to fry chips, but about 20 minutes to boil potatoes. Larger potatoes take even longer to boil.

(a) Why do larger potatoes take longer to cook than small ones?

(b) Why can chips be cooked faster than boiled potatoes?

(c) Why can boiled potatoes be cooked faster in a pressure cooker?

2 Why do gaseous reactions go faster if the pressure of the reacting gases is increased?

3 Which of the following will affect the rate at which a candle burns?

the temperature of the air; the shape of the candle; the air pressure; the length of the wick.

Explain your answer.

State two other factors that will affect the rate at which a candle burns.

4 Temperature and Reaction Rates

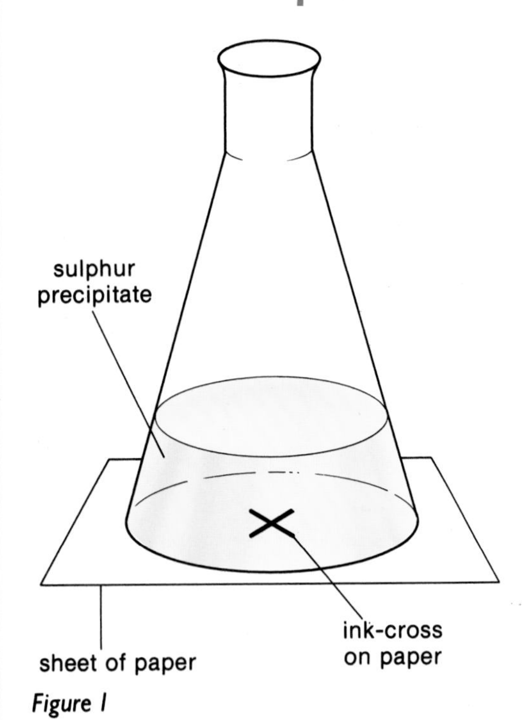

Figure 1

Milk will keep for several days in a cool refrigerator, but it turns sour very quickly if it is left in the sun. This is because *chemical reactions go faster at higher temperatures.*

Studying the effect of temperature on reaction rate

The reaction between sodium thiosulphate solution ($Na_2S_2O_3(aq)$) and dilute hydrochloric acid can be used to study the effect of temperature on reaction rate.

$$Na_2S_2O_3(aq) + 2HCl(aq) \rightarrow 2NaCl(aq) + H_2O(l) + S(s) + SO_2(g)$$

When the reactants are mixed, the solution becomes cloudy because sulphur is precipitated (figure 1). As the precipitate gets thicker, an ink cross on white paper below the flask slowly disappears. We can find the reaction rate by mixing 5 cm^3 of 2.0 M hydrochloric acid with 50 cm^3 of 0.05 M $Na_2S_2O_3(aq)$ and then measuring the time it takes for the cross to disappear.
The table on page 239 shows the results obtained when this reaction was carried out at different temperatures. Figure 2 shows a graph of the time taken for the cross to disappear against temperature.

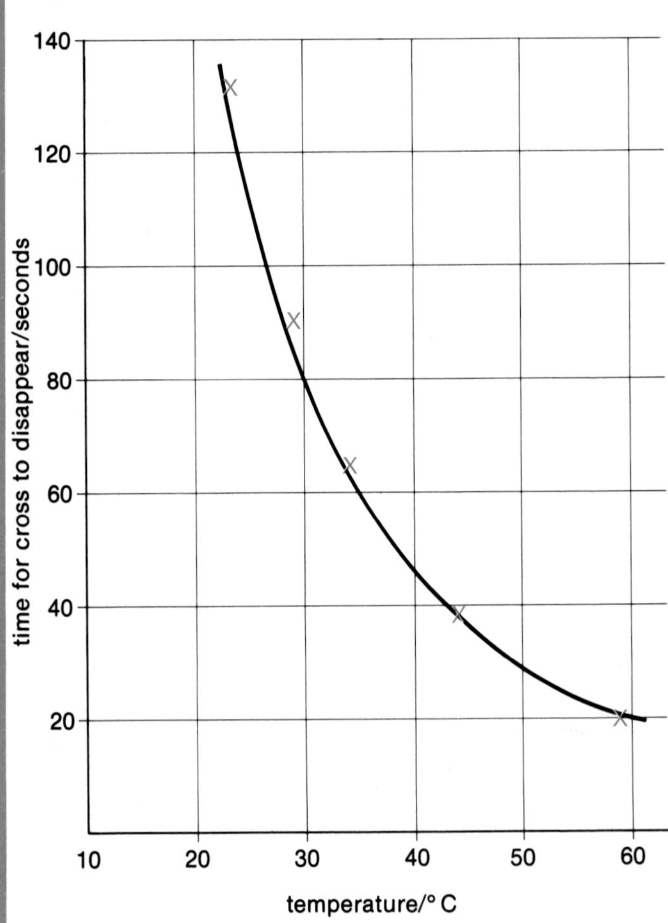

Figure 2
A graph showing the time for the cross to disappear against temperature

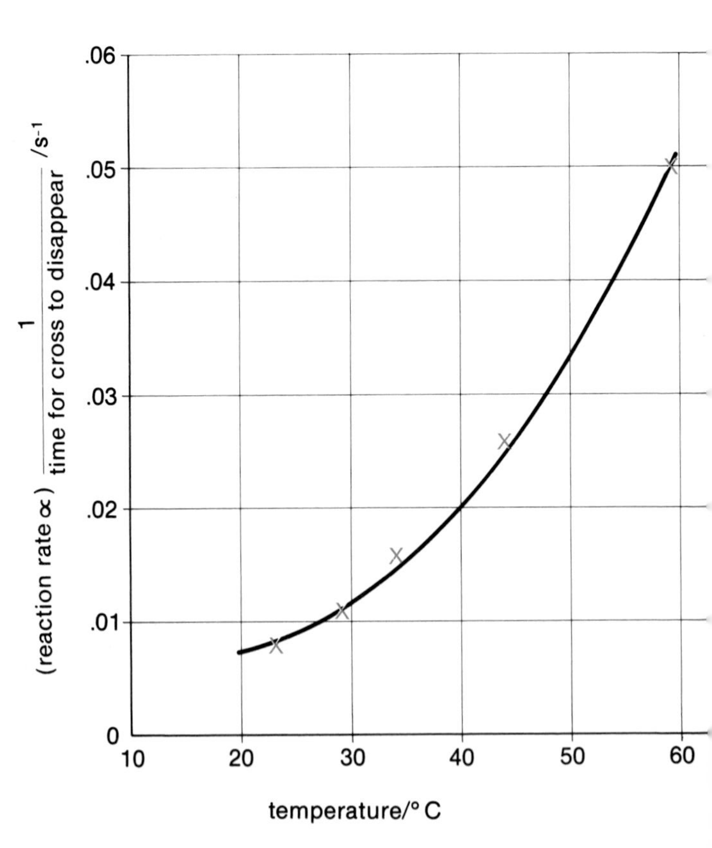

Figure 3
A graph of the reciprocal of the time for the cross to disappear against temperature

Temperature /°C	Time for cross to disappear /s	$\frac{1}{\text{time for cross to disappear}}$ /s^{-1}
23	132	0.0076
29	90	0.0111
34	64	0.0156
44	38	0.0263
59	20	0.0500

The results of an experiment to study the rate of reaction between sodium thiosulphate solution and dilute hydrochloric acid at different temperatures

Notice that:

1 the reaction goes faster at higher temperatures;

2 the reaction rate is about twice as fast if the temperature rises by about 10°C. For example, the cross disappears in 132 seconds at 23°C, but in about half the time (64 seconds) at 34°C.

In this experiment:

$$\text{reaction rate} = \frac{\text{amount of sulphur precipitated}}{\text{time for cross to disappear}}$$

Since the cross disappears at the same thickness of precipitate each time, the amount of sulphur precipitated is the same at each temperature. So,

$$\text{reaction rate} \propto \frac{1}{\text{time for cross to disappear}}$$

Figure 3 shows a graph of this reciprocal against temperature. The graph shows clearly that the reaction rate increases as the temperature increases.

Why does reaction rate increase with temperature?

When particles collide with each other, they do not always react. Sometimes the particles do not have enough energy for bonds to stretch and break during collision so the products cannot form. In some reactions, only the molecules with high energies can react. The same sort of thing happens in a car crash. If the cars collide in slow moving traffic (with low kinetic energy), they hardly dent each other. But if the cars collide at high speed (with high kinetic energy), they get smashed to pieces.

There are two reasons why reactions go faster as the temperature rises.
- *The particles collide more often.*
- *The particles collide with more energy so more of the collisions result in a reaction.*

Fish, meat and soft fruit can be kept for long periods in a deep freeze store where the temperature is about −18°C. (The temperature in a refrigerator is about 5°C.)

Questions

1 What is meant by 'reaction rate'?
2 Why does the rate of a reaction increase as the temperature rises?
3 Give two everyday examples in which temperature affects the rate of a chemical reaction.
4 Design an experiment to study the rate of reaction between magnesium ribbon and dilute sulphuric acid at different concentrations of acid. Draw a diagram of the apparatus you would use, say what you would do and the measurements you would take.
5 Why do plants grow faster in warm, wet weather than in cold, dry weather?

5 Catalysts and Reaction Rates

Hydrogen peroxide solution ($H_2O_2(aq)$) decomposes very slowly at room temperature, but it decomposes very rapidly into water and oxygen when manganese(IV) oxide is added.

$$2H_2O_2(aq) \rightarrow 2H_2O(l) + O_2(g)$$

The manganese(IV) oxide helps the hydrogen peroxide to decompose, but it is not used up during the reaction. The manganese(IV) oxide left at the end weighs exactly the same as that at the start of the reaction. The manganese(IV) oxide has acted as a **catalyst**.

This reaction shows the two important properties of catalysts:
- *they change the rates of chemical reactions;*
- *they are not used up during the reaction.*

Liver and plant tissues contain an enzyme, catalase, which decomposes hydrogen peroxide in living things. As the photograph shows, liver can decompose hydrogen peroxide very rapidly.

Most catalysts are used to speed up reactions, but a few catalysts can be used to slow them down. These substances are called negative catalysts or **inhibitors**. For example, glycerine is sometimes added to hydrogen peroxide as an inhibitor to slow down its rate of decomposition during storage. Hydrogen peroxide is used in industry to bleach textiles, paper and pulp. Catalysts play an important part in the chemical industry. Sulphuric acid (section G, unit 1), petrol (section J, unit 4) and ammonia (section K, unit 7) are all produced by processes involving catalysts.

Many reactions are catalysed by transition metals and their compounds. Margarine is manufactured from vegetable oils, such as palm oil, using a nickel catalyst. The nickel catalyses an addition reaction between the oil (which contains C = C bonds) and hydrogen. The product is a fatty solid which is margarine. By controlling the reaction, the margarine can be made as soft or as hard as required.

Catalysts allow substances to react more easily. They do this by helping bonds to break more easily. So, the particles need less energy to react and the reaction is faster. Catalysts are like motorways. They provide a faster, easier path (route) for the reaction (journey) which needs less energy (petrol) than the uncatalysed reaction (older country roads).

One of the most important discoveries of this century is that all chemical reactions in living things need catalysts. The catalysts in biological processes are called **enzymes**. Enzymes in our bodies catalyse the breakdown of our food and also reactions which synthesize important chemicals like fats, carbohydrates, proteins in our muscles and DNA (deoxyribose nucleic acid) in our genes (parts of chromosomes).

More and more industrial processes are being developed which use enzymes. These processes include the manufacture of fruit juices, beers, yoghurt, vitamins and pharmaceuticals. In some of these processes, enzymes are extracted from living material such as plant extracts, animal tissues, yeast and fungi. Enzyme extracts of this kind are already used in cheese-making, food processing and 'biological' washing powders.

Biological washing powders contain enzymes which break down the chemicals in food, dirt and other stains in clothing

Which way do you prefer to travel? By motorway or along older, winding roads. Motorways are like catalysts. They provide a faster route from the start to the finish. They also require less energy

Questions

1 Explain the words:
catalyst; inhibitor; enzyme.

2 Describe an experiment that you could carry out to show that manganese(IV) oxide is not used up when it catalyses the decomposition of hydrogen peroxide.

3 Design an experiment to collect samples of oxygen by decomposing hydrogen peroxide.

4 How do catalysts speed up reactions?

5 (a) Why do gardeners add fertilizer to the soil?
(b) Are the fertilizers used up by the plants?
(c) Are fertilizers catalysts?

6 In margarine manufacture, nickel catalyses an addition reaction involving hydrogen and vegetable oils.
(a) What is an addition reaction?
(b) What kind of bond must vegetable oils have for this to happen?
(c) How can the vegetable oil be converted into margarine that melts at a higher temperature?

6 Reversible Reactions

Baking a cake is an irreversible reaction

Baking a cake, boiling an egg and burning natural gas are all one-way reactions. When a cake is baked or an egg is boiled, chemical reactions take place in the cake and the egg.

It is impossible to take the cake and turn it back into flour, sugar, water and fat. This also applies to the boiled egg and the carbon dioxide and water produced when natural gas burns.

$$\underset{\text{methane in natural gas}}{CH_4(g)} + 2O_2(g) \rightarrow CO_2(g) + 2H_2O(g)$$

No matter what you do, carbon dioxide and water cannot be turned back into methane and oxygen. Reactions like this which cannot be reversed are called **irreversible reactions**. Most of the chemical reactions that we have studied so far are also irreversible, but there are some processes which can be reversed. For example, ice turns into water on heating,

$$H_2O(s) \xrightarrow{heat} H_2O(l),$$

but the ice reforms if water is cooled.

$$H_2O(l) \xrightarrow{cool} H_2O(s)$$

These two parts of this reversible process can be combined in one equation as:

$$H_2O(s) \underset{cool}{\overset{heat}{\rightleftarrows}} H_2O(l)$$

When blue (hydrated) copper sulphate is heated, it decomposes to white anhydrous copper sulphate and water vapour.

$$\underset{\text{blue}}{CuSO_4.5H_2O(s)} \rightarrow \underset{\text{white}}{CuSO_4(s)} + 5H_2O(g)$$

If water is now added, the change can be reversed and blue hydrated copper sulphate reforms.

$$\underset{\text{white}}{CuSO_4(s)} + 5H_2O(l) \rightarrow \underset{\text{blue}}{CuSO_4.5H_2O(s)}$$

These two processes can be combined in one equation as

$$CuSO_4.5H_2O \underset{\text{mix reactants}}{\overset{\text{heat}}{\rightleftarrows}} CuSO_4 + 5H_2O$$

Ice melts as it warms up in the drink, but if the drink is cooled the ice will reform

Reactions like this, which can be reversed by changing the conditions or adding and removing reagents, are called **reversible reactions**. During a reversible reaction, the reactants are sometimes completely changed to the products. But, in some cases, the reactants are not *completely* converted to the products. For example, if ice and water are kept at 0°C, neither the ice nor the water seems to change. We say the two substances are in **equilibrium**. When two substances are in equilibrium like this, we replace the reversible arrows sign ($\rightleftarrows$) in the equation with the equilibrium arrows sign ($\rightleftharpoons$).

So, at 0°C, $H_2O(s) \rightleftharpoons H_2O(l)$

Another reversible reaction that can come to equilibrium under certain conditions involves ammonia, hydrogen chloride and ammonium chloride. Ammonia and hydrogen chloride will react at room temperature to form a white smoke which is a suspension of solid ammonium chloride (figure 1).

$$NH_3(g) + HCl(g) \rightarrow NH_4Cl(s)$$

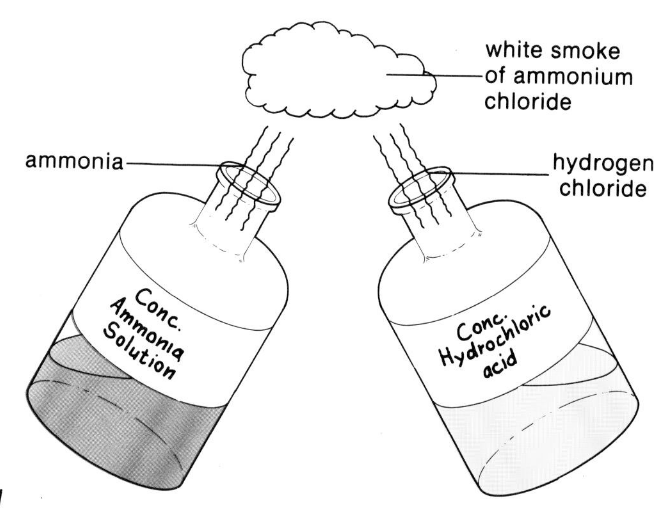

Figure 1

If ammonium chloride is heated, this reaction is reversed. The ammonium chloride decomposes to ammonia and hydrogen chloride.

$$NH_4Cl(s) \rightarrow NH_3(g) + HCl(g)$$

But if ammonium chloride is heated in a sealed container, only part of the solid will decompose. The reactants and products will be in equilibrium (figure 2).

$$NH_4Cl(s) \rightleftharpoons NH_3(g) + HCl(g)$$

At equilibrium this reaction may be well to the left, with most of the ammonium chloride unchanged and very little ammonia and hydrogen chloride present. Or it may be well to the right or at some point in between the two extremes. The position will depend on the amount of ammonium chloride used and the conditions of temperature and pressure in the container.

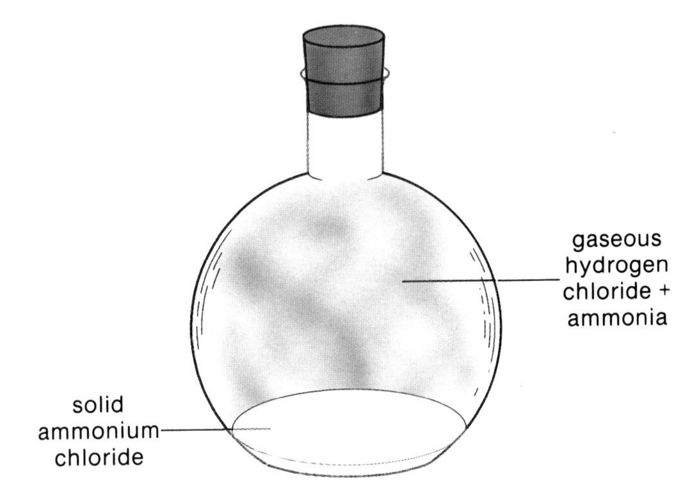

Figure 2

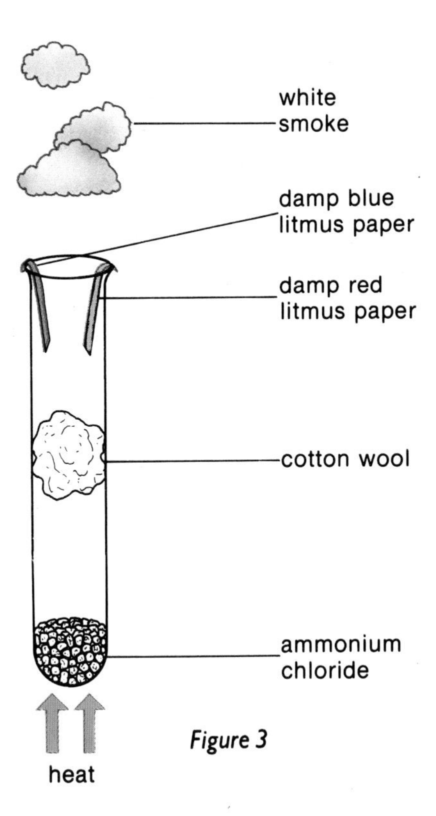

Figure 3

Questions

1 Explain the following:
irreversible reaction; reversible reaction; equilibrium.

2 Figure 3 shows what happens when some ammonium chloride is heated to form ammonia (which is alkaline) and hydrogen chloride (which is acidic). The red litmus turns blue first, then both pieces of litmus paper turn red.

(a) Write an equation for the decomposition of ammonium chloride.

(b) Why do the gases produced separate as they pass up the tube through the cotton wool?

(c) Which gas is detected first and why?

(d) Why does a white smoke form above the tube?

3 When purple hydrated cobalt chloride ($CoCl_2.6H_2O$) is heated, it changes to blue anhydrous cobalt chloride.

(a) Write an equation for this reaction.

(b) How is the reaction reversed?

(c) How is this reaction used as a test for water?

7 Reaction Rates and Industrial Processe

Fritz Haber (1868–1934)

Industrial chemists want to produce materials as cheaply as possible. In order to do this, they chose conditions which

1 increase the reaction rate and

2 use the most economic methods.

One way to speed up a reaction is to increase the concentration of reactants. At the same time, the products must be removed as fast as they form to prevent the reverse reaction happening.

Catalysts are also important in industrial processes. By using a suitable catalyst it is possible to carry out some processes that would otherwise be impossible. Other processes can be carried out at lower temperatures when a catalyst is used and this makes them more economical. Temperature and pressure are also chosen carefully in the manufacture of most chemicals.

The importance of these factors in industrial processes is well illustrated by the Haber process.

The Haber process

During the last century, the populations of Europe and America rose very rapidly. More food and more crops were needed to feed more and more people. So farmers began to use nitrogen compounds as fertilizers (unit 9 of this section). The main source of nitrogen compounds for fertilizers was sodium nitrate from Chile, but by 1900 supplies of this were running out.

Another supply of nitrogen had to be found or many people would starve. The obvious source of nitrogen was the air. But how could this unreactive gas be converted into ammonium salts and nitrates for use as fertilizers? The German chemist, Fritz Haber, solved the problem. In 1904, Haber began studying the reaction between nitrogen and hydrogen. By 1908 he had found the conditions needed to make ammonia (NH_3). Eventually, the Haber process became the most important method of manufacturing ammonia.

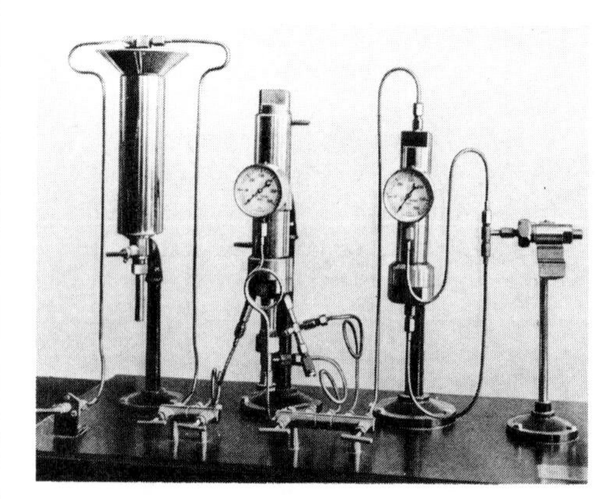

This picture shows the original apparatus used by Fritz Haber to make ammonia

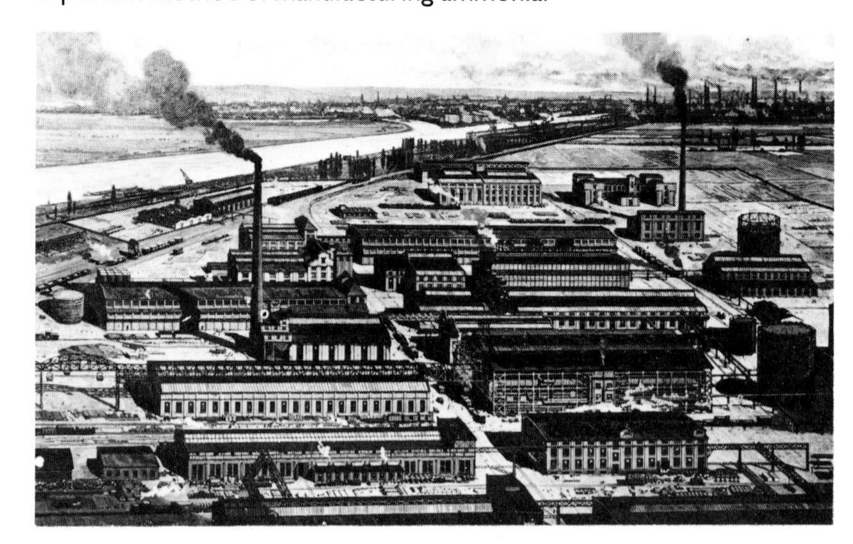

The vast ammonia manufacturing plant at Ludwigshafen, Germany. The plant was opened in 1913, only five years after Haber had found a cheap way of making ammonia

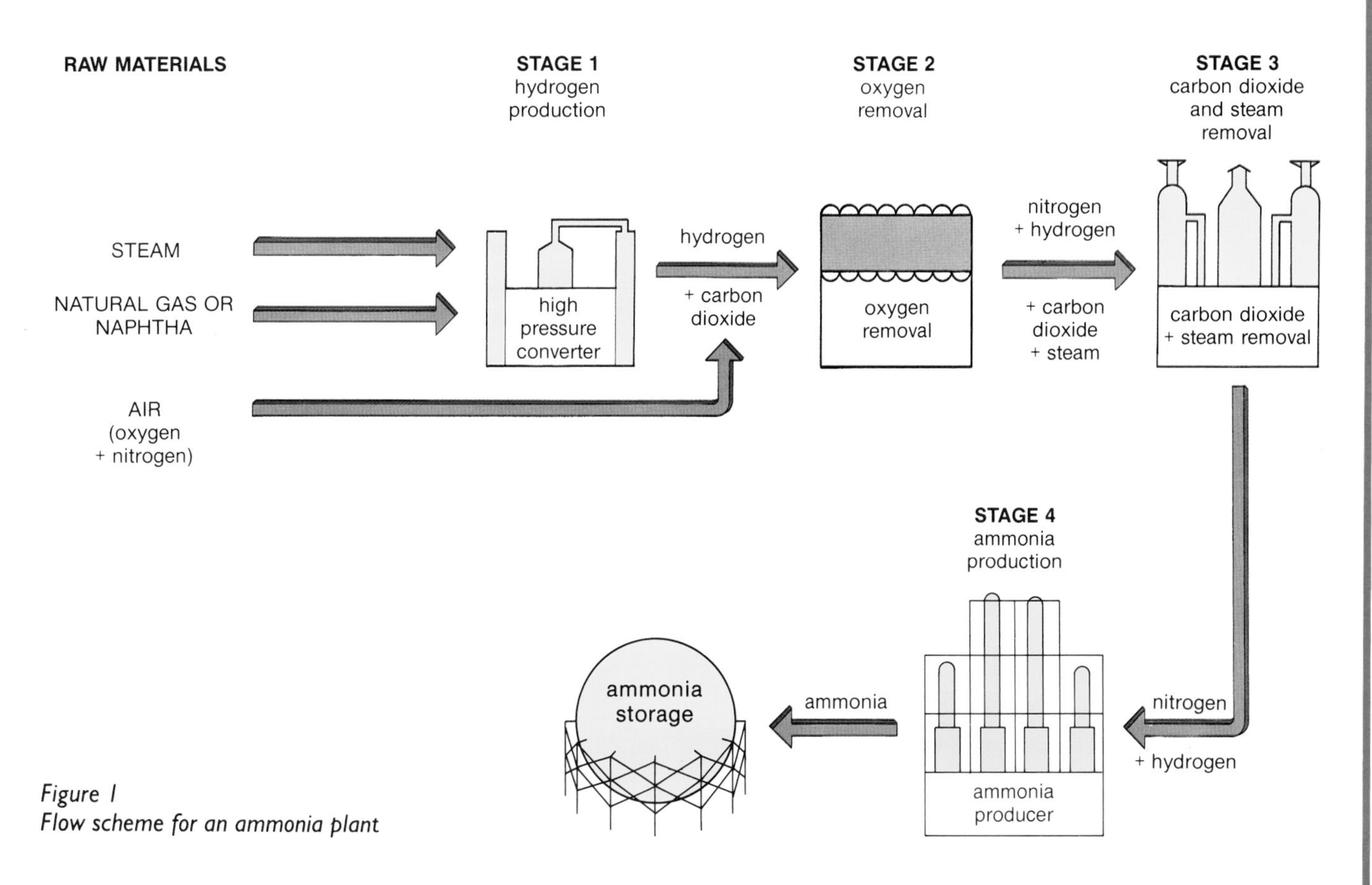

Figure 1
Flow scheme for an ammonia plant

A flow scheme for the Haber process is shown in figure 1.

In Stage 1, steam reacts with natural gas (methane) or with naphtha from crude oil. This produces a mixture of hydrogen and carbon dioxide.

$$CH_4(g) + 2H_2O(g) \xrightarrow[\text{pressure + catalyst}]{\text{high temperature and}} CO_2(g) + 4H_2(g)$$

In Stage 2, air is added to the mixture. Oxygen in the air reacts with some of the hydrogen, forming steam.

$$2H_2(g) + O_2(g) \rightarrow 2H_2O(g)$$

The product gases now contain nitrogen, hydrogen, carbon dioxide and steam.

In Stage 3, the carbon dioxide and steam are removed by passing the gases through concentrated potassium carbonate solution. This leaves a mixture of nitrogen and hydrogen.

Stage 4 is the key reaction in the Haber process. Nitrogen and hydrogen pass over
a catalyst of iron
at a pressure of 150–250 atmospheres and
a temperature of 400°C
This converts 15–35% of the reactants to ammonia.

$$N_2(g) + 3H_2(g) \xrightarrow[\text{400°C, + iron catalyst}]{\text{150–250 atmospheres,}} 2NH_3(g)$$

The hot gases from the converter are cooled to liquefy the ammonia. The unreacted nitrogen and hydrogen are recycled.

Questions

1 What are the main factors that affect the rate of a chemical reaction?

2 (a) Name three industrial processes that use a catalyst. Say what the catalyst is in each case.
(b) Why are catalysts important in industry?

3 At the beginning of this century, Haber synthesized ammonia from nitrogen and hydrogen. Why was this so important?

4 What conditions in the Haber process increase the rate of reaction between nitrogen and hydrogen?

8 Ammonia

Use	Approx. %
Fertilizers	75
Nitric acid	10
Nylon	5
Wood pulp and organic chemicals	10

The main uses of ammonia

Ammonia is an important chemical in industry and agriculture. Most of it is used for fertilizers and nitric acid (see the table). To understand these uses we must look at the properties of ammonia itself, which are listed in figure 1.

Ammonia is very soluble in water because it reacts with water to form a solution containing ammonium ions (NH_4^+) and hydroxide ions (OH^-).

$$NH_3(g) + H_2O(l) \rightleftharpoons NH_4^+(aq) + OH^-(aq)$$

The ammonia solution is a weak electrolyte. It is only partly dissociated into NH_4^+ and OH^- ions. The OH^- ions make the solution alkaline.

Ammonia as a base

Ammonia acts as a base in many reactions. Ammonia molecules pick up H^+ ions to form ammonium ions. So, ammonia reacts with acids to form ammonium salts. This is how fertilizers, such as ammonium nitrate ('Nitram') and ammonium sulphate, are made from ammonia.

$$\underset{\text{ammonia}}{NH_3(g)} + \underset{\text{nitric acid}}{HNO_3(aq)} \rightarrow \underset{\text{ammonium nitrate}}{NH_4NO_3(aq)}$$

Ammonia also reacts with hydrogen chloride to form a white smoke. The white smoke is tiny particles of solid ammonium chloride suspended in the air. This reaction is sometimes used as a test for ammonia.

$$NH_3(g) + HCl(g) \rightarrow NH_4Cl(s)$$

When ammonia dissolves in water, it acts as a base by taking H^+ ions from the water to form NH_4^+ ions.

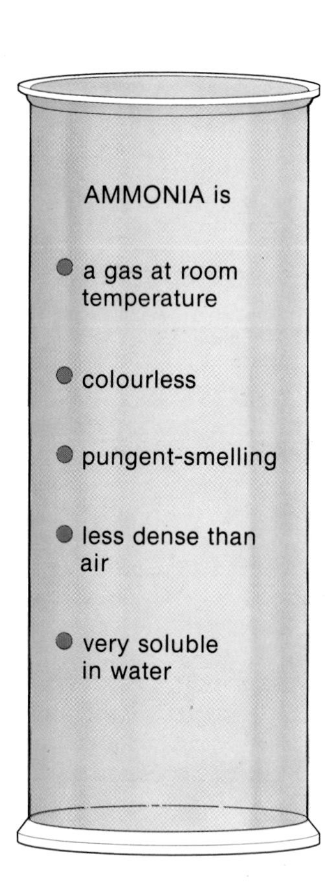

Figure 1
Properties of ammonia

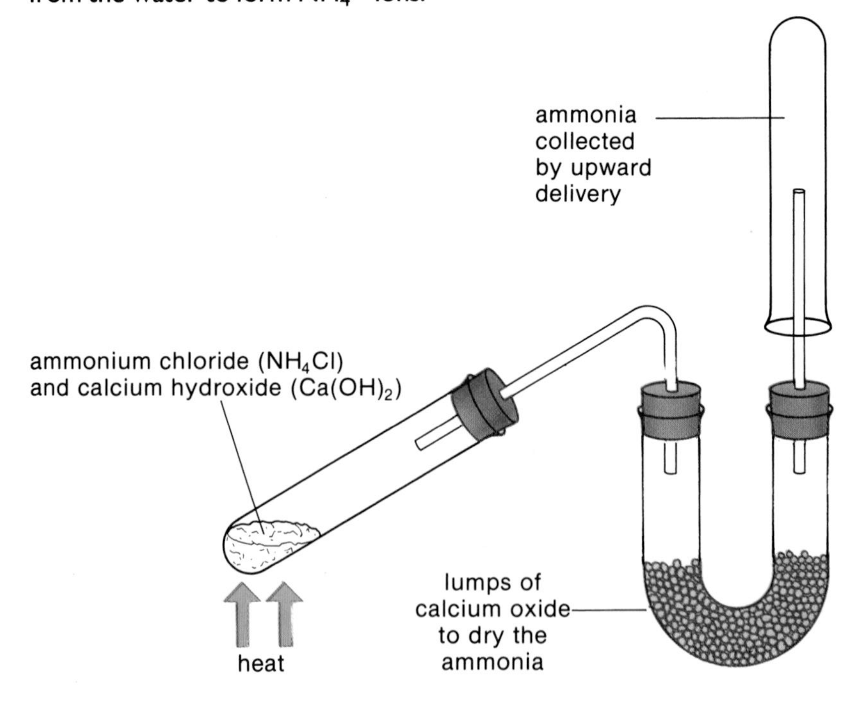

Figure 2
Making ammonia on a small scale

Making ammonia on a small scale

The easiest way to make a small amount of ammonia is to remove H^+ ions from NH_4^+ ions in an ammonium salt, like ammonium chloride (NH_4Cl). Alkalis containing OH^- ions will do this. Figure 2 shows how dry ammonia can be made by heating ammonium chloride and calcium hydroxide.

Ammonia to nitric acid—base to acid

About 10% of ammonia is used to manufacture nitric acid. Nitric acid is used to produce fertilizers such as potassium nitrate, and explosives like TNT (trinitrotoluene) and dynamite.

There are 3 stages in the manufacture of nitric acid.

1 Oxidizing the ammonia to nitrogen oxide (NO) using a platinum alloy catalyst at 900°C.

$$4NH_3 + 5O_2 \rightarrow 4NO + 6H_2O$$

2 Oxidizing the nitrogen oxide to nitrogen dioxide (NO_2) by mixing with air.

$$2NO + O_2 \rightarrow 2NO_2$$

3 Reacting the nitrogen dioxide and oxygen with water to form nitric acid (HNO_3).

$$4NO_2 + O_2 + 2H_2O \rightarrow 4HNO_3$$

The first two stages in this process can be carried out using the apparatus in figure 3. Ammonia evaporates from the solution and reacts with oxygen on the platinum spiral. Nitrogen oxide (NO) is produced and reacts with oxygen to form brown fumes of nitrogen dioxide in the flask.

Nitric acid is a typical mineral acid, like hydrochloric acid and sulphuric acid. The dilute acid shows typical acid reactions with indicators, metals, bases and carbonates. (See section G, unit 3.)

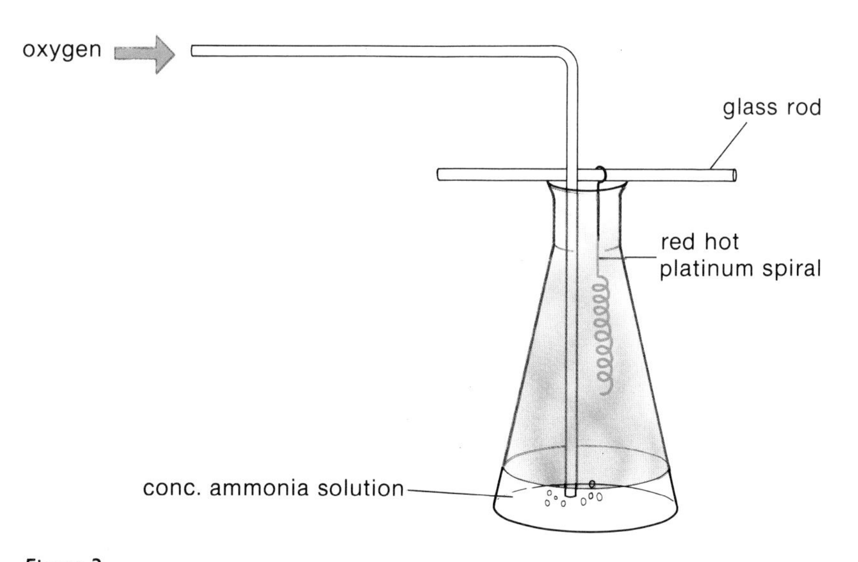

Figure 3

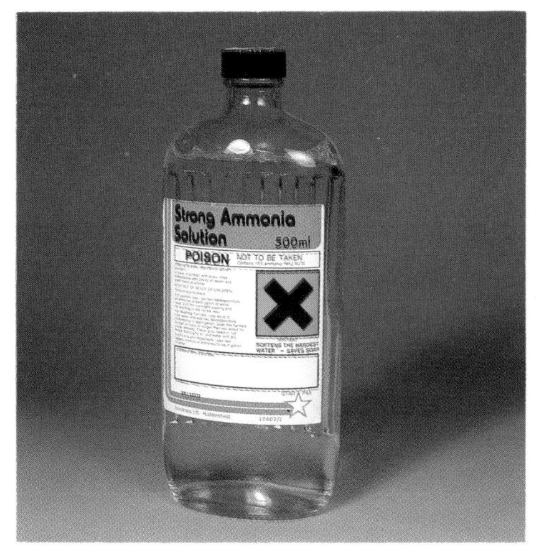

Ammonia solution can be used to clean toilets and sinks

Questions

1 Look at figure 2 and the small-scale preparation of ammonia.
(a) Write an equation for the reaction involved.
(b) Why is it necessary to dry the ammonia?
(c) Why is concentrated H_2SO_4 *not* used to dry ammonia?
(d) Why is ammonia *not* collected over water?
(e) Why can ammonia be collected by upward delivery?

2 *True* or *false?*
Nitric acid
A forms salts called nitrites.
B forms soluble salts with all metals.
C is an important fuel.
D can be oxidized to ammonia.

3 Write equations for the reactions of nitric acid with (i) zinc; (ii) copper(II) oxide; (iii) potassium hydroxide; (iv) sodium carbonate.

9 Fertilizers

A bag of NPK fertilizer. The percentages of N, P and K are indicated at the top of the bag—17%N, 17%P and 17%K

Fertilizer being bagged

Plants need essential elements (**nutrients**) from the soil to grow well. If crops are grown every year on the same piece of land, these nutrients get used up. The soil becomes infertile and plants are stunted with poor seeds and small fruit.

The most important elements for plant growth are nitrogen, phosphorus and potassium. Plants need more of these three elements than other elements, so shortages of them are soon noticed. Table 1 shows the role of these three elements in plant growth and the effects of shortages. So, these three elements must be replaced by adding fertilizers to the soil.

Fertilizers can be used as single compounds such as ammonium nitrate, or as mixtures of compounds containing nitrogen, phosphorus and potassium ('NPK' fertilizers). The proportions of nitrogen, phosphorus and potassium in NPK fertilizers are usually shown as % nitrogen (N), % phosphorus(V) oxide (P_2O_5) and % potassium oxide (K_2O).

Liquid ammonia being injected into the soil as a fertilizer. What are the advantages and disadvantages of using liquid ammonia?

Nutrient	Role in plant growth	Effect of shortage
Nitrogen	Essential for synthesis of proteins and chlorophyll	Plants are stunted, leaves become yellow (due to lack of chlorophyll)
Phosphorus	Essential for synthesis of nucleic acids (DNA)	Plants grow slowly. Small seeds and small fruit
Potassium	Plays a part in synthesis of carbohydrates and proteins	Leaves become yellow and curl inwards

Table 1: the role of three important elements in plant growth

- **Nitrogen fertilizers** are usually nitrates or ammonium salts. Ammonium nitrate ('Nitram'), NH_4NO_3, is the most widely used fertilizer because it is soluble, it can be stored and transported as a solid and it has a high percentage of nitrogen (table 2). The higher the percentage of nitrogen the better because less useless material needs to be stored and transported. Other nitrogen fertilizers are ammonium sulphate, urea and nitrochalk. Nitrochalk is ammonium nitrate crystals coated with calcium carbonate. This provides calcium for the soil as well as nitrogen and it also corrects soil acidity.

Fertilizer	Formula	Mass of one mole	Mass of nitrogen in one mole	% of nitrogen
Ammonium nitrate	NH_4NO_3	80 g	28 g	$\frac{28}{80} \times 100 = 35$
Ammonia	NH_3	17 g	14 g	$\frac{14}{17} \times 100 = 82$
Ammonium sulphate	$(NH_4)_2SO_4$	132 g	28 g	$\frac{28}{132} \times 100 = 21$
Urea	N_2H_4CO	60 g	28 g	$\frac{28}{60} \times 100 = 47$

Table 2: the percentage of nitrogen in different fertilizers

- **Phosphorus fertilizers** are manufactured mainly from phosphate rock containing calcium phosphate. This is insoluble in water, so it must be converted to soluble phosphorus compounds which plants can absorb through their roots. The phosphate rock is reacted with concentrated sulphuric acid. This converts it to 'super-phosphate'—a mixture of soluble calcium dihydrogenphosphate and insoluble calcium sulphate.

Although fertilizers are important in providing an adequate food supply, problems are caused by their over use.

1 They may change the soil pH,

2 they may harm plants and animals in the soil,

3 they allow those elements not required by plants to accumulate in the soil,

4 they get washed out of the soil and lead to the pollution of rivers.

Fertilizers are essential for producing high crop yields year after year

Questions

1 (a) Why should a fertilizer be soluble?
(b) What are the problems in storing, transporting and using fertilizers?
(c) Make a list of the important properties of an ideal fertilizer.

2 Liquid ammonia has been used as a fertilizer in some countries.
(a) What are its advantages?
(b) What are its disadvantages?

3 (a) Why are fertilizers important?
(b) What problems are caused by their over use?

4 (a) Describe how you would make a sample of ammonium sulphate.
(b) Design an experiment to see if ammonium sulphate acts as a fertilizer for peas or beans.

5 (a) Why is NPK fertilizer so called?
(b) Rain water washes fertilizers into streams and rivers. What effect does this have on plants and animals that live in the water?
(c) Nitrochalk can act as a fertilizer and cure soil acidity. Explain why it can do both of these jobs.

10 The Nitrogen Cycle

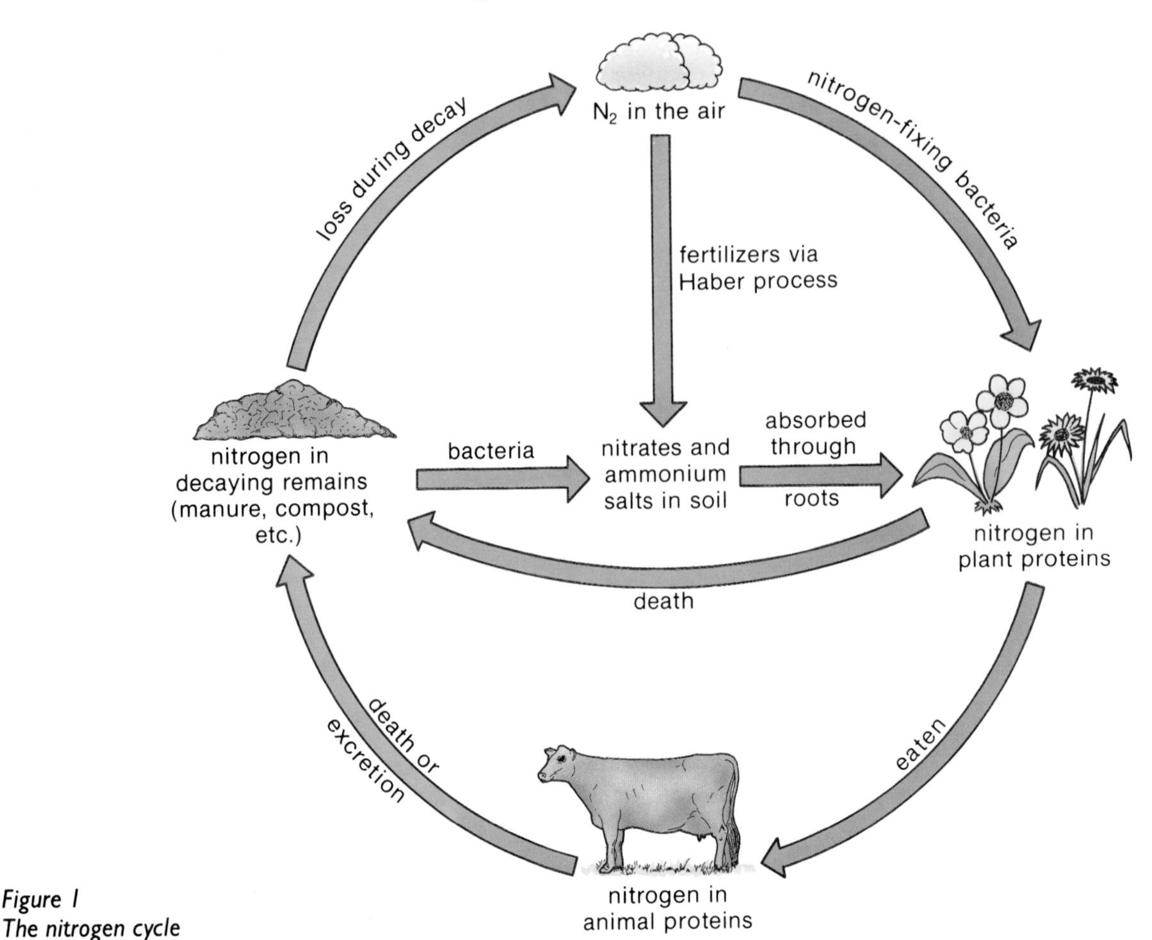

Figure 1
The nitrogen cycle

Nodules on the root of a plant

In the last unit, we looked at man-made inorganic fertilizers like ammonium nitrate and super-phosphate. Large quantities of natural organic fertilizers are also used by farmers and gardeners. The most widely used organic fertilizers are manure and compost which contain decaying matter from animals and plants (see the lower half of figure 1). The nitrogen compounds in manure and compost are decomposed by bacteria to nitrates and ammonium salts. These substances dissolve in rain water and are absorbed through the roots of plants (centre of figure 1). The nitrates and ammonium salts are then used by plants to synthesize chlorophyll and proteins that they need for growth. Animals have to eat plants and other animals in order to get the nitrogen and the proteins which they need.

Manure and compost are excellent fertilizers, but they take time to break down and there is not enough of them for all crops. Because of this, we need large amounts of inorganic fertilizers. Small quantities of the nitrogen that crops need are provided by nitrogen-fixing bacteria in the soil. These bacteria can convert nitrogen from the air into nitrogen compounds that can be used by plants. Some of these nitrogen-fixing bacteria live in nodules on the roots of plants such as peas, beans and clover (top right, figure 1). This way of converting atmospheric nitrogen to nitrogen compounds in plants is sometimes called *natural* nitrogen fixation. The conversion of atmospheric nitrogen to ammonia using the Haber process is described as *industrial* nitrogen fixation.

Although nitrogen is returned to the soil when living matter decays, some of the nitrogen compounds in the manure and compost are decomposed to nitrogen gas which escapes into the air (top left, figure 1). This is another reason why fertilizers are added to the soil in heavily cultivated areas.

The world food problem

There are about 4000 million people in the world. Nearly 3000 million of them are not properly fed and 400 million are starving. These figures suggest that we must increase food production, but the problem is not quite so simple.

- In Europe and North America, there are 'mountains' of surplus food but many poor countries cannot afford to buy it.
- In some poor countries, the rich have plenty to eat.
- Some countries need to export food to earn foreign currency even though some of their population are starving.

Three possible ways of reducing the food problem are as follows.

1 Increasing birth control. It is estimated that the world population will be 6000 million by the year 2000. Many people think that we are facing a population problem and *not* a food problem.

2 Improving farming methods. This includes watering the deserts, preventing soil erosion, developing better varieties of crops, breeding better cattle and using improved pesticides.

3 Finding new food supplies such as farming the sea and growing bacteria on vegetable oils and cellulose to produce food.

During a lightning flash, nitrogen reacts with oxygen in the air to form nitrogen oxides. These react with rain water to produce nitric acid, which increases the nitrogen content of the soil

Part of the European 'grain mountain'. Is it right that huge amounts of food are kept in store whilst thousands of people are starving?

Questions

1 What is meant by (i) natural nitrogen fixation; (ii) industrial nitrogen fixation; (iii) the nitrogen cycle?

2 (a) What is the difference between man-made and natural fertilizers?
(b) What happens to the nitrogen compounds in compost as it decays?
(c) Why do plants require nitrogen?
(d) Why do farmers and gardeners use man-made fertilizers as well as natural fertilizers?

3 What should the more developed nations do to solve the world food problem?

Section K: Study Questions

1 1.0 g of manganese(IV) oxide is added to 50 cm^3 of a solution of hydrogen peroxide. This breaks down to form oxygen and water. The oxygen is collected and its volume measured, at room temperature and pressure, at timed intervals.

Time/minutes	0	1	2	3	4	5	6	7	8
Volume/cm^3	1	20	33	44	52	58	59	60	60

The equation for the reaction is:

$$2H_2O_2 \rightarrow 2H_2O + O_2$$

(a) (i) Plot a graph of volume of oxygen (vertically) against time. Label this curve X.
(ii) Mark on curve X the time at which the rate of reaction is fastest. Label this point Y. Explain why you have chosen this point.
(iii) The experiment is repeated using 1.0 g of the same catalyst, 25.0 cm^3 of the hydrogen peroxide solution and 25.0 cm^3 of water. Sketch the curve obtained from the results of this experiment between the same axes. Label this curve Z. (7)
(b) The speed of a chemical reaction depends on two factors:
1. the rate at which reacting particles collide;
2. the energy possessed by these particles.
Use these factors to explain each of the following:
(i) heating the solution of hydrogen peroxide increases the speed of the reaction;
(ii) 1.0 g of a finely powdered catalyst gives a faster reaction than 1.0 g of small lumps of the same catalyst. (3)
MEG

2 This graph shows the total volume of hydrogen produced in the reaction of magnesium ribbon with excess dilute hydrochloric acid over a period of time.

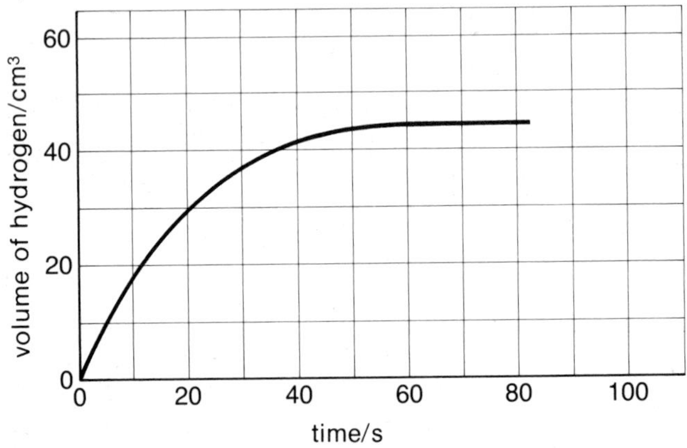

(a) What volume of hydrogen has been produced after 15 s? (1)
(b) How long does it take to produce 28 cm^3 of hydrogen? (1)
(c) Use the graph to work out the volume of hydrogen produced after 100 s. (1)
(d) Sketch *on the graph* the results that you would expect to obtain if the same mass of magnesium was treated with more concentrated acid. (2)
SEG

3 Farmers use large amounts of nitrogen-containing fertilisers. Some of these fertilisers are washed off the farm land by rain into streams and lakes. Also present in streams and lakes are phosphates from domestic detergents. The dissolved fertilisers and phosphates increase the amount of chemicals needed by plants in the water. The surface of the water then becomes covered with algae. Because of this, the plants below the surface die. When the algae decays, the amount of dissolved oxygen in the water is lowered.
(a) Give the chemical name of a fertiliser which contains nitrogen.
(b) Why are fertilisers used on farms?
(c) Name an element, other than nitrogen, which is needed for plant growth.
(d) How do phosphates get into the water system?
(e) What would be the effect of lowering the amount of dissolved oxygen in streams and lakes? (5) **MEG**

5 (a) Describe the industrial production of ammonia by the Haber process, including the formation of the starting materials from North Sea gas. (10)
(b) State what is meant by a fertiliser and give an account of the production of *one* fertiliser from ammonia. (4)
(c) Explain how the following could be proved to be present in a commercial 'NPK compound fertiliser' by simple chemical tests:
(i) potassium ions (ii) ammonium ions
(iii) nitrate ions. (9)
(d) State what is meant by
(i) a pesticide (ii) a herbicide.
What dangers are involved in the use of these compounds? (4)
Total [27] **WJEC**

6 Most of the people in the world do not have enough to eat, so we must think of ways of growing more food. Sometimes chemicals are spread on the ground to provide food for plants.
(a) What is the general name given to chemicals used by farmers to help plants to grow? (1)
(b) Ammonia is an important chemical used as a plant food.
(i) Which *two* elements are combined together to obtain ammonia? (2)
(ii) From which raw materials are these elements obtained to make ammonia gas? (2)
(iii) Unfortunately, these two elements combine together slowly so chemists use a catalyst.
What is a catalyst? (1)
Name the catalyst used in making ammonia. (1)
(iv) Describe *one* other way in which chemists can speed up a reaction. (1)
(c) Spreading chemicals on the land can have bad effects as well as good. Mention *two* bad things that might occur when chemicals are spread on the land. (2)
(d) (i) Give the chemical name of a polymer which can be used to make a bucket. (1)
(ii) State a substance that could have been used for making buckets, before chemists discovered the chemical that you have named in (d)(i). (1)
(iii) Describe *one* advantage of the polymer material. (1)
Total [13] **LEAG**

SECTION L

Atomic Structure and Radioactivity

Scientists working in radiation laboratories must handle radioactive isotopes by remote control from the other side of very thick glass windows

1 Inside Atoms

J. J. Thomson, Professor of Experimental Physics at Cambridge University and winner of the Nobel Prize for Physics in 1906

Less than ninety years ago scientists believed that atoms were solid particles like tiny snooker balls. Since then, experiments have shown that all atoms are made of three types of particles—protons, neutrons and electrons. In this unit we shall study some of the evidence for these particles.

1897: Thomson discovers electrons

In 1897, J. J. Thomson was investigating the way that gases conduct electricity. When he applied 15 000 volts across the electrodes of a tube containing a gas, the glass walls glowed a bright green colour. Rays travelling in straight lines from the cathode hit the glass and made it glow. Thomson called these rays **cathode rays** because they come from the cathode. Experiments with a narrow beam of cathode rays (figure 1) showed that they could be deflected by an electric field. When cathode rays passed between charged plates, they always bent towards the positive plate. This showed that the rays were negatively charged.

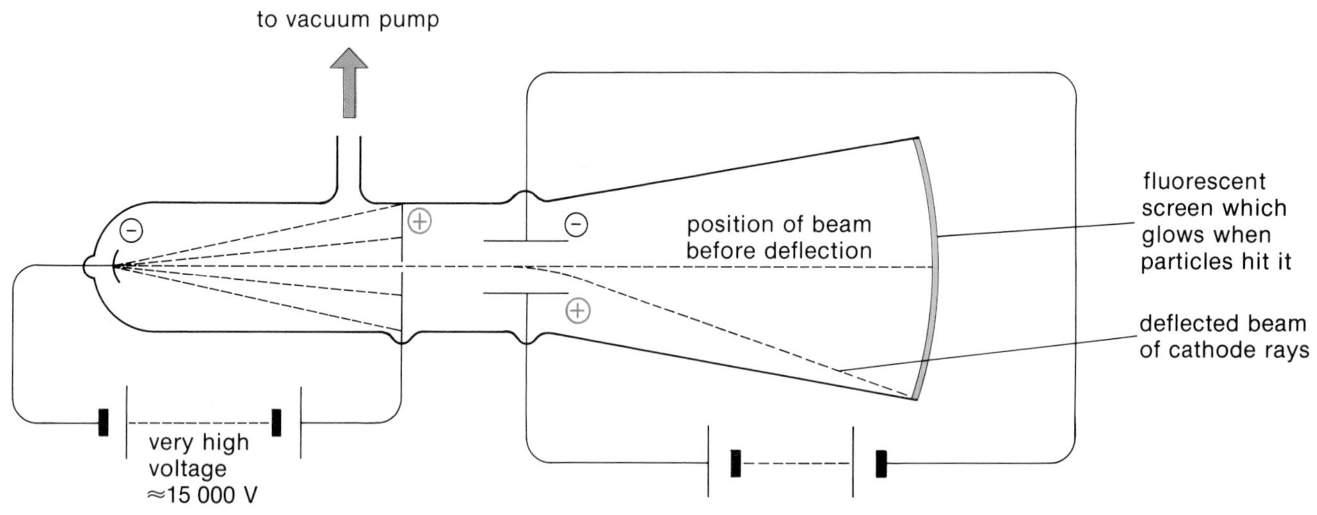

Figure 1
Deflection of cathode rays by an electric field

Further study of the deflection showed that cathode rays consisted of negative particles 1840 times lighter than hydrogen atoms. Thomson called these tiny negative particles **electrons**. The cathode rays were always the same, no matter what gas was present in the tube or what the electrodes were made of. This suggested that the electrons in all substances were identical.

1909: Geiger and Marsden explore the nucleus

Since atoms are neutral, they must contain positive charge to balance the negative charge on their electrons. Geiger and Marsden found a method of probing inside atoms using alpha particles from radioactive substances as 'bullets'. Alpha particles are helium ions, He^{2+}. When alpha particles from radium were fired at thin sheets of metal foil, most of the alpha particles passed straight through the foil. But, some of the alpha particles were deflected by the foil and a few of them even appeared to bounce back from it (figure 2).

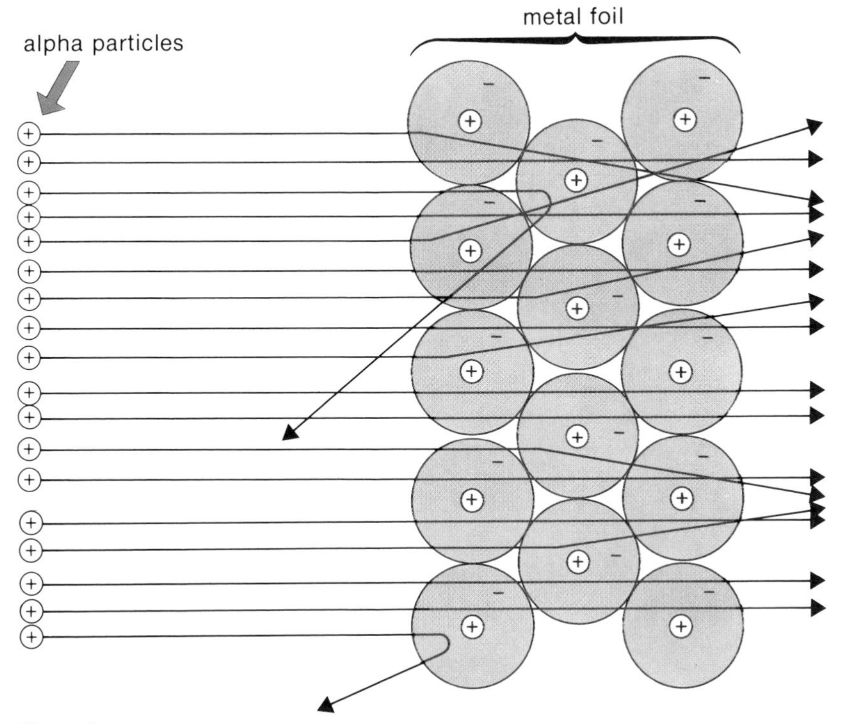

Figure 2
Most alpha-particles pass straight through the foil, some are deflected, but a few rebound from the foil.

Ernest Rutherford who succeeded J. J. Thomson as Professor of Experimental Physics at Cambridge in 1919. Rutherford received the Nobel Prize for Chemistry in 1908

1911: Rutherford explains the structure of atoms

Rutherford explained Geiger and Marsden's results by suggesting that atoms in the foil must have a very small positive **nucleus**. Surrounding this is a much larger region of empty space in which the electrons move. Most of the positive alpha particles pass straight through the large empty space where the electrons are moving. A few alpha particles pass close to the positive nucleus and get deflected. Occasionally an alpha particle approaches a nucleus head-on. When this happens, the positive alpha particle is repelled by the positive nucleus and bounces back. Rutherford suggested that the structure of an atom could be compared to a miniature solar system. Each atom has a positive nucleus, orbited by tiny negative electrons like planets orbiting the Sun. He suggested that the positive charge of the nucleus was provided by positive particles which he called **protons**. Hydrogen, the smallest atom, has one proton in the nucleus, balanced by one orbiting electron. Atoms of helium, the next smallest, contain two protons and two electrons; lithium atoms have three protons and three electrons, and so on.

Questions

1 What are (i) cathode rays; (ii) electrons; (iii) protons?

2 (a) What evidence is there that electrons are (i) negatively charged; (ii) the same in all substances?
(b) What sort of experiments might be carried out to show that electrons are about 2000 times lighter than hydrogen atoms?

3 Why did Geiger and Marsden's experiment suggest that atoms have a small positive nucleus surrounded by a much larger region of empty space?

4 How is the position of an element in the periodic table related to the number of electrons in its atoms?

5 When electrons pass between charged plates they are deflected towards the positive plate. What will happen when alpha particles pass between charged plates?

2 The Structure of Atoms

	Hydrogen atom	Helium atom
Number of protons	1	2
Number of neutrons	0	2
Relative mass	1	4
Relative atomic mass	1	4

Table 1: the relative atomic masses of hydrogen and helium

In spite of Rutherford's success in explaining atomic structure, one big problem remained. Hydrogen atoms contain one proton and helium atoms contain two protons. So the relative atomic mass of helium should be two, since the relative atomic mass of hydrogen is one. Unfortunately, the relative atomic mass of helium is four and *not* two.

James Chadwick, one of Rutherford's colleagues, showed where the extra mass in helium came from. Chadwick discovered that the nuclei of atoms contained *uncharged* particles as well as positively charged protons. Chadwick called these uncharged particles **neutrons**. Further experiments showed that neutrons have the same mass as protons, so Chadwick was able to explain the problem concerning the relative atomic masses of hydrogen and helium (table 1). Hydrogen atoms have one proton, no neutrons and one electron. Since the mass of the electron is almost zero compared to the proton and neutron, a hydrogen atom has a relative mass of one unit. Helium atoms have two protons, two neutrons and two electrons. The two protons and two neutrons give a helium atom a relative mass of four units. Thus, a helium atom is four times as heavy as a hydrogen atom and the relative atomic mass of helium is four.

James Chadwick who discovered neutrons in 1932

Protons, neutrons and electrons

We now know that all atoms are made up from three basic particles—protons, neutrons and electrons. The nuclei of atoms contain protons and neutrons. Both of these particles have a mass about the same as a hydrogen atom. Neutrons have no charge, but protons have a positive charge. Moving around the nucleus are electrons that are negatively charged. The electrons are arranged in layers or **shells** at different distances from the nucleus. The mass of the electron is so small that it can be ignored when working out the total mass of the atom. The positions, masses and charges of these three sub-atomic particles are shown in table 2.

Particle	Position	Mass (relative to a proton)	Charge (relative to that on a proton)
Proton	Nucleus	1	+1
Neutron	Nucleus	1	0
Electron	Shells	$\frac{1}{1840}$	−1

Table 2: properties of the three sub-atomic particles

Different atoms have different numbers of protons, neutrons and electrons. The hydrogen atom is the simplest of all atoms. It has one proton in the nucleus, no neutrons and one electron (figure 1). The next simplest atom is that of helium, with two protons, two neutrons and two electrons. The next, lithium, has three protons, four neutrons and three electrons. Some of the heavier atoms can have large numbers of protons, neutrons and electrons. For example, atoms of uranium have 92 protons, 92 electrons and 143 neutrons. Notice that hydrogen, the simplest atom and the first element in the periodic table, has one proton. Helium, the second element in the periodic table, has two protons. Lithium, the third element in the periodic table, has three protons and so on. Thus, the position of an element in the periodic table tells you how many protons it will have. Furthermore, an atom must always have an equal number of protons and electrons, so that the positive charges (on the protons) balance the negative charges (on the electrons).

If the nucleus of an atom was enlarged to the size of a pea and put on top of Nelson's Column, the electrons furthest away would be on the pavement.

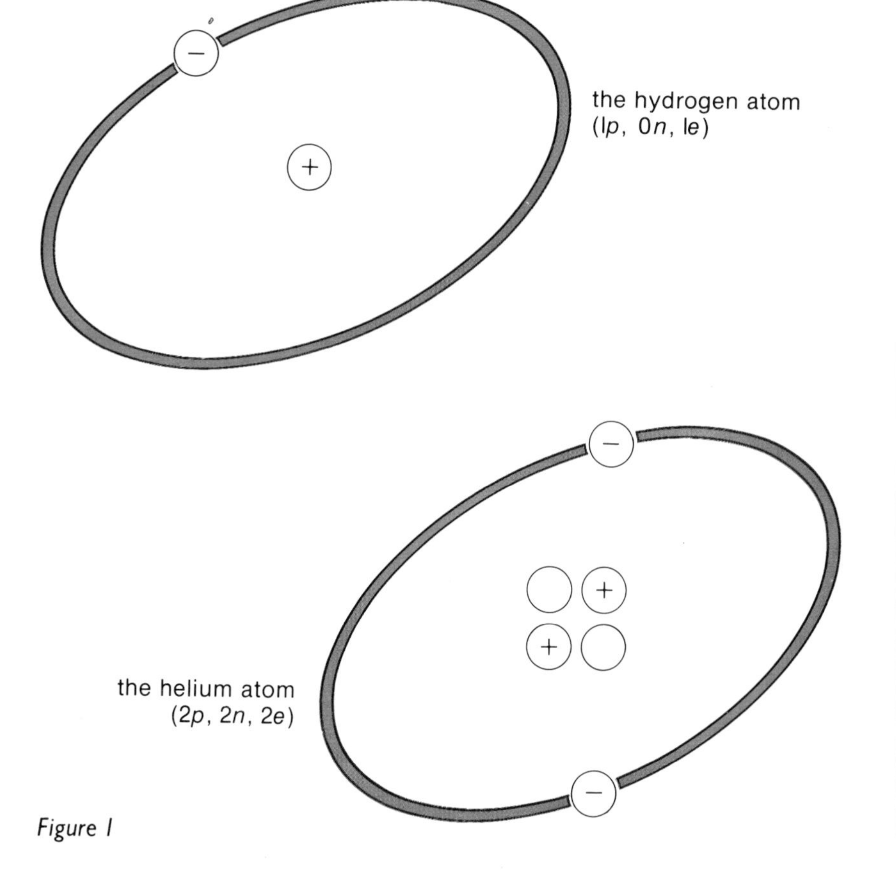

Figure 1

Questions

1 What are the charges, relative masses and positions in an atom of protons, neutrons and electrons?
2 How many protons, neutrons and electrons are there in one (i) H atom; (ii) H^+ ion; (iii) Li atom, (iv) Li^+ ion?
3 Oxygen is the eighth element in the periodic table. How many protons and electrons are there in one (i) O atom, (ii) O^{2-} ion, (iii) O_2 molecule, (iv) H_2O molecule?
4 Lithium atoms have three protons, four neutrons and three electrons. Why is the relative atomic mass of lithium about seven?

3 Atomic Number and Mass Number

Only hydrogen atoms have one proton. Only helium atoms have two protons. Only lithium atoms have three protons, and so on. This shows that the number of protons in an atom decides which element it is. Because of this, scientists have a special name for the number of protons in the nucleus of an atom. They call it the **atomic number** (symbol Z). Thus, hydrogen has an atomic number of one (Z = 1), helium has an atomic number of two (Z = 2), lithium has an atomic number of three (Z = 3) and so on. Aluminium, the thirteenth element in the periodic table with 13 protons and 13 electrons, has an atomic number of 13.

Protons alone do not account for all the mass of an atom. Neutrons in the nucleus also contribute to the mass. Therefore, the *mass* of an atom depends on the number of protons and the number of neutrons added together. This number is called the **mass number** of the atom (symbol A). So,

Atomic number = number of protons.
Mass number = number of protons + number of neutrons.

This photograph shows evidence for the two isotopes in neon, neon-20 and neon-22. Notice that the trace from neon-20 is much more prominent than that from neon-22. What does this tell us about the two isotopes?

Thus, hydrogen atoms (with one proton and no neutrons) have a mass number of one (A = 1). Helium atoms (two protons and two neutrons) have a mass number of four (A = 4) and lithium atoms (three protons and four neutrons) have a mass number of seven (A = 7). We can write the symbol $^{7}_{3}Li$ (figure 1) to show the mass number and the atomic number of a lithium atom. The mass number is written at the *top* and to the left of the symbol. The atomic number is written at the *bottom* and to the left. A lithium ion is written as $^{7}_{3}Li^{+}$. A sodium atom (11 protons and 12 neutrons) is written as $^{23}_{11}Na$. An electron (mass almost zero, charge −1) is shown as $^{0}_{-1}e$.

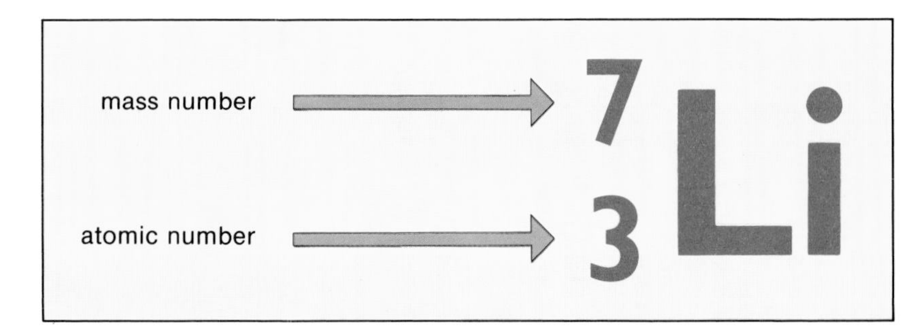

Figure 1

Using the periodic table (section E), we can predict the atomic number of an element, since the elements are arranged in order of atomic number. Therefore, the sixth element in the periodic table has an atomic number of six; the twentieth element an atomic number of 20 and so on.

Isotopes

Many elements have relative atomic masses which are nearly whole numbers. For example, the relative atomic mass of nitrogen is 14.007 and that of sodium is 22.99. This is not surprising, since the mass of an atom depends on the mass of protons and neutrons in its nucleus and the relative mass of both these particles is 1.00.

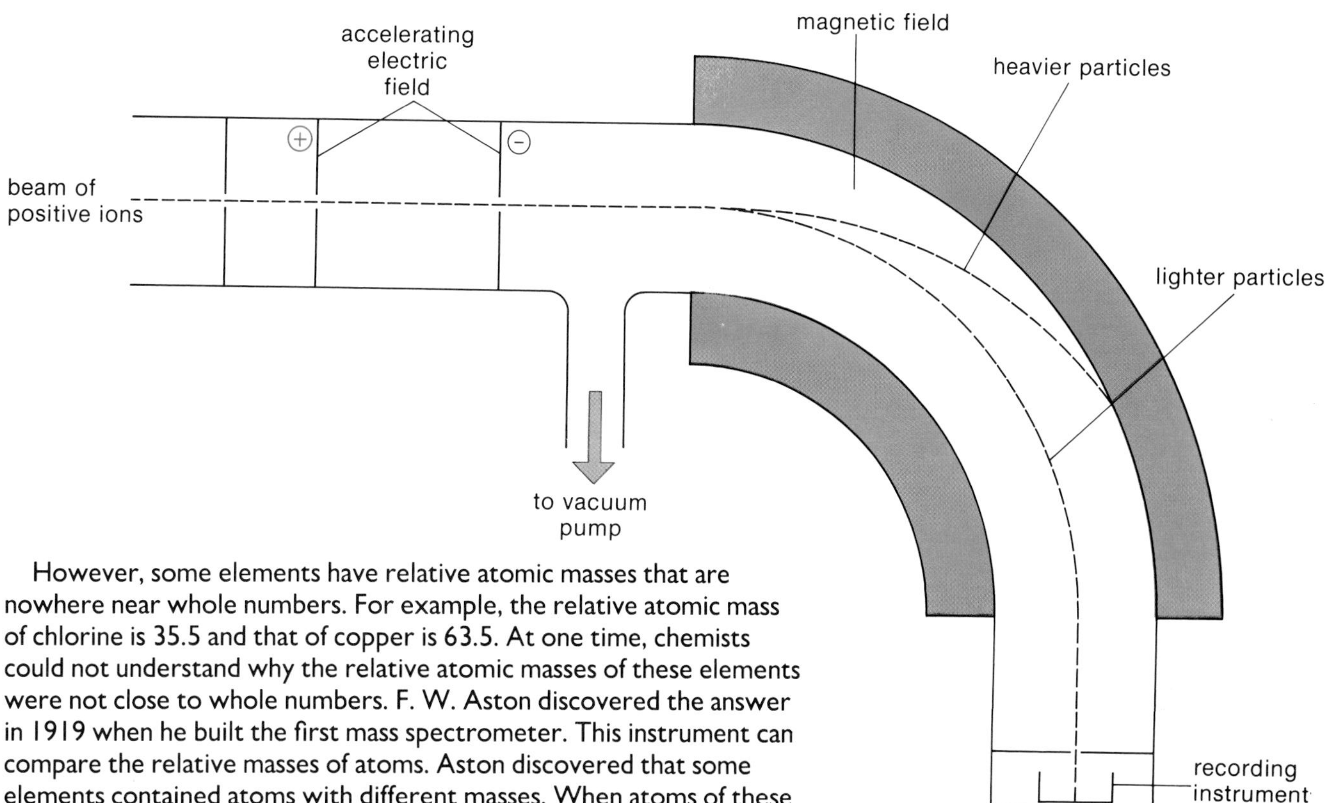

However, some elements have relative atomic masses that are nowhere near whole numbers. For example, the relative atomic mass of chlorine is 35.5 and that of copper is 63.5. At one time, chemists could not understand why the relative atomic masses of these elements were not close to whole numbers. F. W. Aston discovered the answer in 1919 when he built the first mass spectrometer. This instrument can compare the relative masses of atoms. Aston discovered that some elements contained atoms with different masses. When atoms of these elements were ionized and passed through a mass spectrometer, the beam of ions separated into two or more paths (figure 2). This suggested that one element could have atoms with different masses. These *atoms of the same element with different masses are called* **isotopes**. Each isotope has a relative mass close to a whole number, but the average atomic mass for the mixture of isotopes is not always close to a whole number. This is studied further in the next section.

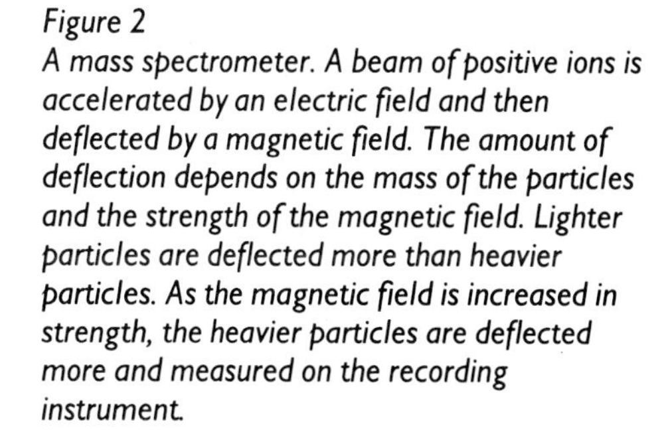

Figure 2
A mass spectrometer. A beam of positive ions is accelerated by an electric field and then deflected by a magnetic field. The amount of deflection depends on the mass of the particles and the strength of the magnetic field. Lighter particles are deflected more than heavier particles. As the magnetic field is increased in strength, the heavier particles are deflected more and measured on the recording instrument.

Aston's original mass spectrometer. Positive ions were accelerated along the cylindrical metal tube at the top of the instrument. The coils of the electromagnet produced a magnetic field which deflected the particles.

Questions

1 Explain the following: (i) *atomic number*; (ii) *mass number*; (iii) *isotope*.

2 (a) What is the atomic number of fluorine?
(b) How many protons, neutrons and electrons are there in one fluorine atom of mass number 19?

3 (a) What do 16, 8, 2− and O mean in the symbol, ${}^{16}_{8}O^{2-}$?
(b) How many protons, neutrons and electrons are there in one ${}^{23}_{11}Na^{+}$ ion?

4 Why do some elements have relative atomic masses which are not close to whole numbers?

4 Isotopes

Uranium ore being treated in giant tanks at the Rossing Mine in Namibia

Isotopes are atoms of the same element with different masses. All the isotopes of one element have the same number of protons. Therefore, they have the same atomic number. Since isotopes have the same number of protons, they must also have the same number of electrons. This gives them the same chemical properties because chemical properties depend upon the number of electrons in an atom and the way in which these electrons are transferred and shared during reactions.

Isotopes do, however, contain different numbers of neutrons. This means **they have the same atomic number but different mass numbers**. For example, neon has two isotopes. Each isotope has 10 protons and 10 electrons and therefore an atomic number of 10. But one of these isotopes has 10 neutrons and the other has 12 neutrons. Their mass numbers are therefore 20 and 22 (figure 1). They are sometimes called neon-20 and neon-22. These 2 isotopes of neon have the same chemical properties because they have the same number of electrons, but they have different physical properties because they have different masses. Samples of $^{20}_{10}Ne$ and $^{22}_{10}Ne$ will have different densities, different melting points and different boiling points. The similarities and differences between isotopes of the same element are summarised in table 1.

Figure 1
The two isotopes of neon

	neon-20 $^{20}_{10}Ne$	neon-22 $^{22}_{10}Ne$
number of protons	10	10
number of electrons	10	10
atomic number	10	10
number of neutrons	10	12
mass number	20	22

Isotopes have the same
Number of protons Number of electrons Atomic number Chemical properties
Isotopes have different
Numbers of neutrons Mass numbers Physical properties

Table 1: the similarities and differences between isotopes of the same element

Chemists can obtain samples of uranium with a higher percentage of uranium-235 because of the different physical properties of uranium-235 and uranium-238. Uranium-235 is needed for use in nuclear reactors. Natural uranium contains only about 0.7% of uranium-235 and 99.3% of uranium-238. The natural uranium is converted to uranium hexafluoride (UF_6) which is very volatile. The uranium hexafluoride is vaporised and allowed to diffuse through a porous solid. Particles of $^{235}UF_6$ are slightly lighter than those of $^{238}UF_6$. So they can move faster and diffuse faster than those of $^{238}UF_6$. After repeated diffusion through the porous solid, the uranium hexafluoride contains 3 or 4% $^{235}UF_6$. This can be converted back to uranium and used as nuclear fuel.

Relative atomic mass

Most elements contain a mixture of isotopes. This explains why their relative atomic masses are *not* whole numbers. The relative atomic mass of an element is the average mass of one atom, taking account of its isotopes and their relative proportions. For example, the mass spectrometer trace in figure 2 shows that chlorine consists of two isotopes with mass numbers of 35 and 37. These isotopes can be written as $^{35}_{17}Cl$ and $^{37}_{17}Cl$.

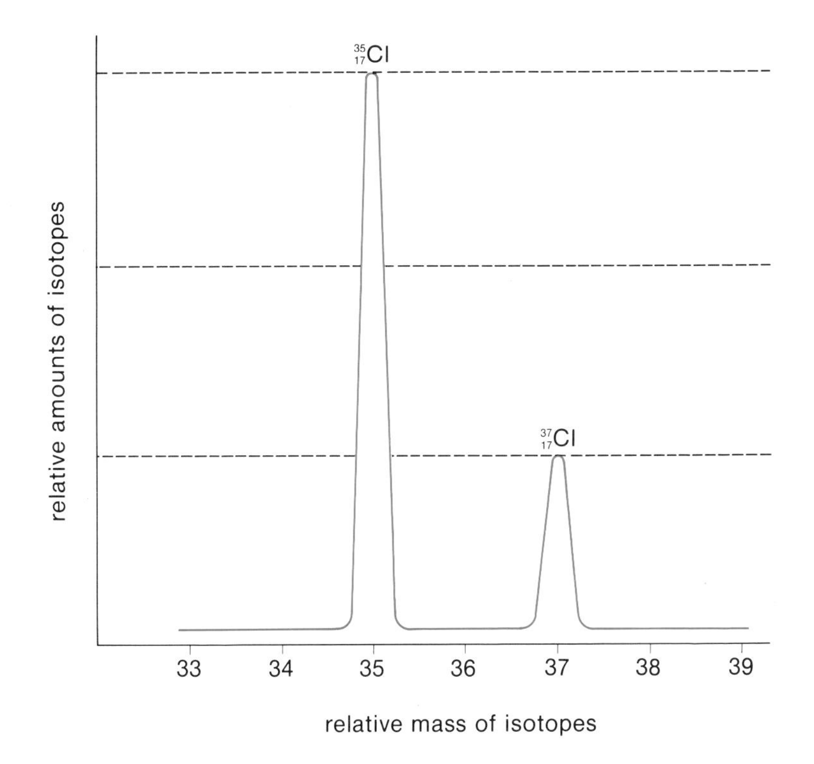

Figure 2
A mass spectrometer trace for chlorine. What are the relative amounts of $^{35}_{17}Cl$ and $^{37}_{17}Cl$?

If chlorine contained 100% $^{35}_{17}Cl$, then its relative atomic mass would be 35. If it contained 100% $^{37}_{17}Cl$, then its relative atomic mass would be 37. A 50:50 mixture of $^{35}_{17}Cl$ and $^{37}_{17}Cl$ would have a relative atomic mass of 36. Figure 2 shows that naturally-occurring chlorine contains three times as much $^{35}_{17}Cl$ as $^{37}_{17}Cl$, i.e. 75% to 25%. This gives a relative atomic mass of 35.5, as shown in table 2.

Percentage of $^{35}_{17}Cl$	100	75	50	25	0
Percentage of $^{37}_{17}Cl$	0	25	50	75	100
Relative atomic mass	35	35.5	36	36.5	37

Table 2: the relative atomic mass of different mixtures of the isotopes of chlorine

Questions

1 What are the important similarities and differences between the isotopes of an element?

2 There are three isotopes of hydrogen with mass numbers of one, two and three. (Naturally-occurring hydrogen is almost 100% $^{1}_{1}H$.) How many protons, neutrons and electrons do each of the three hydrogen isotopes have?

3 Neon has two isotopes, with mass numbers of 20 and 22.

(a) How would you expect the boiling point of $^{20}_{10}Ne$ to compare with that of $^{22}_{10}Ne$? Explain your answer.

(b) Suppose a sample of neon contains equal numbers of the two isotopes. What is the relative atomic mass of neon in this sample?

(c) Neon in the air contains 90% of $^{20}_{10}Ne$ and 10% of $^{22}_{10}Ne$. What is the relative atomic mass of neon in the air?

5 Electron Structures

Chemical reactions involve changes in the number of electrons that atoms have. This led chemists to suggest that the noble gases must have very stable electron structures because they are so unreactive. This means that atoms or ions will have stable electron structures if they have 2 electrons (like helium), 10 electrons (like neon), 18 electrons (like argon), etc.

Chemists believe that electrons occupy the outer parts of atoms in layers or **shells**. The first shell is filled and stable when it contains two electrons like helium. The second shell is filled when it contains eight electrons. So neon with 10 electrons has 2 electrons in the first shell and eight electrons in the second shell. We say that its electron structure is 2, 8. Argon is stable because its first, second and third shells are all filled with 2, 8 and 8 electrons respectively. We write the electron structure of argon as 2, 8, 8.

Figure 1 shows the first 20 elements arranged as in the periodic table. The electron structure of each element is written below its symbol. When the first shell is full at helium, electrons go into the second shell. So the electron structure of lithium is 2, 1; beryllium is 2, 2; boron is 2, 3; etc. When the second shell is full at neon, electrons start to fill the third shell and so on. Using these electron structures, it is possible to explain why elements in the same group have similar properties.

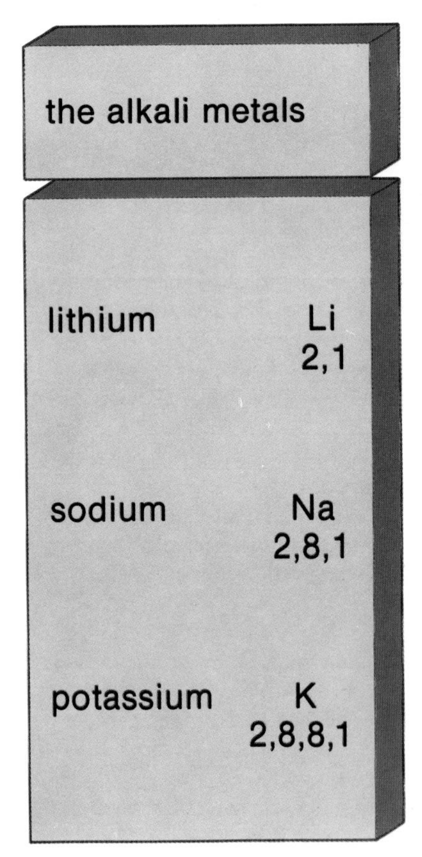

Figure 2

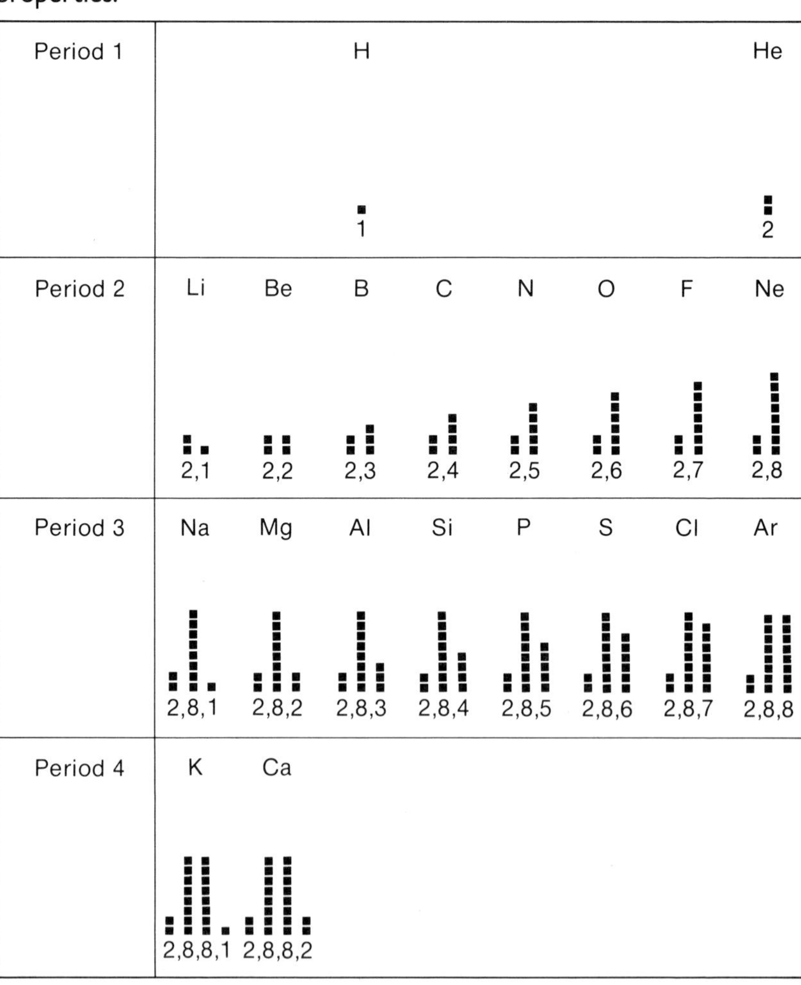

Figure 1

Group I: the alkali metals

Elements in the same group have similar electron structures. Each alkali metal follows a noble gas in the periodic table. So alkali metals have one electron in their outer shell (figure 2). By losing this one electron, their atoms form positive ions (Li^+, Na^+, K^+) with the stable electron structure of a noble gas. For example, Na^+ ions have an electron structure 2, 8 which is like neon. So, all alkali metals have a valency of one and they are very reactive because they are keen to lose the single electron in the outer shell.

Group VII: the halogens

Halogens come just before a noble gas in the periodic table. So, halogen atoms have seven electrons in their outer shells (figure 3). By gaining one electron, they form negative ions (F^-, Cl^-, Br^-) with stable electron structures like the next noble gas. Thus, all halogens have a valency of 1.

Look at the table. It shows the electron structures of atoms and ions for elements in period 3.

Element	**Na**	**Mg**	**Al**	**Si**	**P**	**S**	**Cl**	**Ar**
Electron structure	2,8,1	2,8,2	2,8,3	2,8,4	2,8,5	2,8,6	2,8,7	2,8,8
Electrons in outer shell	1	2	3	4	5	6	7	8
Common ion	Na^+	Mg^{2+}	Al^{3+}	—	—	S^{2-}	Cl^-	—
Electron structure of ion	2,8	2,8	2,8	—	—	2,8,8	2,8,8	—

Electron structures of the atoms and ions of elements in period 3

Notice that:

1 The first three elements in the period (sodium, magnesium and aluminium) *lose* the electrons in their outer shell to form positive ions (Na^+, Mg^{2+}, Al^{3+}) with an electron structure like the previous noble gas.

2 Sulphur and chlorine near the end of the period *gain* electrons to form negative ions (S^{2-} and Cl^-) with an electron structure like the next noble gas, argon.

3 Elements in the middle of the period, like silicon and phosphorus, do not usually form ions. They get stable electron structures in their compounds by *sharing* electrons with other atoms instead of gaining them or losing them. This sharing of electrons results in covalent bonds between atoms. It is the usual type of bonding in non-metal compounds. *So, during most reactions, atoms either lose, gain or share electrons in order to get a stable electron structure like a noble gas.* This idea forms the basis of the electronic theory of chemical bonding.

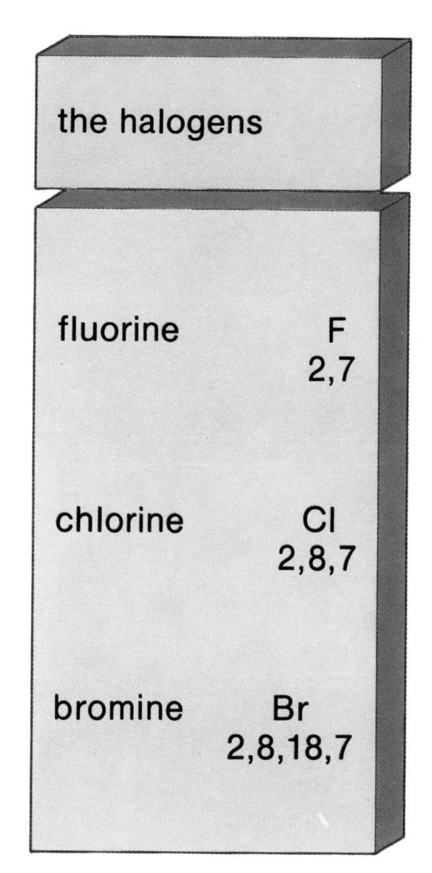

Figure 3

Questions

1 Why do all the alkali metals have similar chemical properties?

2 (a) Write down the electron structures of magnesium and calcium.
(b) How many electrons are there in the outer shell of an atom of an element in Group II?
(c) What charge will stable ions of Group II elements have?
(d) How many electrons are there in the outer shell of oxygen atoms?
(e) What is the charge on stable ions of oxygen?
(f) Why does magnesium react with oxygen to form an ionic compound, $Mg^{2+}O^{2-}$?

3 How many protons, neutrons and electrons will the following have? (i) N; (ii) N^{3-}; (iii) Ca; (iv) Ca^{2+}.

6 Ionic and Covalent Bonds

Ionic bonds: transfer of electrons

Figure 1 shows what happens when sodium chloride (Na^+Cl^-) is formed from sodium and chlorine atoms.

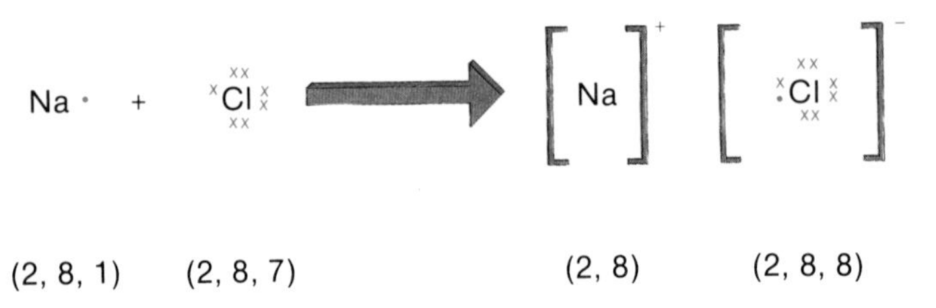

Na·	+	Cl	→	$[Na]^+$	$[Cl]^-$
(2, 8, 1)		(2, 8, 7)		(2, 8)	(2, 8, 8)

Figure 1

The number of electrons in the outer shell of each atom is shown by dots or crosses round its symbol and the full electron structures are shown below the symbols. Each sodium atom loses the one electron in its outer shell to form an Na^+ ion with the same electron structure as neon. The electrons given up by sodium atoms are taken by chlorine atoms. Each chlorine atom gains one electron to form a Cl^- ion with the same electron structure as argon. So, the formation of NaCl involves the *complete transfer* of an electron from a sodium atom to a chlorine atom, forming Na^+ and Cl^- ions.

Ionic (electrovalent) bonds result from the attraction between these oppositely charged ions.

Compounds containing ionic bonds are called ionic compounds. The structure, bonding and properties of ionic compounds are discussed in section D, unit 7 and section H, units 10 and 11.

Figure 2 shows the electron transfer that takes place in the formation of magnesium sulphide and lithium oxide. Transfer of electrons to form ionic bonds is typical of the reactions between metals and non-metals.

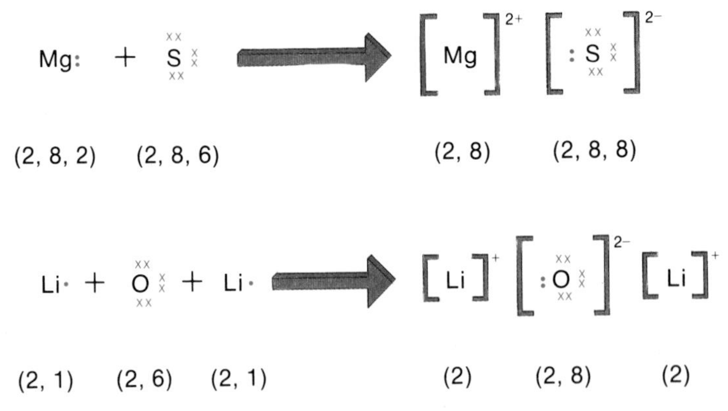

Mg:	+	S			→	$[Mg]^{2+}$	$[S]^{2-}$	
(2, 8, 2)		(2, 8, 6)				(2, 8)	(2, 8, 8)	
Li·	+	O	+	Li·	→	$[Li]^+$	$[O]^{2-}$	$[Li]^+$
(2, 1)		(2, 6)		(2, 1)		(2)	(2, 8)	(2)

Figure 2
Electron transfers in the formation of magnesium sulphide and lithium oxide

Covalent bonds: sharing electrons

A chlorine atom is very unstable. Its outer shell contains only seven electrons. At normal temperatures, chlorine atoms join up in pairs to form Cl_2 molecules. Why is this? If two chlorine atoms come close together, the electrons in their outer shells can overlap so that one pair of electrons is shared by each atom (figure 3). The shared pair of electrons is attracted by the positive nucleus of each atom forming a **covalent bond**. The shared pair contributes to the outer shell of both the chlorine atoms. Circles are used to enclose the electrons in the outer shell of each chlorine atom. So,

> *A covalent bond is formed by the sharing of a pair of electrons between two atoms. Each atom contributes one electron to the bond.*

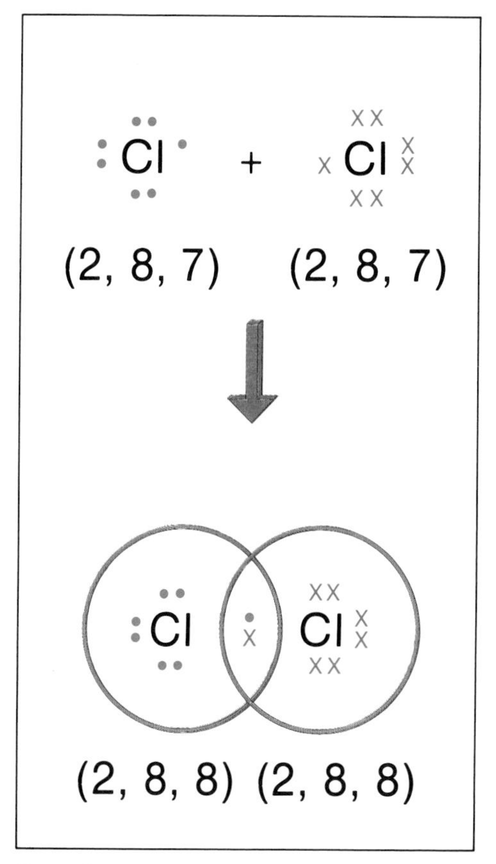

Figure 3

The atoms in molecular compounds are joined by covalent bonds. The structure, bonding and properties of these compounds are discussed in section D, unit 8 and section H, units 6 and 9. The electron structures of some common molecular substances are shown in figure 4. Notice the following points.

1 All the atoms have an electron structure like a noble gas.

2 The electron structures of these molecular compounds can be related to their structural formulas, which show bonds as lines between atoms (e.g. H—O—H for water). The number of lines to an atom equals its valency and each line represents a shared pair of electrons.

3 Double covalent bonds result from the sharing of two pairs of electrons as in oxygen and carbon dioxide. Triple covalent bonds with three pairs of electrons are also known.

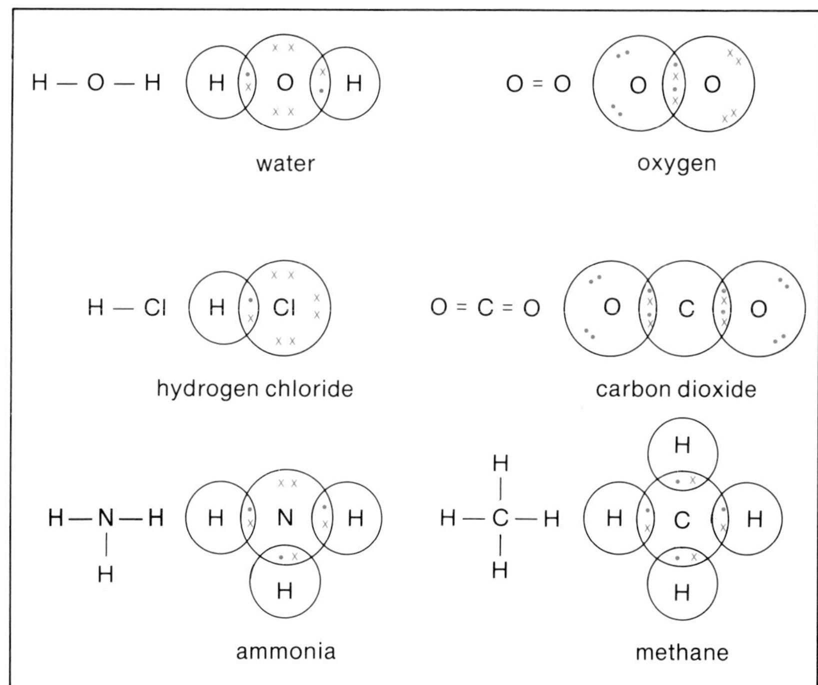

Figure 4
The electron structures of some simple molecular substances

Questions

1 Look at figure 2.
(a) How many electrons do magnesium and sulphur atoms lose or gain in forming magnesium sulphide, MgS?
(b) Why are these numbers of electrons lost or gained?
(c) How many electrons do lithium and oxygen atoms lose or gain in forming lithium oxide, Li_2O?
(d) Which noble gases have electron structures like the ions in lithium oxide?
(e) Why do two lithium atoms react with one oxygen atom in forming lithium oxide?

2 Element *X* has the electron structure 2, 6 and element *Y* has the electron structure 2, 8, 3.
(a) Write an equation similar to those in figure 2 for the formation of a compound between *X* and *Y*.
(b) What is the formula of this compound?

7 Radioactivity

Henri Becquerel—the discoverer of radioactivity

1896: Becquerel discovers radioactivity

In 1896, the Frenchman Henri Becquerel was investigating the reactions of uranium salts. Most of his experiments were carried out in bright sunlight. By chance, Becquerel left some uranium salt *in the dark* and on a photographic plate wrapped in black paper. When he developed the photographic plate, Becquerel was very surprised to see that it had been darkened (fogged) in the area near the uranium salt. Further experiments showed that the uranium salt was giving off some kind of radiation. Becquerel called the process **radioactivity** and described the uranium salts as **radioactive**. He also suggested that the radiation was a form of energy like light. This would explain why it caused a reaction on the photographic plate.

1898: Marie Curie discovers radium

Two of Becquerel's colleagues, Marie and Pierre Curie, decided to examine the radioactivity of uranium salts in more detail. They found that all uranium salts showed radioactivity and affected photographic plates. They also found that radioactive substances would cause the discharge of an electroscope. This meant that the radiation must carry a charge. During their investigations the Curies noticed that pitchblende (impure uranium sulphide) was much more radioactive than they had expected from its uranium content. This led them to think that the pitchblende contained an element more radioactive than uranium.

Marie Curie (1867–1934) and her husband Pierre (1859–1906). The Curies spent nearly four years isolating the radioactive elements radium and polonium from pitchblende. In 1903 they shared the Nobel Physics Prize with Becquerel. Then in 1911, Marie was awarded the Chemistry Prize—the first person to win two Nobel Prizes

In 1898, after months of hard work, Marie Curie isolated two other radioactive elements from pitchblende. She named these elements radium (after the term 'radioactivity') and polonium (after Poland where she was born). Radium was found to be two million times more radioactive than uranium.

The fogging of a photographic plate by radioactive materials has practical applications today. For example, we can follow the way in which a plant leaf takes in carbon dioxide by exposing the leaf to carbon dioxide containing radioactive carbon-14. The leaf is then held against a photographic plate. When the plate is developed, the darkest parts show where the carbon-14 has gone.

Radioactivity can be used to study the uptake and distribution of heavy metals in plants. In the picture the plant has been grown in a solution of a radioactive mercury compound for 48 hours. The radioactive mercury can then be detected in the stem and leaves by laying the plant on photographic film and allowing sufficient exposure time

Before very long, another effect of radioactive substances was discovered. When a screen coated with zinc sulphide is placed near a radium salt and examined with a magnifying lens, tiny flashes of light appear on the surface of the zinc sulphide. The flashes of light are called **scintillations**. When particles hit the zinc sulphide their kinetic energy is converted to light which causes the scintillations. The particles are produced as the radioactive radium atoms break up.

The radiations and particles emitted by radioactive substances were originally detected in one of three ways:

1 the fogging of photographic plates;

2 the discharge of an electroscope;

3 scintillation methods.

Nowadays, the radiations are detected using a Geiger-Müller tube (GM tube) connected to a counter or ratemeter. This is discussed in unit 9 of this section.

Questions

1 Why are photographic plates darkened ('fogged') when they are left in the dark near uranium salts?
2 Why did the Curies think that pitchblende contained an element more radioactive than uranium?
3 Why do radioactive materials cause the discharge of an electroscope?
4 (a) What are scintillations?
(b) What causes the scintillations when a radium salt is placed near zinc sulphide?

8 Nuclear Reactions

Alchemists dreamed of changing base metals to gold.

Radioactive radium compounds are used in the luminous paints on the hands and numbers of clocks and watches

Mixtures of zinc sulphide and a small amount of any radium salt glow in the dark. These mixtures are used in luminous paints on the hands and numbers of clocks and watches. We now know that the glow is caused by **alpha particles** hitting the zinc sulphide in the paint. This stops the alpha particles and their kinetic energy is converted to light energy. The alpha particles are produced as radium atoms split up (disintegrate) of their own accord. All the time, radium atoms are breaking up and losing alpha particles. This spontaneous break up of atoms results in **radioactivity (radioactive decay)**. Alpha particles are helium ions, $^{4}_{2}He^{2+}$—helium atoms that have lost both of their electrons. The mass number of an alpha particle is four and its atomic number is two. Thus, when an atom of $^{226}_{88}Ra$ loses an alpha particle, the fragment left behind will have a mass number which is four less than $^{226}_{88}Ra$ and an atomic number which is two less than $^{226}_{88}Ra$. So, the fragment will have a mass number of 222 and an atomic number of 86. All atoms of atomic number 86 are those of radon, Rn. Thus, the final products are $^{222}_{86}Rn$ and $^{4}_{2}He$. The nuclear decay for radium-226 can be summarised in a nuclear equation as

$$^{226}_{88}Ra \rightarrow \, ^{222}_{86}Rn + \, ^{4}_{2}He$$

Radioactive decay with the loss of an alpha particle is common for large isotopes with an atomic number over 83. These include uranium-238, radium-226 and plutonium-238. These isotopes decay because they are simply too heavy. They lose mass and try to become stable by losing an alpha particle.

Radioactive isotopes with atomic numbers below 83 lose **beta particles** when they decay, rather than alpha particles. Experiments show that beta particles are electrons, ${}_{-1}^{0}e$. Carbon-14 (${}_{6}^{14}C$) decays by losing beta particles to form ${}_{7}^{14}N$. The decay process can be represented as

$$ {}_{6}^{14}C \rightarrow {}_{7}^{14}N + {}_{-1}^{0}e $$

(Mass: upper number 0; Charge: lower number −1)

Notice that the *total* mass and the *total* charge are the same after beta decay as they were before. During beta decay, a neutron in the nucleus of the radioactive atom splits up into a proton and an electron. The proton stays in the nucleus, but the electron is ejected as a beta particle. Thus, the mass number of the remaining fragment stays the same, but its atomic number increases by one (see the table).

	Isotopes involved	**Particle lost**	**Change in mass number**	**Change in atomic number**
Alpha decay	Atomic number >83	${}_{2}^{4}He^{2+}$	−4	−2
Beta decay	Atomic number ≤83	${}_{-1}^{0}e$	0	+1

Comparing alpha decay and beta decay

Chemical reactions and nuclear reactions

Chemical reactions involve changes in the outer parts of atoms, the electrons. During chemical reactions electrons are either

1 *transferred* from one atom to another; or

2 *shared* between two atoms.

Nuclear reactions, on the other hand, involve changes in the central parts of atoms, the nucleus. During nuclear reactions, one element may be converted to another element by

1 *radioactive decay;*

2 *atomic fission; or*

3 *atomic fusion* (unit 12).

The first scientist to show that one element could be converted to another was Rutherford. In the 1920s, Rutherford and his colleagues turned sodium into magnesium and aluminium into silicon. At first, the changes from one element to another involved only elements with low mass numbers. However, in 1940, American chemists managed to build up heavier elements from uranium, the heaviest natural element. For example, neptunium-239 and plutonium-239 were obtained by bombarding uranium-238 with neutrons. By 1972, thirteen new elements had been synthesised, all of which come after uranium in the periodic table. All these new elements are radioactive and most of them disintegrate rapidly.

Questions

1 Explain the following words, (i) *radioactivity;* (ii) *alpha particles;* (iii) *beta particles;* (iv) *nuclear reactions.*

2 Uranium-238 (${}_{92}^{238}U$) decays to give an alpha particle and a new atom.
(a) What is the mass number of the new atom?
(b) What is the atomic number of the new atom?
(c) What is the symbol of the new atom?
(d) Write an equation for the decay of U-238.

3 Write nuclear equations for the beta-decay of ${}_{19}^{43}K$ and ${}_{11}^{24}Na$.

4 Look up the names and symbols of the elements after uranium in the periodic table. Most of these elements were first synthesised in the USA, the other two or three in Russia.
(a) Name four elements that you think were first synthesised in the USA.
(b) Name two elements that you think were first synthesised in Russia.
(c) Suggest why the particular name was chosen for four of these elements.
(d) Why do you think these elements are radioactive?

5 How do nuclear reactions and chemical reactions differ?

9 Detecting Radioactivity

When radioactive substances decay they emit alpha particles ($^{4}_{2}He^{2+}$ ions) and beta particles (electrons). At the same time, they usually give off **gamma rays**. Gamma rays are electromagnetic waves, like light. But, they contain so much energy that they can damage cells and kill organisms. As a radioactive atom gives off gamma rays, it loses energy and becomes more stable.

When alpha particles are emitted by radioactive substances, they travel a few centimetres in air and they can be deflected by electric and magnetic fields. Beta particles are much lighter than alpha particles. They travel several metres in air and are deflected much further by electric and magnetic fields. Gamma rays are unaffected by electric and magnetic fields, but they travel a long way in air and penetrate bricks and metal sheets. The properties of alpha particles, beta particles and gamma rays are summarised in the table and in figure 1.

Radiation	Nature	Effect of electric and magnetic fields	Penetration of		
			paper	thin aluminium	thick lead
Alpha	Helium nuclei, $^{4}_{2}He^{2+}$	small deflection	X	X	X
Beta	electrons, $^{0}_{-1}e$	large deflection	√	X	X
Gamma	electromagnetic waves	no deflection	√	√	X

The nature and properties of alpha particles, beta particles and gamma rays

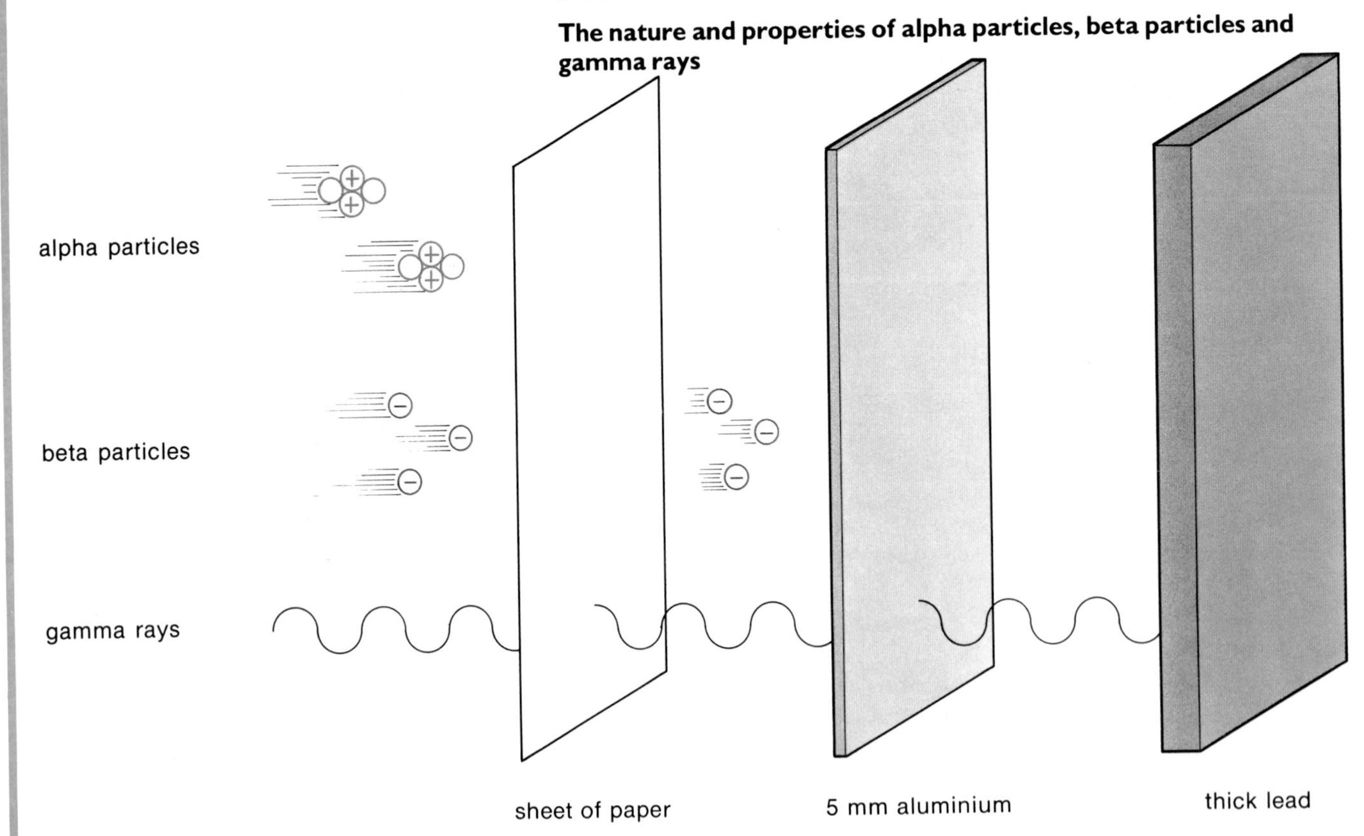

Figure 1
The relative penetrating power of alpha particles, beta particles and gamma rays

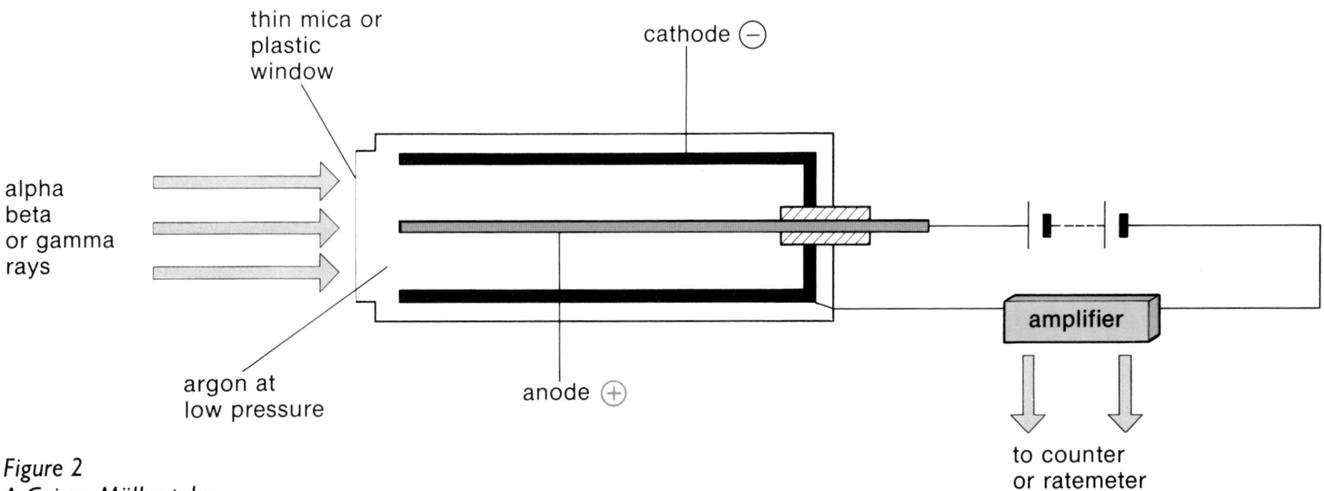

Figure 2
A Geiger-Müller tube

The particles and rays emitted by radioactive isotopes can be detected using a **Geiger-Müller tube** connected to a counter or ratemeter (figure 2). If alpha, beta or gamma rays enter the tube, they ionize the argon inside.

$$Ar(g) \rightarrow Ar^+(g) + e^-$$

The Ar^+ ions and electrons from the ionization are attracted to the electrodes in the tube and a tiny current flows in the circuit. This current is amplified and recorded on a counter or ratemeter. When radioactive substances decay, no-one knows exactly why a particular atom splits up at a particular moment. The process is entirely random. Experiments show, however, that each radioactive isotope decays at its own rate, and this rate of decay is always the same. It is not affected by changes in temperature, changes in concentration or by other atoms combined with the radioactive isotope. This is an important difference between nuclear reactions and chemical reactions.

Half life

The rate of decay of a radioactive isotope is shown by its half life. *This is the time it takes for half of the atoms of the isotope to decay.* Half lives can vary from a few milliseconds to several million years. The shorter the half life, the faster the isotope decays and the more unstable it is. The longer the half life, the slower the decay process and the more stable the isotope. Uranium-238 with a half life of 4500 million years is 'almost stable', but polonium-234, with a half iife of only 0.15 milliseconds is quite the reverse.

The radioactive isotope iodine-131 has been used by doctors to measure the uptake of iodine by the thyroid, an important gland in the human body. Iodine-131 has a half life of eight days. This means that, starting with 1 g of iodine-131, only ½ g remains eight days later. After another eight days, the half gram will have decayed to ¼ g and eight days after that (i.e. 24 days from the start) only ⅛ g would remain. In fact, only tiny quantities of iodine-131 are used in this treatment. By measuring the amount of radioactivity in the thyroid gland, doctors can tell how much of the iodine-131 has been taken up by the gland.

Questions

1 Explain the following:
gamma rays; Geiger-Müller tube; half life.

2 Suppose that you had 8 g of strontium-90, with a half life of 28 years.
(a) How much would be left after (i) 28 years; (ii) 56 years; (iii) 112 years?
(b) When will its rate of decay be half as great as it was at the start? Explain your answer.

3 Radon-222 decays by alpha particle emission, with a half life of 3.8 days. 8.88 g of Ra-222 was allowed to decay for 7.6 days.
(a) Calculate the mass of radon-222 left after 7.6 days.
(b) How many moles of radon-222 decay in 7.6 days?
(c) Calculate the number of alpha particles emitted in 7.6 days. (Avogadro constant = 6×10^{23} mol^{-1}.)

4 (a) What do you understand by the term 'radioactive element'?
(b) Name one radioactive element which occurs naturally.
(c) Which part of an atom is responsible for radioactivity?
(d) Which of the particles or rays emitted by radioactive substances (i) is most penetrating; (ii) contains positive particles; (iii) is not deflected by a magnetic field?

10 Measuring a Half Life

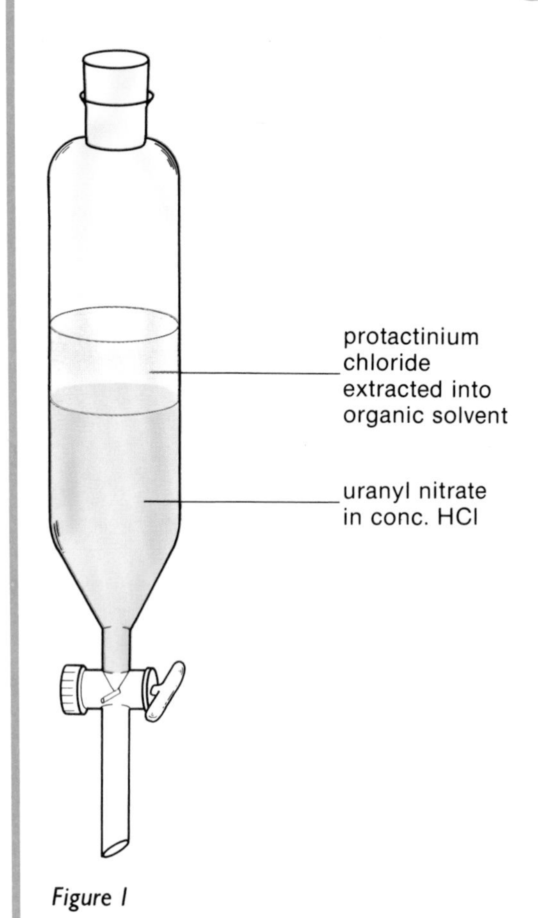

Figure 1

When a radioactive element decays, a new ('daughter') element begins to form. As the parent element decays, more 'daughter' forms. For example, as uranium-238 loses alpha particles it forms thorium-234.

$$^{238}_{92}\text{U} \rightarrow {}^{234}_{90}\text{Th} + {}^{4}_{2}\text{He}$$

uranium thorium helium

As $^{238}_{92}$U decays, the amount of $^{234}_{90}$Th increases. This daughter product ($^{234}_{90}$Th) is also radioactive, and it decays to a 'grand-daughter' product (protactinium-234).

$$^{234}_{90}\text{Th} \rightarrow {}^{234}_{91}\text{Pa} + {}^{0}_{-1}\text{e}$$

Further decays happen until a stable isotope is formed. Many of the heavy elements in the periodic table decay like this in a series of steps until they form a stable isotope of lead.

In the following experiment, we can study the decay of protactinium-234, which is the grand-daughter product from uranium-238. The $^{234}_{91}$Pa can be obtained from uranyl nitrate which contains radioactive uranium-238.

Concentrated hydrochloric acid is added to a concentrated solution of uranyl nitrate. This converts any protactinium to protactinium chloride. The mixture is then shaken with an organic solvent that extracts the protactinium chloride (figure 1). The organic layer is transferred to a modified GM tube for use with liquids and the radioactivity due to protactinium-234 is measured (figure 2). The number of counts shown on the counter is proportional to the amount of radiation entering the GM tube. The results in the table show the number of counts in 10 second periods, at an interval of 40 seconds.

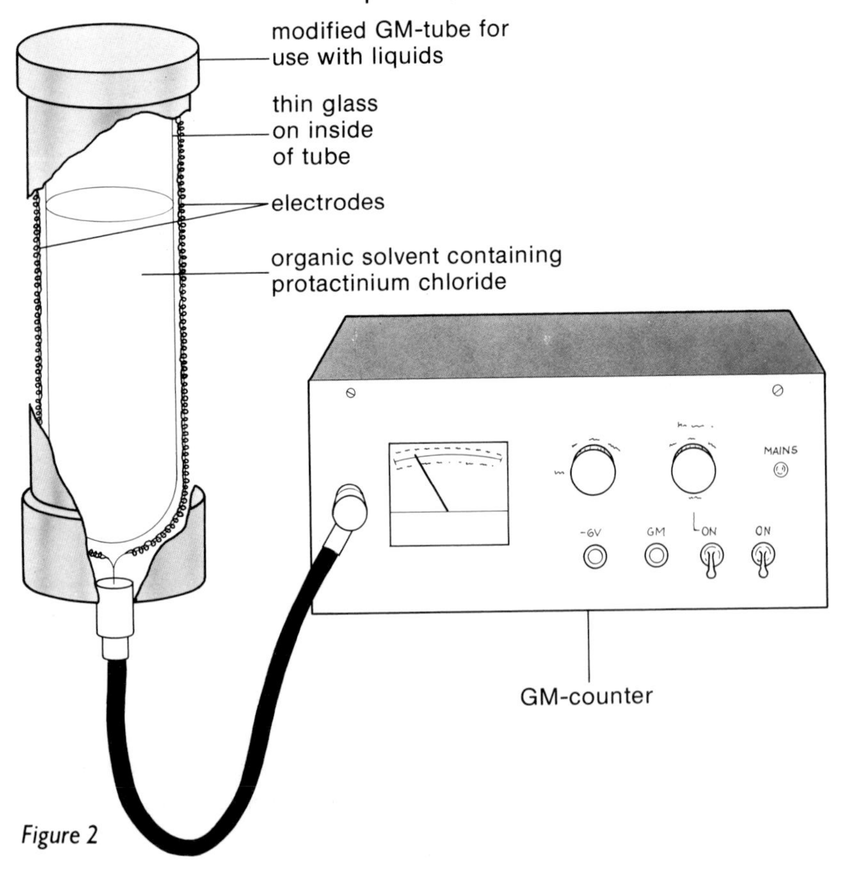

Figure 2

These results have been plotted on a graph in figure 3.

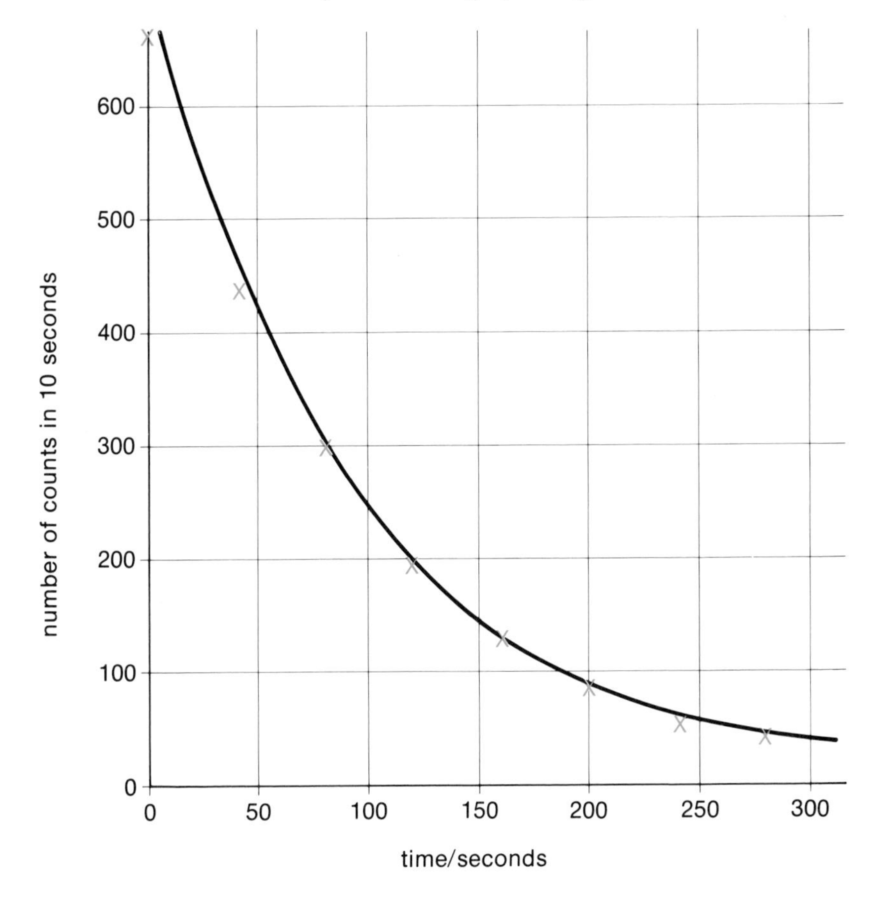

Time at start of 10 second count period/s	Number of counts in 10 seconds
0	660
40	437
80	300
120	195
160	130
200	87
240	58
280	40

The results of an experiment to study the decay of protactinium-234

Figure 3

1 How long does it take for the number of counts to fall from (i) 600 to 300; (ii) 400 to 200; (iii) 200 to 100?
2 What is the average time, from question 1, for the count rate to fall by half?
3 When the count rate has fallen by half, half of the Pa-234 atoms have decayed. What is the half life of Pa-234?

Accurate experiments show that the half life of protactinium-234 is 71 seconds. The symbol for half life is $t_{1/2}$.

Students under 16 years of age are not allowed to carry out experiments with radioactive materials. If your teacher demonstrates experiments like the one in this section, it is important to follow standard radioactive precautions. For relatively weak isotopes like Pa-234 these are:

1 Carry out experiments in a tissue-lined tray.

2 Wear polythene gloves.

3 Never pipette liquids by mouth. Always use a safety filler.

4 Keep all contaminated waste separate.

These precautions do *not* protect the experimenter from radiation, but they do prevent contamination of the laboratory and the experimenter's hands.

Questions

1 Write nuclear equations for the beta decay of the following isotopes: ${}^{212}_{82}Pb$; ${}^{131}_{53}I$; ${}^{234}_{91}Pa$.
2 Write nuclear equations for the alpha decay of the following isotopes: ${}^{212}_{84}Po$; ${}^{237}_{93}Np$.
3 0.1 g of ${}^{59}_{26}Fe$ decays by beta decay. ($t_{1/2}$ = 30 days) How much is left after 90 days?
4 What precautions should be taken in experiments with radioactive isotopes?
5 $t_{1/2}$ for ${}^{131}_{53}I$ is eight days. Draw a graph showing the percentage of ${}^{131}_{53}I$ remaining over a period of 32 days.
6 Why is radioactive waste *not* destroyed by incineration (burning)?

11 Using Radioactive Isotopes

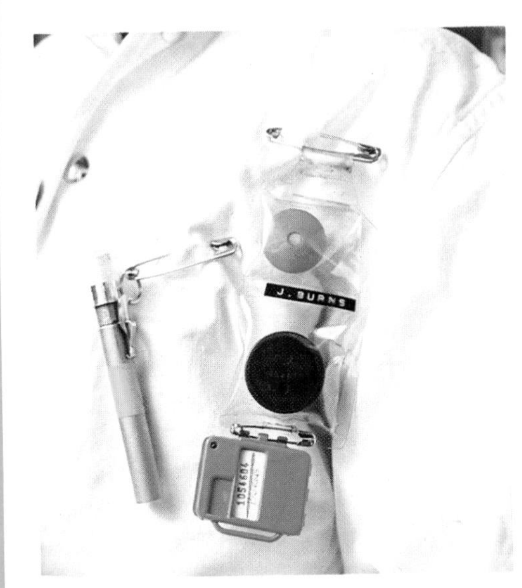

A technician wearing a blue radiation-level badge

Radioactive isotopes are widely used in industry and medicine. People who work with radioactive materials must wear a radiation badge which measures the radiation to which they have been exposed. Gamma rays are the most dangerous form of radiation. They can cause changes in the structure of chemicals in our bodies and kill living cells. Scientists and technicians who use dangerous isotopes must be protected by lead or concrete shields and handle radioactive materials by remote control.

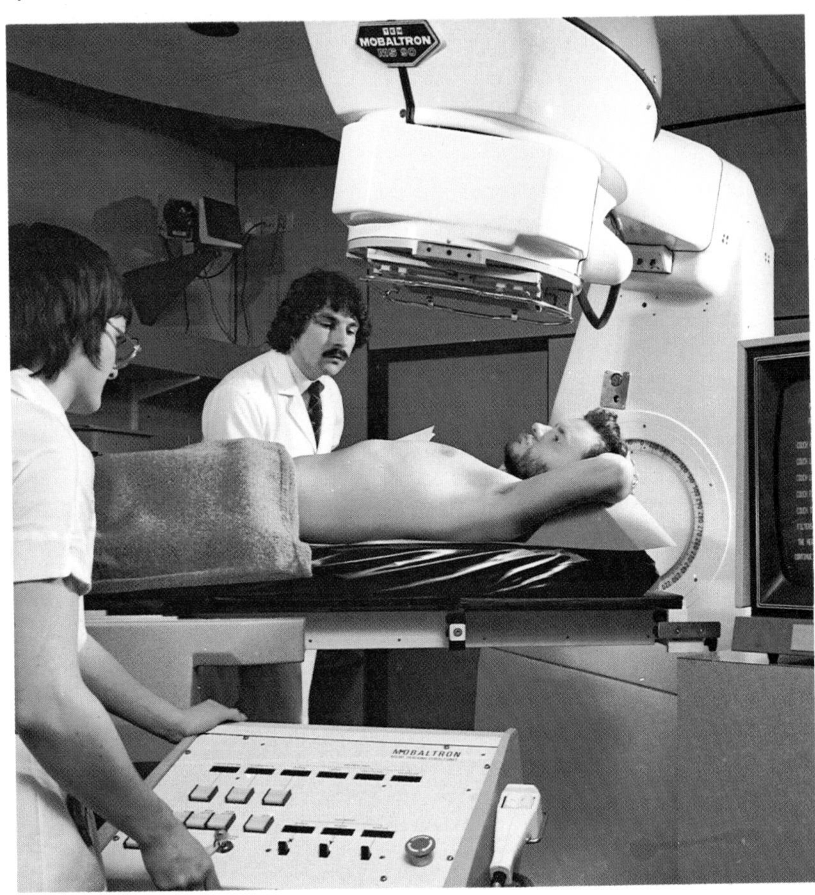

A patient being treated by radiation from cobalt-60

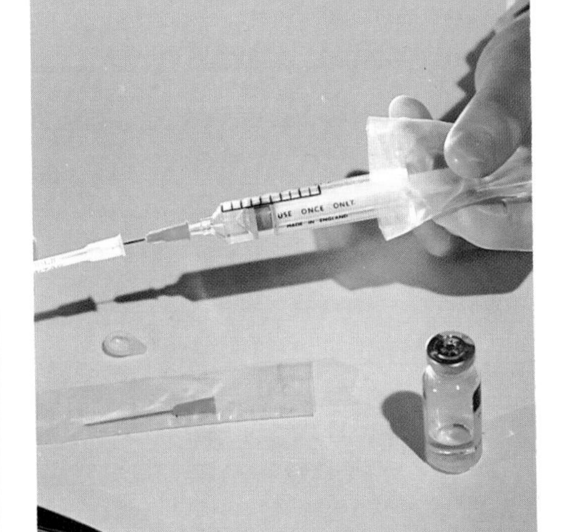

Syringes and other medical items are sterilized by gamma radiation

Medical uses

When a living cell is exposed to radiation, the structure of chemicals (genes) in the nucleus may be changed. This can cause the cell to die. Penetrating gamma rays from cobalt-60 ($^{60}_{27}Co$) are used to kill cancer cells and to treat growths inside the body. Cancer cells on the surface of the body, as in skin cancer, can be treated with less penetrating beta rays. This is done by strapping a plastic sheet containing $^{32}_{15}P$ (radioactive phosphorus) or $^{90}_{38}Sr$ (radioactive strontium) on the affected area. Cancer can be fatal, but doctors are becoming more and more successful at curing less serious cases. Medical items, such as dressings and syringes, are often sealed in polythene bags and sterilized by gamma rays. This method of sterilization is much easier than the old method which used steam. Intense doses of gamma radiation sterilize the articles by killing any bacteria on them.

Tracer studies

Radioactive isotopes are easy to detect. Therefore they are used to *trace* what happens to different substances in chemical, physical and biological processes. Radioactive isotopes are usually mixed with non-radioactive atoms of the substance under investigation. For example, the way that plants take in and use phosphates can be studied using a fertilizer containing $^{32}_{15}P$. Tracer studies using $^{14}_{6}C$ have helped in the study of photosynthesis and protein synthesis. The activity of the thyroid gland can be studied by measuring the uptake of iodine-131 (unit 9 of this section).

Archaeological uses

The common isotope of carbon is carbon-12. The carbon in living things is therefore mainly carbon-12 with a small constant percentage of *radioactive* carbon-14. This gets into living things from the carbon-14 which is part of the carbon dioxide in the air. When an animal or plant dies, the carbon-14 in it continues to decay. However, the replacement of decayed carbon-14 from food and carbon dioxide stops. The amount of carbon-14 left in the remains of the animal or plant can be measured. Then, knowing the half life of carbon-14, it is possible to work out how long it is since the animal or plant died.

'Carbon dating' has been used to check the age of ancient documents and the bones of ape men. This method was used to show that the famous Dead Sea Scrolls were in fact 2000 years old and that the skull of the Piltdown Man was a fake.

Part of the Dead Sea Scrolls. Radiocarbon dating shows that the Dead Sea Scrolls were probably authentic

Questions

1 What precautions are taken to ensure that scientists and technicians who work with radioactive isotopes are not exposed to dangerous radiations?

2 Suppose that carbon-14 makes up $x\%$ of the carbon in living things and that the half life of carbon-14 is 5700 years.

(a) How long will it take for the percentage of carbon-14 to fall from:

(i) $x\%$ to $\frac{1}{2}x\%$, (ii) $\frac{1}{2}x\%$ to $\frac{1}{4}x\%$?

(b) How old is an object which has $\frac{1}{8}x\%$ of its carbon as carbon-14?

3 Why are radioactive isotopes important?

4 In some countries, gamma radiation is used to prevent food going off.

(a) What will gamma radiation do to bacteria to prevent the food going off?

(b) This process is not yet used in the UK. Why do you think that some people are opposed to its use?

12 Nuclear Energy

Enormous amounts of energy can be obtained from nuclear reactions, but this discovery was made almost by chance. In 1938, the German scientists Hahn and Strassmann were trying to make a new element by bombarding uranium with neutrons. Instead of producing a new element, the neutron caused the uranium nucleus to break up violently into two smaller nuclei. Two or three separate neutrons were also released together with large quantities of energy (figure 1).

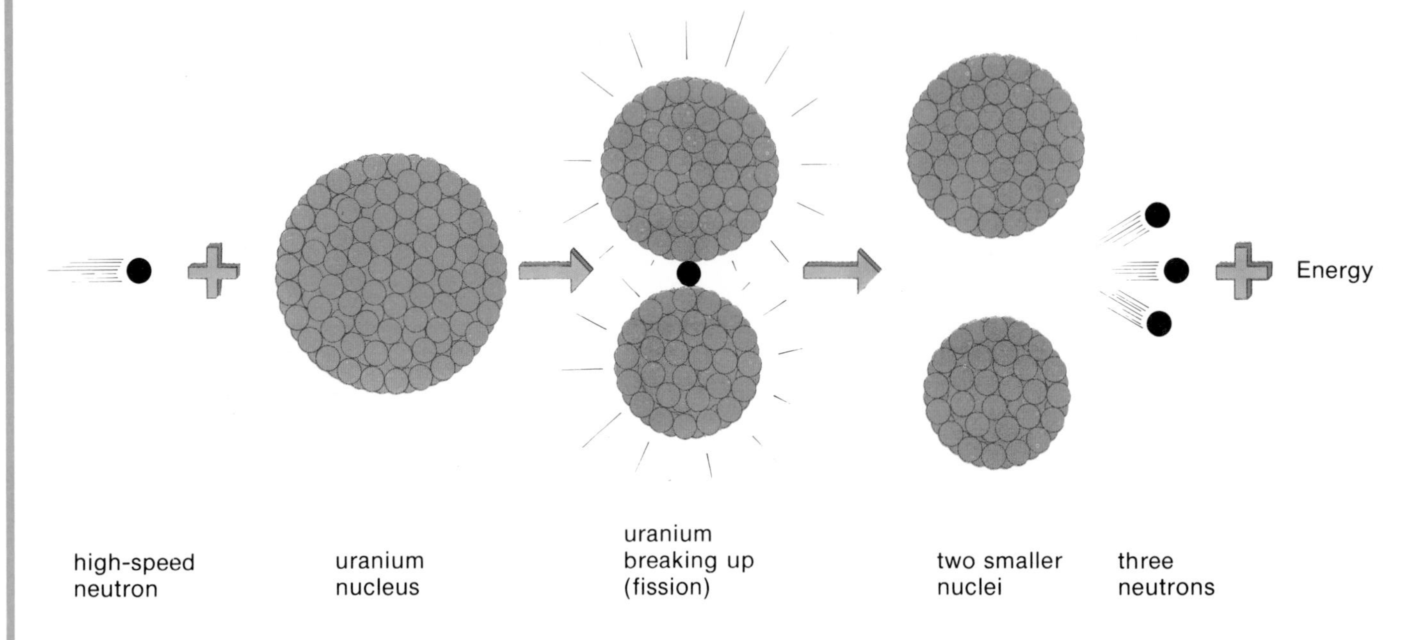

Figure 1
The fission of a nucleus of uranium-235

Notice how this nuclear reaction differs from radioactive decay. First, it does not happen of its own accord like radioactive decay. It only happens when the uranium is bombarded by neutrons. Second, it involves the break up of one large nucleus into two fragments of roughly the same size. During radioactive decay, the products are one large fragment and one very small fragment (either an alpha particle or an electron). The special name **nuclear fission** (atomic fission) is used to describe the splitting of an atom into two fragments of roughly the same size.

Atomic bombs and atomic reactors

Natural uranium contains two isotopes, $^{235}_{92}U$ and $^{238}_{92}U$. Only 0.7% is uranium-235. Experiments showed that only uranium-235 took part in nuclear fission during neutron bombardment. This led scientists to realize that if the uranium sample contained a larger fraction of uranium-235, the neutrons released during fission would split more uranium nuclei and cause a chain reaction.

Figure 2 shows how an exploding chain reaction occurs in uranium-235. One reaction releases three neutrons, three neutrons release nine, then 27, 81 and so on. Each time more and more energy is produced as more and more uranium-235 atoms undergo fission.

In an atomic bomb, the fission of *enriched* uranium-235 happens in an *uncontrolled* manner and enormous amounts of energy and radiation are released. In an atomic reactor, a *mixture* of uranium-235 and uranium-238 is used to give a *controlled* chain reaction. In this case, the

Atomic bomb explosions release vast amounts of energy

heat produced is used to generate electricity. Figure 3 shows a simplified diagram of a gas-cooled nuclear reactor. At the centre of the reactor, rods of 'enriched' uranium containing 3% of uranium-235 are stacked inside a large block of graphite. Neutrons released by the fission of the uranium are slowed down by collision with carbon atoms in the graphite.

The temperature of the reactor is controlled by moveable rods of boron or cadmium which absorb neutrons. The deeper these rods are inserted, the more neutrons are absorbed and the slower the reaction becomes. By carefully adjusting the neutron-absorbing rods, a controlled chain reaction can be obtained. Heat is produced steadily and taken away by carbon dioxide gas circulating through the reactor. The hot gas is then used to make steam which drives turbines and generates electricity. In the USA, most reactors use water as a coolant in place of carbon dioxide.

Recently, reactors have been developed that use plutonium as the fuel and liquid sodium as the coolant. Unlike uranium, which reacts best with slow neutrons, plutonium can use *fast* neutrons. This means that the reactor does not need a graphite block to slow them down. These reactors are called fast reactors. The first power station to use a fast reactor for generating electricity was the one at Dounreay in Scotland.

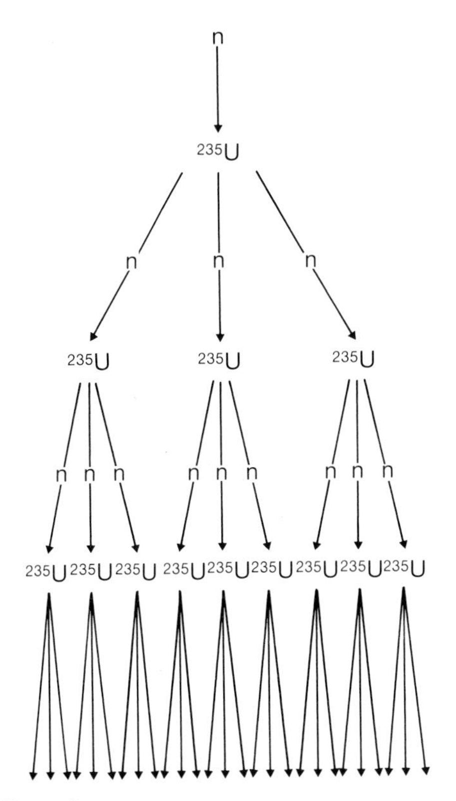

Figure 2
A chain reaction in uranium-235

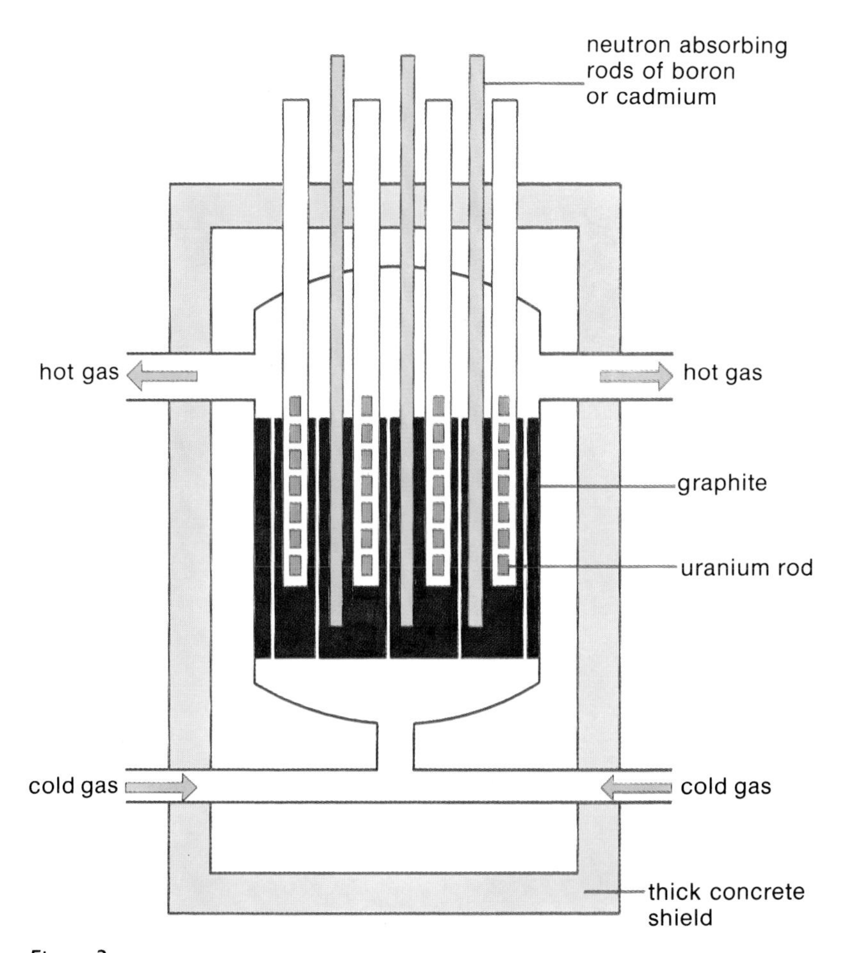

Figure 3
A simplified diagram of a gas-cooled nuclear reactor

Questions

1 What do you understand by the following:
nuclear fission; nuclear energy; chain reaction; fast reactor?

2 What are the differences between nuclear fission and radioactive decay?

3 In what ways are nuclear reactors similar to nuclear bombs? In what ways are they different?

4 This question is about nuclear reactors.

(a) Why are the uranium rods sunk in a graphite block?

(b) What is the function of the boron or cadmium rods?

(c) How is the rate of nuclear fission controlled?

(d) Why is the reactor surrounded by thick concrete?

(e) How is the energy from nuclear fission converted into electricity?

(f) Recently, some reactors have used liquid sodium as the coolant. Mention one advantage and one disadvantage of using sodium.

13 The Nuclear Debate

Hiroshima—a few weeks after the atomic bomb exploded in August 1945. The long wooden building in the foreground was, of course, erected after the blast

Hardly a week passes without some mention of nuclear technology in the newspapers or on television. A lot of the publicity concerns the risks and dangers involved. These include pollution by radioactive waste from nuclear power stations and, of course, the horrors of nuclear war.

The first atomic reactor was built in America in 1942 by the Italian scientist, Enrico Fermi. In October 1956, the first industrial nuclear power station was opened at Calder Hall, now Sellafield, in Cumbria (UK).

Progress was marred in 1945 when the first nuclear (atomic) bomb was dropped on Hiroshima in Japan. Then, in 1953, a hydrogen bomb was tested in the Pacific Ocean. This was even more destructive than an atomic bomb. Many people were very worried about the possibility of a war. Large numbers of people joined CND (the Campaign for Nuclear Disarmament). During the last ten years, the USA and the USSR have continued to develop even more powerful nuclear weapons. This has led to even more support for CND and the anti-nuclear movement.

As more nuclear reactors are built, there is also an increasing danger of pollution by radioactive waste. After use, the spent fuel is still very radioactive. Some of the waste material will remain dangerous for hundreds of years. Several methods of disposing of the radioactive wastes have been used. These include dumping the waste in deep mines or solidifying it in glass or concrete and then burying it below the sea bed.

CND marchers carrying banners and CND slogans

Although the risks from radioactive wastes are very real, many people are more concerned about the link between nuclear power stations and nuclear weapons. For example, terrorists might steal uranium or plutonium and use them to make bombs. There is also the risk that terrorists might capture a nuclear power station and use it as a bargaining device by threatening to destroy the reactor.

Many people feel that the problems created by nuclear technology outweigh the advantages. Other people say that we must learn to live with the risks if we are to maintain our industrial activity, preserve our standards of welfare and improve the quality of life for those in the Third World.

Any discussions on nuclear technology are always linked to energy supplies. The world consumption of coal, oil and gas is likely to increase rapidly during the next few decades. This is particularly so in developing countries. We must find another large source of energy and the main possibility seems to be nuclear power. Solar energy, tidal energy, wave energy and wind energy may help in certain areas, but none of them could provide the huge amounts of energy that we need.

At present, about 18% of electricity in the UK is generated from nuclear energy. This is expected to rise to about 70% by the end of the century. Unfortunately, the reserves of uranium are limited and many experts think that there is not enough for the next century. Eventually, we may be forced to build more fast reactors that use plutonium rather than uranium. These can extract 50 to 60 times more energy from the fuel than ordinary uranium reactors.

Some radioactive waste used to be sealed in drums and then dumped at sea. There has been a worldwide agreement to stop this. Instead waste is to be buried deep underground on land.

Dounreay Power Station—the first power station to use a fast reactor for generating electricity

Questions

1 Why did the CND movement start?
2 What are the main dangers associated with nuclear technology?
3 (a) What methods are used to dispose of waste from nuclear power stations?
(b) Why are such thorough methods of disposal needed?
4 Why do some people think that we will have to rely on nuclear energy in the future?
5 Is nuclear power the answer to our energy problems or are the risks too great? What do *you* think?

Section L: Study Questions

1 The table below gives information about four elements, *V*, *W*, *X* and *Y*, which are in the same group in the periodic table.

Element	Atomic Number	Melting point/°C	Boiling point/°C
V	9	−220	−188
W	17	−101	−33
X	35	−7	58
Y	53	114	183

(a) Use the melting and boiling points to give the letters of all the elements in the table which, at atmospheric pressure and at room temperature, are:
(i) solids (ii) liquids (iii) gases. (3)
(b) Describe what happens to the particles of a solid when it melts to form a liquid. (3)
(c) The next element in this group after *Y* is *Z*.
(i) Would you expect *Z* to exist under room conditions as a solid, liquid or gas? State your reasons.
(ii) How many electrons will there be in the outer shell of a *Z* atom?
(iii) Using the symbol *Z*, write down the formula of the ion of *Z*. (4)
(d) *X* reacts with a solution of the sodium salt of *Z* according to the equation

$$X_2 + 2NaZ \rightarrow Z_2 + 2NaX.$$

If 4.00 g of X_2 produce 10.5 g of Z_2, find the relative atomic mass of *Z*. (Relative atomic mass of *X* = 80.) (3)

Total [13] **LEAG**

2 Statements about pollution are listed below.
A The china clay industry produces large unsightly spoil heaps.
B The waste products from nuclear reactors are radioactive and the disposal of these products presents long term difficulties.
C 'Acid rain', produced by the large scale burning of fossil fuels, is causing a reduction in the trout population of some Scottish lochs.
D Phosphates, from domestic sewage, and nitrogenous fertilisers cause the uncontrolled growth of algae in lakes and streams.

(a) (i) What is a spoil heap? (1)
(ii) Name one other industry that produces large spoil heaps. (1)
(iii) Why are spoil heaps unsightly? (1)
(b) What are the problems associated with the disposal of radioactive materials? (2)
(c) (i) Explain how the burning of fossil fuels causes acidity in rain. (3)
(ii) Give one other problem caused by this type of pollution. (1)
(d) (i) How do nitrogenous fertilisers get into lakes and streams? (1)
(ii) Which domestic product is the major source of the phosphates? (1)
(iii) Why, other than the appearance of the lake, are extensive surface growths of algae undesirable? (1)

Total [12] **NEA**

3 (a) Name an element in:
(i) the same group of the periodic table as chlorine; (1)
(ii) the same period of the periodic table as carbon. (1)
(b) Write down or draw a diagram of the electronic structure of:
(i) an atom of carbon (atomic number = 6);
(ii) an atom of chlorine (atomic number = 17). (2)
(c) Draw a diagram to show the electron arrangement in a molecule of tetrachloromethane (CCl_4). Only the outer shells of electrons need to be shown in your diagram. (2)

SEG

4 This question is about the 'Holy Shroud of Turin'. This is the cloth which, some people believe, covered the body of Jesus Christ after his crucifixion.
(a) How does 'carbon dating' work?
(b) How could carbon dating be used to check whether the 'Holy Shroud of Turin' is genuine or whether it is a fake?
(c) Do you think it is right that religious relics are dated by radioactive methods?

5 The ion X^{2-} contains 54 electrons.
(a) To which group of the periodic table does the element *X* belong? (1)
(b) Write the formula of the compound between caesium and X. (Caesium has symbol Cs and is in Group 1 of the periodic table.) (1)
(c) How many protons are present in the nucleus of an atom of *X*? (1)

SEG

6 The following passage was written to support the building of more nuclear power stations.

'World energy requirements will rise rapidly in the next decade. By the year 2000, it is estimated that the average world citizen will consume twice as much energy as the present average world citizen in the USA. This energy can only be obtained by building 4000 nuclear fission reactors using the inexhaustible supplies of nuclear fuel.

We must stop using fossil fuels to produce energy because they are irreplaceable as raw materials for the chemical industry. Their use also damages the environment. Atmospheric pollution is caused by burning coal and oil, oil drilling and transport causes marine pollution and coal mining damages the environment.'

(a) Criticise the arguments used in the passage.
(b) Suggest four alternative sources of energy not considered in the passage.
(c) Discuss three important arguments against increasing the number of nuclear power stations.

7 Natural boron contains 20% boron-10 and 80% boron-11. Boron is the fifth element in the periodic table.
(a) What is the atomic number of boron?
(b) What are the mass numbers of the two boron isotopes?
(c) How many protons, neutrons and electrons does one atom of boron-11 possess?
(d) What is the relative atomic mass of boron?

Index

Index

Relative Atomic Masses

Element	Symbol	A_r	Element	Symbol	A_r	Element	Symbol	A_r
Aluminium	Al	26.9	Gold	Au	197.0	Rubidium	Rb	85.5
Antimony	Sb	121.8	Helium	He	4.0	Scandium	Sc	45.0
Argon	Ar	39.9	Hydrogen	H	1.0	Selenium	Se	79.0
Arsenic	As	74.9	Iodine	I	126.9	Silicon	Si	28.1
Barium	Ba	137.3	Iridium	Ir	192.2	Silver	Ag	107.9
Beryllium	Be	9.0	Iron	Fe	55.8	Sodium	Na	23.0
Bismuth	Bi	209.0	Krypton	Kr	83.8	Strontium	Sr	87.6
Boron	B	10.8	Lead	Pb	207.2	Sulphur	S	32.1
Bromine	Br	79.9	Lithium	Li	6.9	Tellurium	Te	127.6
Cadmium	Cd	112.4	Magnesium	Mg	24.3	Thorium	Th	232.0
Caesium	Cs	132.9	Manganese	Mn	54.9	Tin	Sn	118.7
Calcium	Ca	40.1	Mercury	Hg	200.6	Titanium	Ti	47.9
Carbon	C	12.0	Molybdenum	Mo	95.9	Tungsten	W	183.9
Chlorine	Cl	35.5	Neon	Ne	20.2	Uranium	U	238.0
Chromium	Cr	52.0	Nickel	Ni	58.7	Vanadium	V	50.9
Cobalt	Co	58.9	Nitrogen	N	14.0	Xenon	Xe	131.3
Copper	Cu	63.5	Oxygen	O	16.0	Zinc	Zn	65.4
Fluorine	F	19.0	Phosphorus	P	31.0			
Gallium	Ga	69.7	Platinum	Pt	195.1			
Germanium	Ge	72.6	Potassium	K	39.1			

Acknowledgements

The following companies, institutions and individuals have given permission to reproduce photographs in this book:

Ace Photo Agency (241, bottom), Malcolm Aird (124), D. Allan BAS (36, top), Allsport (186, top), Heather Angel/Biophotos (201), Banking Information Service (56, middle), Professor J. Barber, Imperial College, London (267), BASF (244, top, middle and bottom), BICC (83), Bilpot Limited (63, left), Biophoto Associates (104, middle; 250), Syd Bishop & Sons Limited (232, bottom), Nic R. Brawn (200), Paul Brierley (69; 75; 95), Brighton Borough Council (61, middle), Brighton Palace Pier (187, bottom), Britain on View (207, top; 208; 209, top; 210, right; 231, bottom), British Airways (10, top right; 115), British Alcan (71; 82; 181, top), BBC Stills (229), British and European Sales Limited (2, top), British Geological Survey (206; 209, bottom), BP Oil (61, right; 220, bottom; 187), British Rail (237, bottom), British Steel 101; 127, middle; 130; 131; 133), Bulmer Limited (226, top), Burmah-Castrol (UK) Limited (41, top right), Calor Gas Limited (218 two pictures), Cambridge University Library (254; 255; 256), Camera Press Limited (2, bottom; 42; 129; 278, bottom), Canadian Embassy (199, left), Simon-Carves Limited (143, top), Cavendish Laboratory, Cambridge University (258), CEGB (33; 65; 68, top; 189; 199, right), Chanel (48), Chloride Automotive Batteries Limited (89, left; 146, top), Chubb (204, right), Corning (233, top), Sean Crampton (26, top), Marie Curie Memorial Foundation (5, top; 9, top left), CVG Internacional CA (43), De Beers Industrial Diamond Division (178, top left), DOE (30, middle and bottom), Diamond Information Centre (179; 231, top), Distiller's Company (175; 205), Dunton Green Fruiterers (146, bottom), ECON Group Limited (187, top), The Electricity Council (66, top), Express Milk (88, top), Farmer's Weekly (145, left; 186, right; 248, bottom; 251, top), Ford Motor Company Limited (84, top; 233, bottom), Friends of the Earth (3, top; 144, top left), Geological Museum (104, left; 120; 158, top and bottom; 163 (three pictures); 164 (two pictures); 184; 196, top), GeoScience Features (132; 166), GHI Studios (247), Girl Guides Association (236, top), Goddard and Gibbs Studios, London (10, top left), Goodyear (170, bottom left), GLC (3, bottom; 181, bottom), Sally and Richard Greenhill (50, middle and left), Greenpeace (34; 276), Robert Harding (7), John Hillelson (17; 99 top), Hulton (28, bottom), ICI (4, top; 81; 108; 110, 142; 153; 248, top right and left; 249), IMI Range Limited (118), Israel Museum, Jerusalem (275), Chris James (51, right), Japanese Ministry of Foreign Affairs (31, middle), Key Terrain Limited (225), Martin Leach (32), Lego (UK) Limited (6, top), Manchester City Council (45), Sir Robert McAlpine Limited (207, bottom; 214), Alfred McAlpine PLC (51, middle), Meteorological Office (103, bottom), Museo Prado, MAS (28, top left), NCB (Cover; 197), National Gallery (268, top), NHPA (11; 16, top; 25; 35, top; 37; 121, middle; 188; 251, bottom), Norsk Hydro Polymers (111, right), NASA (29; 92, bottom; 190), National Diving Museum (103, right), Nuffield-Chelsea Curriculum Trust (122, top), Ohmeda (237, top), Polysius Limited (63, right), Popperfoto (9, middle; 111, bottom; 278, top), Preussicher Kulturbesitz (93), Rex Features (232, bottom; 266, bottom), RHS/Smith (145, bottom left), Rio Tinto Zinc (260), Ronan Picture Library (167), Royal Society (6, bottom; 76, top), RSPB (170, bottom right), Salter Clark Associates (66, bottom), Jerry Sampson (31, right; 144, right; 232, top), Science Museum (9, top right; 52, bottom left; 60; 74; 94; 102; 259; 266 top), Science Photo Library (26, bottom; 38; 46, bottom; 52; 121; 152, bottom), Scot Rail (39; 41, middle; 119), J. Scott and Son Limited (77), Scottish Tourist Board (126, left), Dr Seakins (21), Shell (1; 5, bottom; 198; 213; 214; 219, top; 221), South Yorkshire County Council (196), Sulphur Institute (143, bottom; 170), Susan Grigg Agency Limited (141), Syndication International (23; 30, top; 35, middle), Tate and Lyle (12), John Taylor and Company (8), Tesco Stores Limited (224; 242, bottom), John Topham (40), UKAEA (9, top middle; 87; 99, bottom; 253; 274, top, middle and bottom; 279, top and bottom), Van de Burghs and Jergens (223, left), Wessex Water (36, middle), West Kent Cold Store (239), Whitbread (18; 226, bottom), Woodmansterne (10, bottom; 80), Yorkshire Museum (117; 126, middle).

We are grateful to the following examining bodies for permission to reproduce questions from specimen GCSE papers and from recent joint 16+ examinations in Chemisty.
The London and East Anglian Group (LEAG), The Midland Examining Group (MEG), The Southern Examining Group (SEG), The Welsh Joint Education Committee (WJEC), The Northern Examining Association (Associated Lancashire Schools Examining Board, Joint Matriculation Board, North Regional Examinations Board, North West Regional Examinations Board, Yorkshire and Humberside Regional Examinations Board) (NEA).

British Library Cataloguing in Publication Data
Hill, Graham, 1942-
Chemistry counts.
1. Chemistry
I. Title
540 QD33

ISBN 0-340-37631-7

First published 1986
Ninth impression 1991

Printed in Hong Kong for Hodder and Stoughton Educational a division of Hodder and Stoughton Ltd, Mill Road, Dunton Green, Sevenoaks, Kent by Colorcraft